PROBABILITY AND STOCHASTICS SERIES

Edited by Richard Durrett and Mark Pinsky

Probability and
Stochastics Series

Real and Stochastic Analysis

Recent Advances

Edited by
M.M. Rao
Department of Mathematics
University of California
Riverside, CA

CRC Press
Boca Raton   New York

Acquiring Editor: *Tim Pletscher*
Senior Project Editor: *Susan Fox*
Cover Design: *Denise Craig*
Prepress: *Gary Bennett*

Library of Congress Cataloging-in-Publication Data

Real and stochastic analysis : recent advances / edited by M.M. Rao.
    p.   cm. — (Probability and stochastics)
    Includes bibliographical references and index.
    ISBN 0-8493-8078-2 (alk. paper)
    1. Stochastic analysis.   I. Rao, M. M. (Malempati Madhusudana),
1929-   . II. Series: Probability and stochastics series.
QA274.2.R423   1997
519.2—dc21   97-143
                                                   CIP

Contributors

René A. Carmona, Department of Mathematics, University of California at Irvine, Irvine, CA 92717
(Current address: Statistics and Operations Research Program, E- Quad, Princeton University, Princeton, NJ 08544)

Michael L. Green, Department of Mathematics, Baylor University, Waco, TX 76798

V. Haynatzka, Department of Statistics and Applied Probability, University of California at Santa Barbara, Santa Barbara, CA 93106

G. Haynatzki, Department of Mathematics and Statistics, McMaster University, Hamilton, ON L8S 4K1, Canada

S. T. Rachev, Department of Statistics and Applied Probability, University of California at Santa Barbara, Santa Barbara, CA 93106

Randall J. Swift, Department of Mathematics, Western Kentucky University, Bowling Green, KY 42101

N. N. Vakhania, Muskhelishvile Institute of Computational Mathematics, Georgian Academy of Sciences, Tbilisi- 93, Georgia (FSU)

V. Tarieladze, Muskhelishvile Institute of Computational Mathematics, Georgian Academy of Sciences, Tbilisi- 93, Georgia (FSU)

Preface

One of the main purposes of this monograph is to highlight the key role played by abstract analysis in simplifying and solving some fundamental problems in stochastic theory. This effort is analogous to a similar one published over a decade ago. The presentation is in a research-expository style, with essentially complete details, to make the treatment self-contained so as to be accessible to both graduate students seeking dissertation topics and other researchers desiring to work in this area. The aim is to give a unified and general account for a selected set of topics covering a large part of stochastic analysis. A central thread running through all the articles here is employment of functional analytic methods.

The work presented in the following six chapters is not only a unified treatment, but each one contains a substantial amount of new material appearing for the first time. They are devoted to both the random processes and fields, Gaussian as well as more general classes, with some serious applications and also several indications of them at several places. A more detailed synopsis of each chapter appears in the Introduction and Overview that follows immediately. All chapters have been reviewed.

It is hoped that these results will stimulate further research in these areas. In preparing this volume, I received considerable assistance from Dr. Y. Kakihara and Ms. Jan Patterson, as well as the authors. This is much appreciated. I also wish to thank CRC Press for their enthusiastic cooperation in publishing this book on schedule.

M. M. Rao

Contents

Introduction and overview

M. M. Rao

The material of the following chapters, presented in a research-expository style, consists of recent advances in some key areas of stochastic analysis wherein real (= functional) analysis methods and ideas play a prominent role. The topics discussed are detailed with numerous references bringing the reader to current research activity in the subject, and at the same time pointing out several problems that lead to promising investigations. This is particularly helpful for graduate students as well as other researchers who would pursue work in the areas covered in the following chapters. These appear in stochastic theory, and moreover interesting new functional analysis problems are suggested by the former. The current work complements the studies of a previous volume, published a decade ago, and concentrates on areas mostly considered since that time and includes certain topics that could not be treated then. Some important applications are also presented now and they motivate new areas of potential interest. Let us discuss the material of the chapters in some detail giving an overview of the topics to potential readers.

1. The first chapter, written by Carmona, is on stochastic modeling useful to several important problems, of interest in applications, along with an account of the underlying theory. Thus a comprehensive model that can be specialized to describe several stochastic flows, including Brownian, Jacobian and Manhatten flows as well as the mass transport, is discussed. This general model is defined by the following (nonlinear first order) stochastic differential equation:

$$dX_t = \overrightarrow{v}(t, X_t)dt + \sqrt{2\kappa}dB_t, \tag{1}$$

where $\{B_t, \geq 0\}$ is a d-dimensional Brownian motion, $\kappa \geq 0$, and $\{\overrightarrow{v}(t, x), t \geq 0, x \in \mathbb{R}^d\}$ is itself a Gaussian random field on $\mathbb{R}^+ \times \mathbb{R}^d$ of the following type. For each $t \geq 0$, $\overrightarrow{v}(t, \cdot) : \mathbb{R}^d \to \mathbb{R}^d$ is stationary and for each $x \in \mathbb{R}^d$, $\overrightarrow{v}(\cdot, x)$ has an extension to $\mathbb{R} \to \mathbb{R}^d$, which is again stationary so that the covariance function of $\overrightarrow{v}$, say Γ, satisfies:

$$\Gamma(s, x; t, y) = \tilde{\Gamma}(s - t, x - y), \tag{2}$$

a $d \times d$-positive definite matrix. The values of interest for the first variable are $s, t \geq 0$, to be identified as time and $x, y \in \mathbb{R}^d$ as space variables. Here $\tilde{\Gamma}(\cdot, \cdot)$ is assumed integrable on $\mathbb{R} \times \mathbb{R}^d$ so that it has (a.e. Leb.) a spectral density. Under various specializations, the author discusses the solutions of

(1). It is applied, with $\kappa = 0$ resulting in a Gaussian velocity field, to shear Brownian and Manhatten flows, and their transport properties. If $\vec{v}(t, x) = -ax$ and $2\kappa = b^2$, the solution of (1) is an Ornstein-Uhlenbeck process. This is described first for the one-dimensional case and then extended to finite dimensions with $2\kappa = B$ as a positive definite matrix and $a = A$ a square matrix. This motivates a study of an infinite dimensional O.U. process with B as a positive definite operator on the state space of the B_t-process (a Hilbert space $\mathcal{H}$) and A as a linear operator on it. If A is a partial differential operator, then one gets a representation of the solution of the stochastic PDE (or SPDE) as:

$$\langle x, X_t \rangle = \langle x, X_0 \rangle - \int_0^t \langle Ax, X_s \rangle ds + W(\chi_{[0,t)} x), \qquad (3)$$

for each $x \in dom(A)$, $W(\cdot)$ being a Brownian motion process.

Several applications of the solution process are considered. Numerical simulations are also discussed. Applications to mass transport are given using the theory of ODE (especially utilizing the Lyapounov exponents – indeed, for the ODE in Banach spaces Lyapounov and Bohl exponents play a key role as seen from Dalecky and Krein (1974), p.116 ff). A discussion of SPDE in this context is included, along with diffusion approximations, and a good collection of related papers are in the bibliography. Several unsolved problems are also pointed out at various places. In fact the paper is based on a series of lectures given by the author recently, and the freshness of the exposition is maintained.

2. One of the important reasons to separate the index of the process X in the above work as $\mathbb{R}^+ \times \mathbb{R}^d$ is to treat the first component as time and all the stochastic differentials (or integrals) are defined relative to this one-dimensional parameter, and the existing theory suffices. If it is multidimensional, there are several new problems and multiple stochastic integration and its properties are needed. The second chapter, written by Green, addresses this problem if the index is a two-dimensional parameter, and hence planar stochastic integrals are studied. Such integrals relative to a Brownian sheet have been developed in a fundamental paper by Cairoli and Walsh (1975), and for an extended analysis in SDEs these have to be obtained for more general integrators than Brownian sheets that include, for instance, (sub)martingales. To take into account all such cases, Green develops a planar integration relative to quasimartingales, using a generalized boundedness principle, originally due to Bochner (1955). An extension of the work of Cairoli and Walsh presents several problems involving conditions that are automatic in the Brownian case, but must be suitably formulated to have the desired generality.

The extended Bochner boundedness principle may be stated as follows. Let $X : [a, b] \to L^p(P)$, $p \geq 0$ be a (random) function, $\mathcal{O} \subset \mathcal{B}([a, b]) \otimes \Sigma$ a

σ-subalgebra, $\varphi_i : \mathbb{R} \to \mathbb{R}^+, i = 1, 2$, increasing functions, and $\alpha : \mathcal{O} \to \bar{\mathbb{R}}^+$ a σ-finite measure. Then X is said to be L^{φ_1, φ_2}-*bounded relative to* $\mathcal{O}$ and α, if there exists some constant $K(= K_{\varphi_1, \varphi_2, \mathcal{O}, \alpha} > 0)$ such that for each simple function $f = \sum_{i=0}^{n-1} a_i \chi_{(t_i, t_{i+1}]}$, $a \leq t_0 < t_1 < \cdots < t_n \leq b$ one has:

$$E(\varphi_2(\tau(f))) \leq K \int_{\Omega \times [a,b]} \varphi_1(f) d\alpha, \ \tau(f) = \sum_{i=0}^{n-1} a_i(X(t_{i+1}) - X(t_i)). \quad (4)$$

If $\varphi_1(x) = \varphi_2(x) = x^2$, then X is termed $L^{2,2}$-bounded. For a Brownian motion X, one can take $\mathcal{O} = \mathcal{B}([a, b]) \otimes (\emptyset, \Omega)$, and $\alpha = Leb. \otimes P$, so that X is $L^{2,2}$-bounded (with $K = 1$ and equality in (3)). It can also be shown that a square integrable submartingale is $L^{2,2}$-bounded with $\mathcal{O}$, a predictable σ-subalgebra and α a suitable σ-finite measure on $\mathcal{O}$, for (4). Thus an L^{φ_1, φ_2}-bounded X qualifies to be a stochastic integrator and a dominated convergence theorem is valid. Green considers planar integrals relative to such general integrators, namely quasimartingales, and extends most of the Cairoli-Walsh theory under a suitable condition called "cross-term domination". This is satisfied not only for Brownian sheets, but also for square integrable two parameter martingales as well as many quasimartingales. Then he develops stochastic line and surface integrals, a Fubini type theorem, and moreover a (stochastic) Green theorem. Further a definition of stochastic partial differential is given. These results will be of considerable interest in a study of SPDEs in accordance with Walsh's (1984) account. All the basic work with complete details for the latter study is thus contained in this somewhat long article.

3. A different approach to stochastic modeling via probability metrics is the subject of Chapter 3 by Rachev, Haynatzka and Haynatzki. This is particularly useful for studies of large sample behavior and weak convergence of processes. These metrics are of two kinds, namely those depending on distributions of single random variables and those depending on joint distributions of two or more random variables. Typical examples are the classical Lévy and Fréchet metrics, i.e., if X, Y are the random variables with distributions F_X, F_Y then the Lévy metric is:

$$L(X, Y) = \inf\{\varepsilon > 0 : F_X(x - \varepsilon) - \varepsilon \leq F_Y(x) \leq F_X(x + \varepsilon), x \in \mathbb{R}\}, \quad (5)$$

and the Fréchet or convergence in probability metric is:

$$d(X, Y) = \int_\Omega \frac{|X - Y|}{1 + |X - Y|} dP \ (= \int_{\mathbb{R}^2} \frac{|x - y|}{1 + |x - y|} dF_{X,Y}(x, y)). \quad (6)$$

These two cases are generalized to wide classes of probability (semi)metrics. Keeping the applications in mind the triangle inequality is weakened. Thus if $\rho : S \times S \to \mathbb{R}^+$ is a mapping, let

$$(i)\rho(x,y) = 0 \Leftrightarrow (\Leftarrow)x = y, (ii)\rho(x,y) = \rho(y,x),$$
$$(iii)\rho(x,y) \leq k[\rho(x,z) + \rho(y,z)], \forall x,y,z \in S, \text{ and some } k \geq 1.$$

This ρ is termed a *quasi-(semi)metric*. Further one can use some ideas of abstract analysis and define several classes of such metrics. For instance, if $\varphi : \mathbb{R} \to \mathbb{R}^+$ is a generalized Young function satisfying a Δ_2-condition for large values, then define for any random variables X, Y with values in a metric space (S, d):

$$\rho_\varphi(X,Y) = \int_\Omega \varphi(d(X,Y))dP \tag{7}$$

or

$$\tilde{\rho}_\varphi(X,Y) = \inf\{\varepsilon > 0 : \varphi(F_X(x) - F_Y(x + \varepsilon)) < \varepsilon,$$
$$\varphi(F_Y(x) - F_X(x + \varepsilon)) < \varepsilon, \ x \in \mathbb{R}\}. \tag{8}$$

All these and their generalizations play important roles in the (weak) convergence of stochastic processes. Analysis with many of these (quasi)metrics has been extensively carried out by many people, including V. M. Zolotarev and especially Rachev himself together with several collaborators. In this paper first a very readable account of probability metrics, their numerous applications, and an extensive bibliography are included. Also the authors make a novel application of these metrics in obtaining some limit theorems for two problems in the spread of AIDS. This is done first for discrete time, and then the results are extended to continuous time by an approximation where the model now consists of a system of stochastic differential equations depending on a parameter N, for a fixed number (here 4) of communities. Then as $N \to \infty$, they establish that the number of infectives in each class converges to the unique solution of a Liouville type SDE. All the assumptions of the model and the underlying methodology are presented. There is an extensive bibliography on related work.

4. Chapter 4, written by me, is devoted exclusively to higher order SDEs. First order equations starting with the Langevin's, have been generalized in the literature to nonlinear equations using the full development of the Itô calculus and its extension to square integrable martingales. However, the second and higher order equations, which similarly start with the motion of a simple harmonic oscillator, have not received a corresponding treatment. The linear constant coefficient case has been studied by Dym (1966) who noted that, with white noise process as driving force, they exhibit special

features that are not seen in the first order case. He discussed the existence
of solutions of an equation of the form

$$a_0 \frac{d^n X(t)}{dt^n} + a_1 \frac{d^{n-1} X(t)}{dt^{n-1}} + \cdots + a_n X(t) = \varepsilon(t), \qquad (9)$$

with a_i as constants, written symbolically, but may be interpreted rigorously
in integrated form as:

$$\int_{\mathbb{R}} \varphi(t)(L_n X(t)) dt = \int_{\mathbb{R}} dB(t), \qquad (10)$$

where $L_n = a_0 \frac{d^n}{dt_n} + \cdots + a_n$ is a differential opretor, φ a smooth function
of compact support, and $dB(t) = \varepsilon(t) dt, B(t)$ being Brownian motion. The
existence of a distributional solution can be established and shown to be
Markovian. The sample path analysis uses this property through semi-
group theory, but now the associated infinitesimal generator of the latter is
found to be a degenerate elliptic differential operator. This is a characteristic
feature of the higher order cases.

To consider the corresponding nonlinear problem, it is necessary to define
the derivative of X in the sense of mean and present conditions in order
to interpret it in the pointwise sense as well. Only then the higher order
SDEs can be studied. This and a generalization of the linear time dependent
coefficient case (i.e., the a_i of (9) are $a_i(t)$ now) are treated in some detail in
this chapter. Considering the second order case, for simplicity, the nonlinear
form of (9) is studied:

$$d\dot{X}(t) = q(t, X(t), \dot{X}(t)) dt + \sigma(t, X(t), \dot{X}(t)) dB(t), \qquad (11)$$

where q, σ are coefficients satisfying certain Lipschitz type conditions. The
desired solution is an absolutely continuous process, for the driving force
$\{B(t), t \geq 0\}$ which is now taken as an $L^{2,2}$-bounded process in the sense
of (3) above. In this form it is shown that (11) has a unique absolutely
continuous solution, and if moreover the $B(\cdot)$-process has independent increments then the solution is also Markovian. Next the $B(\cdot)$ is specialized to
Brownian motion and an analysis of the associated semigroup is presented.
This time one has to consider weighted continuous function space as the
domain of this semigroup which however is not strongly continuous. The
infinitesimal operator is again a degenerate elliptic operator with coefficients
as functions of t, x, y. Analysis of these PDEs is difficult and many problems are pointed out for future investigations. A few results on the path
behavior are then given. This work leads to multiparameter analogs (the
solutions being random fields) and the SPDEs wherein the ideas and results
of Chapter 2 can be utilized. But much of this remains to be explored.

5. An analysis of random processes and fields, which need not be Brownian but are square integrable, is considered in Chapter 5 by Swift. Most of the work here is related to several generalizations of (weakly) stationary processes and fields using the Fourier analysis methods and ideas. The central class is what is known as the harmonizable family. Thus a family $\{X_t, t \in \mathbb{R}^n\}$ is a harmonizable random field if it can be expressed as:

$$X_t = \int_{\mathbb{R}^n} e^{t \cdot \lambda} dZ(\lambda), \tag{12}$$

where $Z(\cdot)$ is a random measure on (the Borel σ-ring of) $\mathbb{R}^n$ with values in $L^2(P)$. Here one takes $E(X_t) = 0 = E(Z(\lambda))$ for simplicity. If $F(\lambda, \lambda') = E(Z(\lambda)\bar{Z}(\lambda'))$, then F defines a bimeasure. If F has finite Vitali variation, then X_t is called *strongly harmonizable*, and *weakly harmonizable* field otherwise. If moreover the covariance $r(\mathbf{s}, \mathbf{t}) = E(X_\mathbf{s}\bar{X}_\mathbf{t})$ is unchanged under rotations then X_t is *isotropic*. It is also possible to consider the increments of the process (field) to have the harmonizability and isotropy properties. In fact most of the questions investigated in the stationary case can be asked for the harmonizable families (cf., e.g., Yaglom (1987)). It should be noted that the covariance representation

$$r(\mathbf{s}, \mathbf{t}) = \iint_{\mathbb{R}^n \times \mathbb{R}^n} e^{i\mathbf{s} \cdot \lambda - i\mathbf{t} \cdot \lambda'} dF(\lambda, \lambda') \tag{13}$$

is in the Lebesgue sense only when F has finite Vitali variation. Otherwise one has to use a weaker Morse-Transue integration, suitably modified. The former is used in the strongly harmonizable case, and the latter for the weakly harmonizable ones. Now the structure of both these classes is discussed in detail in this chapter. Also treated are periodically correlated processes, local continuity, and almost periodicity of the sample paths. A recent general account of the basic theory of such processes, in multidimensions, is given by Kakihara (1997), and the current chapter essentially complements it. A large part on fields generalizes the work in Yadrenko (1983).

The analyticity of harmonizable random fields, and sampling theorems as well as the Cramér classes are treated here. Also discussed are integral (spectral) representations of fields with m^{th}-order increments of this class. Further harmonizable spacially isotropic random fields are treated. Some multidimensional extensions are briefly described. Several problems in the area that await future investigation are pointed out at various places. The account gives a comprehensive view of the area along with an extended bibliography for related studies.

6. The final chapter is devoted to the Gaussian dichotomy problem by Vakhania and Tarieladze. There are several proofs of the dichotomy theorem

in the literature. Here the authors present a simpler argument than before, avoiding any usage of conditioning which appears at least in invoking the martingale convergence before. An interesting aspect now is that the result is reduced to Kakutani's (1948) theorem on infinite product measures. The authors reformulated this result, for uncountable sets of probability measures, to use it in their dichotomy proof. Some refinements of others' works are also obtained from these ideas.

The authors then proceeded to present simple proofs of the assertions of the following type for measures on general locally convex topological vector spaces (LCTVSs). For instance, let $\mathcal{X}$ be a real LCTVS and $G \subset \mathcal{X}^*$ be a set of continuous linear functionals that separate points of $\mathcal{X}$. If $\sigma(G)$ is the smallest σ-algebra of $\mathcal{X}^*$ relative to which all elements of G are measurable, then a pair of Gaussian measures on $\sigma(G)$ (i.e., $\mu \circ f^{-1}, \nu \circ f^{-1}$ are Gaussian probabilities on $\mathbb{R}$ for each $f \in G$) satisfy the dichotomy theorem. Thereafter conditions in terms of mean and covariance operators for equivalence of μ, ν are presented. This work gives a fresh approach to an old problem, and harvests some consequences.

It is thus evident that real analysis methods play a fundamental role in all the works presented here. Moreover, there are places where the stochastic theory raises new questions of abstract analysis such as for SPDEs, not strongly continuous semigroup study, and degenerate elliptic operator theory itself. One expects that this interaction will help advance both subject areas as well as their applications.

Bibliography

[1] S. Bochner (1956), "Stationarity, boundedness, almost periodicity of random valued functions," *Proc. Third Berkely Symp. Math. Statistist. and Probab.*, **2**, 7–27.

[2] R. Cairoli and J. B. Walsh (1975), "Stochastic integrals in the plane," *Acta Math.*, **134**, 111–183.

[3] Ju. L. Dalecky and M. G. Krein (1974), *Stability of Solutions of Differential Equations in Banach Spaces*, Am. Math. Soc., Providence, RI (translation).

[4] H. Dym (1966), "Stationary measures for the flow of a linear differential equation driven by white noise," *Trans. Am. Math. Soc.*, **123**, 130–164.

[5] Y. Kakihara (1997), *Multidimensional Second Order Stochastic Processes*, World Scientific Publishers, Singapore, (in press).

[6] S. Kakutani (1948), "On equivalence of infinite product measures," *Ann. Math.*, **49**, 214–224.

[7] J. B. Walsh (1984), "An introduction to stochastic partial differential equations," *Springer Lect. Notes in Math.*, **1180**, 268–439.

[8] M. I. Yadrenko (1983), *Spectral Theory of Random Fields*, Optimization Software Inc., New York.

[9] A. M. Yaglom (1987), *Correlation Theory of Stationary and Related Random Functions: Basic Results*, Springer-Verlag, New York.

Chapter 1

Transport Properties of Gaussian Velocity Fields

René A. Carmona[1]

Abstract

The purpose of these lecture notes is to describe several mathematical problems which arise in the study of the statistical properties of the solutions of the equation:

$$dX_t = \vec{v}(t, X_t)dt + \sqrt{2\kappa}dB_t$$

when $\{\vec{v}(t, \mathbf{x}); \ t \geq 0, \mathbf{x} \in \mathbb{R}^d\}$ is a mean zero stationary and homogeneous Gaussian field and $\{B_t; \ t \geq 0\}$ a process of Brownian motion. We are mostly interested in velocity fields with spectra of the Kolmogorov type. The study is motivated by problems of transport of passive tracer particles at the surface of a two-dimensional medium. We are mostly concerned with the mathematical analysis of problems from oceanography and we think of the surface of the ocean as a physical medium to which our modeling efforts could apply.

1.1 Introduction

The purpose of these notes is to present in a more or less informal manner a set of mathematical problems which arise in the study of the statistical properties of the solutions of the equation:

---
[1]Partially supported by ONR N00014-91-1010

$$dX_t = \vec{v}(t, X_t)dt + \sqrt{2\kappa}dB_t \qquad (1.1)$$

when $\{\vec{v}(t, \mathbf{x}); \ t \geq 0, \mathbf{x} \in \mathbb{R}^{\mathbf{d}}\}$ is a mean zero stationary and homogeneous Gaussian field, $\kappa \geq 0$ and $\{B_t; \ t \geq 0\}$ is a d-dimensional process of Brownian motion. Except for the last section, in which we discuss stochastic partial differential equations (SPDE for short), we shall restrict ourselves to the case $\kappa = 0$. Moreover, most of our efforts will be devoted to the case of velocity fields with spectra of the Kolmogorov type. This assumption is motivated by problems of fluid mechanics. Instead of considering velocity fields which are solutions of the Navier-Stokes equation, we use the dynamical approach and assume from the beginning that $\{\vec{v}(t, \mathbf{x})\}$ is a stationary and homogeneous random field. According to Kolmogorov's theory of well-developed turbulence (i.e., for systems with high Reynolds numbers), this assumption is well founded. See, for example, Chapters 6 and 7 in [36] for an excellent account of this theory in a modern perspective. We shall add the assumption that the velocity field is Gaussian. As proven by the results of many wind velocity measurements, this is a very reasonable assumption.

Our main concern is the analysis of the transport of passive tracers at the surface of a two-dimensional medium. We are mostly interested in the mathematical modeling of problems from oceanography and we think of the surface of the ocean as a physical medium to which our modeling efforts could apply. For this reason we shall sometimes use the terminology *drifters* or *floats* for the passive tracers.

Transport properties of time-independent velocity fields have been studied both from a theoretical point of view and via computer simulations. The results have been reported in many publications. See, for example, the recent works [34] or [35]. The latter cannot be compared to ours because the time-independence of the velocity field drastically changes the nature of the simulation algorithms and the typical properties of the tracers. At this stage it is important to emphasize the differences between the two approaches. It is very often the case that the terminology *disordered systems* or *random media* is used for models in which the randomness is autonomous (i.e., time independent). The analysis of such systems is most naturally done by first studying the properties of the randomness and exhibiting properties of the random parameters of the model, which almost surely hold. Then the mathematical analysis is performed in a very classical manner, by fixing the values of the environment and then studying the system as if it were deterministic. This approach is rarely possible for time-dependent random models. Indeed, probabilistic-like arguments are needed throughout the analysis.

Closer to our point of view are the theoretical results obtained and the numerical simulations performed in the case of Brownian flows. See [47], [7],

[62], and [25]. The main difference with the models we want to study is the independence of the time increments. Indeed, this assumption does not hold in the case of time-dependent velocity fields with a Kolmogorov spectrum. Brownian models are based on velocity fields which are "white noise" in the time variable while the fields which we are considering are superpositions of Ornstein-Uhlenbeck processes in time. So our velocity fields are more regular in time. This seems to be a desirable property except for the fact that we shall have to face the difficulties caused by the absence of the tools of Ito's stochastic calculus. In fact, the Brownian model can be considered as a diffusion approximation limit of some of the more realistic models which we consider here. See [4] and/or [14] or [54].

These notes are organized as follows. Section 1.2 contains the details of the theoretical models which we consider for the velocity field. We restrict ourselves to stationary (in time) and homogeneous (in space) mean zero Gaussian velocity fields. We shall concentrate most of our efforts on isotropic fields having a spectrum of the Kolmogorov type. The other examples which we present can be regarded as testbeds for the results we are interested in and the methods of attack which we propose. Section 1.3 is devoted to the review of relevant facts about infinite-dimensional Ornstein-Uhlenbeck (O-U for short) processes. We show how the velocity fields with a Kolmogorov spectrum can be viewed as infinite-dimensional O-U processes. This point of view proves to be very useful for the numerical simulations. The transport properties which we are interested in are presented in Section 1.4. We use illustrations from the results of numerical simulations performed on a MASsively PARallel machine. Section 1.5 explains how the numerical simulations have been done. We review some of the explanations given in [15] concerning the program which we developed to analyze the transport properties of the O-U velocity fields. The mathematical conjectures suggested by the numerical simulations are proved in Section 1.6 in the particular case of Brownian flows. Finally, Section 1.7 contains a short discussion of some of the many open problems concerning the transport properties of Ornstein-Uhlenbeck velocity fields. We end these lecture notes with a short review of some of the connections of the analyses of stochastic flows as presented in these notes to the theory of some parabolic SPDEs. It is clear that we cannot do justice to the field of SPDEs in such a short section, but we thought that it was important to recast the theory of transport properties of passive scalars to the analysis of the time evolution of concentrations. Again, it is important to emphasize that our point of view has strongly been biased toward the problems of potential applications to oceanographic modeling.

The lectures on which this paper is based were given at the Rennes SMF School on Random Media, October 10-12, 1994. Since then, significant progress has been made and some of the questions raised during the lectures have been answered. The original version of the lecture notes was updated during the summer of 1996 to account for the existence of new results and the bibliography has been updated accordingly.

1.2 Gaussian Velocity Fields with Kolmogorov Spectra

We present the assumptions of a commonly used mathematical model of a random velocity field. We assume that $\{\vec{v}(t, \mathbf{x}); \ t \geq 0, \mathbf{x} \in \mathbb{R}^d\}$ is a mean zero Gaussian vector field with values in $\mathbb{R}^d$. The hypothesis on the mean is chosen because we are merely interested in the random fluctuations around a mean deterministic motion.

We assume that the velocity field is homogeneous in space. By this we mean that its distribution is invariant under the shifts of the space variable $\mathbf{x}$. In other words, for each fixed $\mathbf{y} \in \mathbb{R}^d$:

$$\{\vec{v}(t, \mathbf{x} + \mathbf{y}); \ t \geq 0, \mathbf{x} \in \mathbb{R}^d\} \qquad \text{and} \qquad \{\vec{v}(t, \mathbf{x}); \ t \geq 0, \mathbf{x} \in \mathbb{R}^d\}$$

have the same distribution. We also assume that the random velocity field is stationary, i.e., its distribution is invariant under the shifts of the time variable t. This means that for each fixed $s \geq 0$:

$$\{\vec{v}(s + t, \mathbf{x}); \ t \geq 0, \mathbf{x} \in \mathbb{R}^d\} \qquad \text{and} \qquad \{\vec{v}(t, \mathbf{x}); \ t \geq 0, \mathbf{x} \in \mathbb{R}^d\}$$

have the same distribution. These three assumptions (Gaussian character, homogeneity and stationarity) are quite common in the random mathematical models of turbulent fluids. See [61], [6], and [50] for example.

The distribution of the random field is completely characterized by its covariance:

$$\mathbb{E}\{\vec{v}(s, \mathbf{x}) \otimes \vec{v}(t, \mathbf{y})\} = \Gamma(t - s, \mathbf{y} - \mathbf{x}), \qquad s, t \geq 0, \quad \mathbf{x}, \mathbf{y} \in \mathbb{R}^d. \quad (1.2)$$

If Φ is a random variable (or more generally a random vector) over the probability space of the velocity field, we use the notation $\mathbb{E}\{\Phi\}$ for its expectation. Notice that $\Gamma(t, \mathbf{x})$ is a $d \times d$ matrix. Its entries are given by the covariances of the components of $\vec{v}$:

$$\Gamma_{j,\ell}(t, \mathbf{x}) = \mathbb{E}\{v_j(0, 0) \otimes v_\ell(t, \mathbf{x})\}, \qquad j, \ell = 1, \cdots, d.$$

1.2.1   Spectral Representation

Because of its positive definiteness (as a function of t and $\mathbf{x}$), the matrix $\Gamma(t, \mathbf{x})$ is the Fourier transform of a nonnegative definite matrix measure $\nu(d\omega, d\mathbf{k})$. We use ω to denote the frequency variable associated to the time variable t and we use the variable $\mathbf{k}$ for the wave number variable associated to the space variable $\mathbf{x}$. This notation may be surprising to some probabilists but we decided to use it because it is a standard convention in fluid dynamics. We also assume that the entries of the covariance matrix are integrable. This guarantees the existence of densities $E_{j,\ell}(\omega, k)$ for the entries $\nu_{j,\ell}(d\omega, dk)$ of the matrix measure $\nu(d\omega, dk)$. In other words we have:

$$\Gamma(t, x) = \int_{I\!R} \int_{I\!R^d} e^{i(\omega t + \mathbf{k}\cdot\mathbf{x})} E(\omega, \mathbf{k}) d\omega \, d\mathbf{k} \qquad (1.3)$$

for some matrix $E(\omega, \mathbf{k}) = [\mathbf{E}_{j,\ell}(\omega, \mathbf{k})]_{j,\ell=1,\cdots,d}$. Further assumptions on the spatial characteristics of the velocity field, such as *incompressibility* and *isotropy* can be used to specify further the covariance $\Gamma(t, \mathbf{x})$. This is usually done in terms of the form of its Fourier transform $E(\omega, \mathbf{k})$. See for example [6], [50], [54], [61], and/or [4]. The assumptions of isotropy and incompressibility essentially determine the form of the spatial part of the covariance. But it might also be desirable to build the structure of the time dependence and the time correlation into the model. In particular one may want to make the velocity field Markovian in time. In this case we get:

$$E_{j,\ell}(\omega, \mathbf{k}) = \frac{\beta(|\mathbf{k}|)}{\omega^2 + \beta(|\mathbf{k}|)^2} \mathcal{E}(|\mathbf{k}|) \frac{1}{|\mathbf{k}|^{d-1}} \left(\delta_{j,\ell} - \frac{k_j k_\ell}{|\mathbf{k}|^2} \right) \qquad (1.4)$$

One of the major assumptions of the 1941 Kolmogorov theory of fully developed turbulence (i.e., theory applying to systems with high Reynolds numbers) is a form of statistical scale invariance of the system. For this reason, the function $\mathcal{E}(\nabla)$ should be a power function. But such a function would be such that the spectral density cannot be integrable at infinity and near the origin simultaneously. This lack of integrability would force the velocity field to be some sort of generalized field whose sample realizations would be Schwartz distributions instead of functions. Since the scale invariance of the Kolmogorov theory is limited to the so-called inertial range, the function $\mathcal{E}(\nabla)$ is usually chosen to be zero near the origin (i.e., for $r \leq r_0$ for some $r_0 > 0$) and near infinity (i.e., for $r \geq r_1$ for some $r_1 < \infty$) and to behave like a power, say $\mathcal{E}(\nabla) = \nabla^{\infty - \epsilon}$ in between. The value of the parameter ϵ controls the spatial correlation in the velocity field. In the classical Kolmogorov's theory of turbulence, the numbers r_0 and r_1 are related to the so-called integral and dissipation scales, respectively.

The function $\beta(r)$ is usually chosen of the form:

$$\beta(r) = ar^z$$

for some $z > 0$. In this way, the parameter z controls the time correlation in the velocity field. As a consequence, a Kolmogorov spectrum can be characterized by the two parameters z and ϵ. Finally, we recall the spectral representation formula:

$$\vec{v}(t, x) = \int_{I\!R} \int_{I\!R^d} e^{i(\omega t + \mathbf{k} \cdot \mathbf{x})} E(\omega, \mathbf{k})^{1/2} \mathbf{W}(d\omega, d\mathbf{k}), \qquad (1.5)$$

which writes the random velocity field in terms of a complex vector white noise measure (in the L^2-sense) $W(d\omega, d\mathbf{k})$. We shall not make use of this formula in the proofs provided in these lectures. But we shall use it to introduce the Ornstein-Uhlenbeck models and to justify the form of the numerical simulations whose results are reported below.

1.2.2 Special Cases

We now introduce special cases which will be used in the sequel for illustration purposes. By dropping some of the requirements in the derivation of the Kolmogorov spectrum or by concentrating on some specific asymptotic regimes, it is possible to justify the introduction of simpler mathematical models. The analyses of these models, though simpler, will shed light on the nature of the transport properties of the more general stochastic flows.

Shear Flows

The analysis of shear flows leads to what is presumably the simplest possible example of an incompressible velocity field. This model is essentially one-dimensional. Let us assume that $d = 2$ and let us consider the velocity field $\{\vec{v}(t, \mathbf{x})\}$ defined by:

$$\vec{v}(t, \mathbf{x}) = \begin{bmatrix} 0 \\ v(t, x_1) \end{bmatrix}, \qquad t \geq 0, \quad \mathbf{x} = (x_1, x_2) \in \mathbb{R}^2. \qquad (1.6)$$

where $\{v(t, x);\ t \geq 0,\ x \in \mathbf{R}\}$ is a real-valued mean zero stationary and homogeneous Gaussian field. Obviously, the divergence of this vector field is 0 so the incompressibility condition is satisfied. On the other hand, the isotropy condition is not satisfied. If needed (see for example the series of works of Avellaneda and Majda starting from [4]), one can also assume that the spectrum of this field is of the Kolmogorov type in the sense that the spectral density $E(\omega, k)$ is given by a formula of the form:

$$E(\omega, k) = \frac{\beta(|k|)}{\omega^2 + \beta(|k|)^2} \mathcal{E}(|k|), \qquad \omega \in \mathbf{R}, k \in \mathbf{R}, \qquad (1.7)$$

and where the functions $\mathcal{E}(\nabla)$ and $\beta(r)$ are as before. If we restrict our study to nonviscous models for which $\kappa = 0$, i.e., we did not include a Brownian motion term in the equation of motion (1.1), the latter can be integrated explicitly. We get:

$$x_1(t) = x_{1,0}, \qquad \text{and} \qquad x_2(t) = x_{2,0} + \int_0^t v(s, x_{1,0})ds, \quad (1.8)$$

if we use the notation $X_t = (x_1(t), x_2(t))$. We shall use the explicit form of the flow given by (1.8) to illustrate some of the points discussed later. It is important to warn the reader that our use of the shear flow might be deceiving because of its simplicity. Indeed, these models are no longer trivial in the presence of a diffusion term, i.e., when the equation of motion is of the form:

$$dX_t = \vec{v}(t, X_t)dt + \sqrt{2\kappa}dB_t$$

for some standard Wiener process $\{B_t\}_{t \geq 0}$ and a positive constant $\kappa > 0$. In this case, Avellaneda and Majda have shown that the asymptotic analysis of this model was very rich. They study the possible renormalizations of the model when the variance of the Wiener process and the inertial range cut-off r_0 tend to 0 simultaneously. We shall review some of their results in the last section.

Manhattan Flows

We introduce this model as a testbed for some of the consequences of the last of the isotropy condition. The shear flows discussed above represent too drastic a departure from the models with rich invariance properties which we are interested in. As an intermediate alternative we propose to use the velocity field:

$$\vec{v}(t, \mathbf{x}) = \begin{bmatrix} v_1(t, x_2) \\ v_2(t, x_1) \end{bmatrix}, \qquad t \geq 0, \quad \mathbf{x} = (x_1, x_2) \in \mathbb{R}^2. \quad (1.9)$$

for some real-valued mean zero stationary homogeneous Gaussian and independent fields $\{v_1(t, x); \ t \geq 0, \ x \in \mathbb{R}\}$ and $\{v_2(t, x); \ t \geq 0, \ x \in \mathbb{R}\}$. Again, the incompressibility condition is obviously satisfied and the group of invariance is now richer. Unfortunately, the equations of motion do not lead to explicit solutions any longer, and the mathematical analysis is still challenging despite the simplicity of the model.

The terminology Manhattan flow is borrowed from the study [3] of a similar model in the time-independent case. But the flow lines (and the behavior of the solutions) are so different in the time-independent case that this justification is very weak and a better name should be found!

Brownian Flows

We now return to the general d-dimensional case, but we consider a limiting regime in the spirit of the diffusion approximation.

Let us assume, for example, that the spectrum of the velocity field is of the Kolmogorov type described above and that the inertial range r_0 is greater than 1. In other words, we assume that the support of the function $\mathcal{E}(\nabla)$ is contained in the interval $(1, \infty)$. We suspect that this assumption can be satisfied in all cases by an appropriate choice of physical units and *nondimensionalization* of the equation. But our ignorance of the "unit ritual" prevents us from being more assertive at this stage. In any case, as mathematicians, we can always make this assumption to justify the study of a specific model. Indeed, if we let the parameter z tend to ∞, then the velocity field converges toward a Gaussian field which has the distribution of a white noise in time. Using the techniques of Chapter 5 of [45] (see also [14]), it is presumably possible to show that the corresponding flow converges toward the stochastic flow generated by the limiting velocity field. Consequently, we shall study the transport properties of this limiting flow as a warm-up, but also – let's be quite candid – because it is one of the very few models for which one can give complete answers to all the mathematical questions in which we are interested.

We shall say that we are considering a white velocity field whenever the covariance of $\{\vec{v}(t, \mathbf{x})\}$ is of the form:

$$\Gamma_{j,\ell}(t, \mathbf{x}) = \delta_0(t)\Gamma_{j,\ell}^{(0)}(\mathbf{x})$$

where $\Gamma^{(0)} = [\Gamma_{j,\ell}^{(0)}]_{j,\ell=1,\cdots,d}$ is the covariance tensor of a mean zero homogeneous Gaussian field on $\mathbf{R}^d$ having desirable invariance properties (such as isotropy and/or incompressibility, for example). It is clear that this definition is only formal because of the delta function in the covariance. A precise mathematical meaning can be given by first defining the antiderivative, which is usually introduced as a Brownian motion. This will be done in Section 1.6 but, for the time being, our definition is precise enough to justify the terminology of Brownian flows.

1.3 Infinite-Dimensional Ornstein-Uhlenbeck Processes

The main thrust of our approach is to view the random velocity fields with Kolmogorov spectra as infinite-dimensional Ornstein-Uhlenbeck (O-U for short) processes. Before we review the properties of the latter, we justify our point of view by revisiting the formulas (1.3) and (1.4) for the spectral representation of the covariance tensor of the field. Performing the integration with respect to ω in (1.3) gives:

$$\Gamma(t, x) = \int_{I\!R^d} e^{ik \cdot x} E(k) \, dk$$

with:

$$E_{j,\ell}(k) = e^{-|k|^\alpha |t|} \mathcal{E}(|k|) \frac{1}{|k|^{d-1}} \left(\delta_{j,\ell} - \frac{k_j k_\ell}{|k|^2} \right).$$

In other words, the covariance is in the Fourier domain the superposition of the covariances of O-U processes and it is natural to conjecture that the spectral representation (1.5) can be rewritten in the form:

$$\vec{v}(t, x) = \int_{I\!R^d} e^{ik \cdot x} \tilde{E}(k)^{1/2} \xi_t(dk). \tag{1.10}$$

where the $\xi_t(dk)$ for each fixed t are, orthogonal (independent) increment L^2-measures in the Fourier domain and for each fixed infinitesimal dk are, O-U processes in the time variable.

Given this motivation, we now review the definitions and the properties of O-U processes. Let us begin with recalling what is usually understood by an O-U process in finite dimensions.

1.3.1   Finite-Dimensional O-U Processes

A one-dimensional O-U process is usually defined as the solution of the stochastic differential equation:

$$d\xi_t = -a\xi_t dt + \sqrt{b}\, dw_t \tag{1.11}$$

where the numbers a and b are positive and where $\{w_t\}_{t \geq 0}$ is a one-dimensional Wiener process. The initial condition, say ξ_0, is (implicitly) assumed to be independent of the driving Wiener process. The solution can be written explicitly in the form:

$$\xi_t = e^{-ta}\xi_0 + \sqrt{b} \int_0^t e^{-(t-s)a} dw_s, \qquad 0 \leq t < +\infty. \tag{1.12}$$

The mean and the covariance of the solution are given by the formulas:

$$\mathbb{E}\{\xi_t\} = e^{-ta}\mathbb{E}\{\xi_0\} \tag{1.13}$$

and

$$\text{Cov}\{\xi_s, \xi_t\} = [\text{Var}\{\xi_0\} + \frac{b}{2a}(e^{2(s \wedge t)a} - 1)]e^{-(s+t)a}. \tag{1.14}$$

$\{\xi_t\}$ is a Gaussian process whenever ξ_0 is Gaussian. The solution process is a Markov process. It has a unique invariant measure, the normal distribution $N(0, b/2a)$. Started with this distribution, the Markov process becomes a stationary ergodic Markov (mean zero) Gaussian process. Its covariance is then given by the formula:

$$\mathbf{E}\{\xi_s \xi_t\} = \frac{b}{2a} e^{-a|t-s|}. \tag{1.15}$$

Let us now review the concept of an n-dimensional (n finite) O-U process, say $\{X_t;\ t \geq 0\}$. As we can see by comparing formulas (1.11 to 1.15) to formulas (1.16 to 1.20) below, in order to go from the definition of a one-dimensional O-U process to the definition of the n-dimensional analog, we need the following substitutions:

$$
\begin{array}{lcl}
a > 0 & \longrightarrow & A \ n \times n \text{ symmetric p. d. matrix} \\
b > 0 & \longrightarrow & B \ n \times n \text{ symmetric p. d. matrix} \\
\{w_t\}_{t \geq 0} \ 1 - \text{dim. Wiener Process} & \longrightarrow & \{W_t\}_{t \geq 0} \ n - \text{dim. Wiener Process}
\end{array}
$$

Indeed, an n-dimensional O-U process can be defined as the solution of a system of stochastic equations written in a vector form as:

$$dX_t = -AX_t dt + \sqrt{B} dW_t \tag{1.16}$$

where $\{W_t\}_{t \geq 0}$ is an n-dimensional Wiener process. More generally, the stochastic system could be driven by an r-dimensional Wiener process, in which case the matrix $\sqrt{B}$ would have to be replaced by a $d \times r$ matrix C and the role of the matrix B in the formulas given below would have to be played by the matrix C^*C. We shall refrain from using this generalization which we will not need.

As before, the solution of the stochastic system can be written explicitly:

$$X_t = e^{-tA} X_0 + \int_0^t e^{-(t-s)A} \sqrt{B} dW_s, \qquad 0 \leq t < +\infty, \tag{1.17}$$

and, if X_0 is assumed to be independent of the driving Wiener process, the mean and the covariance are given by the formulas:

$$\mathbf{E}\{X_t\} = e^{-tA} \mathbf{E}\{X_0\} \tag{1.18}$$

and

$$\text{Cov}\{X_s, X_t\} = e^{-sA} \text{Var}\{X_0\} e^{-tA} + \int_0^{s \wedge t} e^{-(s-u)A} B e^{-(t-u)A} \, du. \tag{1.19}$$

As before, $\{X_t\}$ is a Gaussian process as soon as the initial condition X_0 is Gaussian. We shall always restrict ourself to this case. The process is a strong Markov process. The positivity and the nondegeneracy of the drift matrix A and the dispersion matrix B imply the existence of a unique invariant probability measure, say μ. This invariant measure is the n-dimensional normal distribution $N(0, \Sigma)$ given by the following formula:

$$\mu(dx) = \frac{1}{Z(\Sigma)} e^{-<\Sigma^{-1}x,x>/2} dx,$$

where $Z(\Sigma)$ is a normalizing constant which guarantees that the total mass of μ is one and the matrix Σ is defined as:

$$\Sigma = \int_0^\infty e^{-sA} B e^{-sA}\, ds.$$

Notice that the fact that Σ solves:

$$\Sigma A + A\Sigma = B$$

is the crucial property of the matrix Σ. Started from the invariant measure μ, the process $\{X_t\}_{t\geq 0}$ becomes a stationary mean zero (Markov) Gaussian process. Its distribution is determined by its covariance function:

$$\Gamma(s,t) = \mathbb{E}_\mu\{X_s X_t\} = \begin{cases} \Sigma e^{-(t-s)A} & \text{whenever} \quad 0 \leq s \leq t \\ e^{-(s-t)A}\Sigma & \text{whenever} \quad 0 \leq t \leq s. \end{cases} \quad (1.20)$$

One sees that the formulas can become cumbersome because of the possibility that the matrices A and Σ do not commute. Fortunately, the situation which we consider later deals only with the case of commuting matrices A and B. For this reason we rewrite the above formulas in this special case. In particular, the definition of Σ gives:

$$\Sigma = B \int_0^\infty e^{-2sA}\, ds = B(2A)^{-1}$$

and the equality (1.20) between matrices can be rewritten in terms of the entries of the matrices. More precisely, if $x, y \in \mathbb{R}^n$ we have:

$$< x, \Gamma(s,t)y > = \mathbb{E}_\mu\{< x, X_s >< y, X_t >\} \quad \begin{aligned} &= < \sqrt{\Sigma}x, \sqrt{\Sigma}e^{-|t-s|A}y > \\ &= < x, e^{-|t-s|A}y >_\Sigma \end{aligned}$$

if we use the notation:

$$< \cdot, \cdot >_D = < \sqrt{D}\cdot, \sqrt{D}\cdot > = < D\cdot, \cdot >$$

for the inner product in $\mathbb{R}^n$ naturally associated with a positive definite matrix D. The Markov process $\{X_t\}_{t\geq 0}$ is symmetric in the sense of the theory of symmetric Markov processes developed by Fukushima and his followers. Indeed, a simple integration by parts shows that it is the process associated to the Dirichlet form of the measure μ, i.e., to the quadratic form:

$$Q(f,g) = \int_{\mathbb{R}^n} < \nabla f(x), \overline{\nabla g(x)} >_B \ \mu_\Sigma(dx)$$

defined on the subspace Q of $L^2(\mathbb{R}^n, \mu_\Sigma(dx))$ comprising the absolutely continuous functions whose first derivatives (in the sense of distributions) are still in the space $L^2(\mathbb{R}^n, \mu_\Sigma(dx))$.

1.3.2 Infinite-Dimensional O-U Processes

We now consider the problem of the definition of O-U processes in an infinite-dimensional setup. The three approaches discussed in the finite-dimensional case are possible but, because of our background, we shall have a definite bias for the Gaussian process approach.

- The Euclidean space $\mathbb{R}^n$ is replaced by a (possibly infinite-dimensional) Hilbert space, say H.

- The $n \times n$ dispersion matrix B is replaced by a (possibly unbounded) positive self-adjoint operator which we will still denote by B. We shall need the Hilbert space H_B obtained by completing the domain of the operator $B^{1/2}$ with the inner product:

$$< x, y >_B = < B^{1/2}x, B^{1/2}y >$$

where the inner product in the above right-hand side is the inner product of H.

- The $n \times n$ drift coefficient matrix A is replaced by a (possibly unbounded) positive self-adjoint operator which we will still denote by A.

- The role played by $\sqrt{B}dW_t$ is now played by a H-valued (cylindrical) Wiener process with covariance given by the operator B. The appropriate mathematical object is a linear function, say W_B, from the tensor product $L^2([0,\infty), dt) \hat{\otimes}_2 H_B$ into a Gaussian subspace of $L^2(\Omega, \mathcal{F}, \mathbf{P})$ where $(\Omega, \mathcal{F}, \mathbf{P})$ is the complete probability space we work with. If $x \in H_B$ and $t \geq 0$, then $W_B(1_{[0,t)}(\cdot)x)$ should play the same role as $< x, W_t >$ in the finite-dimensional case.

- **Solution of a SPDE**

In the same way the finite-dimensional O-U processes were defined as solutions to some specific stochastic differential equations, the infinite-dimensional O-U processes can be introduced as solutions of stochastic partial differential equations. Formally we try to solve:

$$\frac{dV(t)}{dt} = -AV(t) + \frac{dW_B}{dt}$$

which can be regarded as a SPDE when A is a partial differential operator. Notice that the dispersion matrix coefficient B is now included in the infinite-dimensional analog of the driving Wiener process. This equation can be given a rigorous meaning by considering the integral equation:

$$< x, V(t) > = < x, V(0) > - \int_0^t < Ax, V(s) > ds + W(\mathbf{1}_{[0,t)}(\cdot)x)$$

which makes sense when x belongs to the domain $\mathcal{D}(\mathcal{A})$ of the operator A. This integral equation can be solved under some restrictive conditions on the data (H, A, B, W_B). See for example [30], [60], or [2] and the references therein.

• Symmetric Process Associated to a Dirichlet Form

The fact that the Markov process $\{X_t\}_{t\geq 0}$ can be characterized as the symmetric process associated to the Dirichlet form of the measure μ can be used in the infinite-dimensional setting as well. As before, let us assume that K is a real separable Banach space containing the Hilbert space H_Σ (recall that the self-adjoint operator Σ is defined as $B(2A)^{-1}$) and such that the canonical cylindrical measure of H_Σ extends into a σ-additive probability measure on K. Notice that this countably additive extension is the measure which we have denoted by μ_Σ so far. The abstract theory of Dirichlet forms gives a construction, starting from the Dirichlet form

$$Q_\mu(f, g) = \int_K < \nabla f(x), \overline{\nabla g(x)} >_B \ \mu_\Sigma(dx)$$

associated to the measure μ_Σ, of a symmetric strong Markov process $\{X_t\}_{t\geq 0}$. This construction is carried out when the dispersion operator B is the identity in the review article [57], but it can be adapted to apply to the general case.

• Gaussian Process

Infinite-dimensional O-U processes can be advantageously constructed as Gaussian processes. See [2]. The construction of the process starting from the origin, say $\{V_0(t)\}_{t\geq 0}$, can be formulated as the definition of a Banach space K and the construction of a mean zero K-valued Gaussian process with continuous sample paths and with covariance function given by:

$$\mathbf{E}\{f(V_0(s))g(V_0(t))\} = \int_0^{s \wedge t} < e^{-(s-u)A}f, e^{-(t-u)A}g >_B du$$

for all the elements f and g of the dual K^* of K. The process $V_v(t)$ starting from a point v of K is then defined by the formula:

$$V_v(t) = e^{-tA}v + V_0(t).$$

It is plain to check that the measure μ_Σ introduced earlier is invariant for this process and that the desired process is the process started from this invariant measure. In other words, the stationary O-U process we are looking for is given as the mean zero (continuous) stationary Gaussian process with covariance:

$$\mathbf{E}\{f(V_0(s))g(V_0(t))\} = < e^{-|t-s|A}f, g >_B .$$

It is clear from the discussion above that, whatever the approach we choose, the construction of the infinite-dimensional O-U process depends upon the construction of the invariant measure μ_Σ. As explained above, this construction relies on the choice of an appropriate (infinite-dimensional) space, say K, on which the cylindrical Gaussian measure μ_Σ can be made countably additive.

One of the main problems discussed below is the positivity of the upper Lyapunov exponent of the Jacobian flow generated by isotropic incompressible velocity fields with Kolmogorov spectra. We shall see that one of the main obstructions to the generalization of existing proofs in the finite-dimensional case is the lack of absolute continuity of various measures. One of the main reasons is that, if one denotes by $P_t(v, dv')$ the transition probability of the Markov process $\{V(t)\}_{t \geq 0}$, then:

$$P_t(v, dv') << \mu_\Sigma(dv')$$

holds only in very restrictive particular cases!!

The works we quoted in this section are only pointing to a small number of elements in a very extensive bibliography on infinite-dimensional O-U processes. Nevertheless, there is still need for work on the understanding of some of the properties of these processes. This should be clear from the discussion of the next section and the difficulties we shall encounter later in the analysis of the Lyapunov exponents of the Kolmogorov flows.

1.3.3 Kolmogorov Velocity Fields as O-U Processes

We show in this section how the velocity fields introduced in Section 1.2 can be recast in the framework of infinite-dimensional O-U processes.

Since the distribution of the velocity field is given in terms of its spectral characteristics, the various spaces introduced above in the general setting of infinite-dimensional O-U processes are best identified in terms of the Fourier transforms of their elements. We start with $H = L^2(\mathbf{R}^d, dx)$ and we identify this Hilbert space with the Hilbert space $L^2(\mathbf{R}^d, d\mathbf{k})$ by means of the unitary Fourier transform. In this way, the dispersion operator B can be identified with the operator of multiplication by the function $\mathcal{E}(|\mathbf{k}|)^{-1/2}$. For this reason the space H_B can be identified (still modulo Fourier transform) with the weighted L^2-space $L^2(\mathbf{R}^d, \mathcal{E}(|\mathbf{k}|)^{-1}d\mathbf{k})$. Notice that the elements of this space have Fourier transforms which have to vanish outside the set where the function $\mathcal{E}$ vanishes. This implies that the elements of H_B are analytic functions of exponential type. This does not say much (yet) about the space K on which the measure μ_Σ can be made countably additive, and by the same token, of the almost sure sample path properties of the random elements $V(t)$ when regarded as functions of the space variable $\mathbf{x}$. In fact, the abstract form of the Karhunen-Loeve expansion of Gaussian processes implies that a typical element $V(t)$ must be the sum of infinite series of random elements of H_B, the convergence being locally uniform. This implies that the $V(t)$s are also analytic functions.

We now identify the operator A on the Hilbert space H. Because A should be the operator of multiplication by the function $|\mathbf{k}|^z$ in the Fourier domain (i.e., modulo the unitary equivalence given by the Fourier transform), it is clear that A is the operator $(-\Delta)^{z/2}$ given by the functional calculus of self-adjoint operators.

Consequently, the evolution semigroup operators $e^{-|t-s|A}$ which appear often in the previous discussion are merely the operators of multiplication by the function $e^{-|t-s||\mathbf{k}|^z}$.

1.4 Transport Problems

Throughout this section we consider a mean zero stationary and homogeneous Gaussian field $\{\vec{\mathbf{v}}(t, \mathbf{x}); \ t \geq 0, \ \mathbf{x} \in \mathbf{R}^d\}$ with a continuous bounded and integrable covariance $\Gamma(t, \mathbf{x})$. For each sample realization of the field we consider the equation of motion:

$$dX_t = \vec{\mathbf{v}}(t, X_t)dt.$$

For each "initial condition" $X_0 = \mathbf{x} \in \mathbf{R}^d$ there exists a unique solution. We denote by $X_t(\mathbf{x}) = \varphi_{0,t}(\mathbf{x})$ its value at time t. $\varphi_{0,t}(\cdot)$ is a diffeomorphism of $\mathbf{R}^d$ for each $t \geq 0$. For $s, t \geq 0$ we set:

$$\varphi_{s,t} = \varphi_{0,t} \circ \varphi_{0,s}^{-1}$$

and more generally

$$\varphi_{s,t} = \varphi_{t,s}^{-1}$$

for all $s, t \in \mathbb{R}$, and with this proviso we have:

$$\varphi_{s,u} = \varphi_{s,t} \circ \varphi_{t,u} \tag{1.21}$$

for all real numbers s, t, and u.

1.4.1 One-Point Motions and Lagrangian Observations

The one-point motions of the system are defined as the $\mathbb{R}^d$-valued stochastic processes $\{\varphi_{0,t}(\mathbf{x})\}_{t \geq 0}$ when the initial position $\mathbf{x}$ is specified.

Let us consider, for example, the case of the shear flow introduced earlier. In this case $d = 2$ and as before we use the notation $\mathbf{x} = (\mathbf{x}_1, \mathbf{x}_2)$. Let us denote by $\varphi_{0,t}^1(\mathbf{x})$ and $\varphi_{0,t}^2(\mathbf{x})$ the two components of $\varphi_{0,t}(\mathbf{x})$. We saw that:

$$\varphi_{0,t}^1(\mathbf{x}) \equiv \mathbf{x}_1 \qquad \text{and} \qquad \varphi_{0,t}^2(\mathbf{x}) = \mathbf{x}_2 + \int_0^t \mathbf{v}(s, \mathbf{x}_1) ds.$$

In other words, the first component remains constantly equal to its initial value while the second component is the antiderivative of a scalar O-U process. More precisely, it is a (real-valued) Gaussian process with mean

$$\mathbb{E}\{\varphi_{0,t}^2(\mathbf{x})\} = \mathbf{x}_2$$

and covariance:

$$\gamma(s,t) = \text{Cov}\{\varphi_{0,s}^2(\mathbf{x})\varphi_{0,t}^2(\mathbf{x})\} = \int_0^s \int_0^t \Gamma(t' - s', 0) ds' dt'.$$

We shall determine later the laws of the one-point motions in the case of white velocity fields.

We now consider the problem of the measure of an observable quantity along a one-point motion trajectory. Let us consider for example a (possibly vector-valued or even matrix-valued) function $A(t, \mathbf{x})$. Notice that $A(t, \mathbf{x})$ can be random **and** dependent upon the field $\{\vec{\mathbf{v}}(t, \mathbf{x}); \ t \geq 0, \ \mathbf{x} \in \mathbb{R}^d\}$ as long as it is stationary and homogeneous. If one thinks of A as an observable, a Lagrangian measurement of this observable is given by the stochastic process $\{A(t, \varphi_{0,t}(\mathbf{x}))\}_{t \geq 0}$ obtained by measuring A along a one-point motion trajectory. A measure theoretic argument due to Zirbel [62] shows that such a stochastic process is always stationary. Weaker forms of this general fact had been known for quite some time (see for example [50], [37], and [56]), but the general form of this remark will be of crucial importance in the sequel where we use it in the case of the observable $A(t, \mathbf{x}) = \nabla \vec{v}(\mathbf{t}, \mathbf{x})$.

We shall discuss several natural examples of these Lagrangian observations.

1.4.2 Separation Process

Let us consider two (different) starting points, $\mathbf{x}$ and $\mathbf{y}$, and let us study the time evolution of the distance separating the positions of two passive tracer particles starting at time $t = 0$ from $\mathbf{x}$ and $\mathbf{y}$, respectively. We want to understand the nature of the process $\{\rho_t\}$ defined by:

$$\rho_t = |\varphi_{0,t}(\mathbf{x}) - \varphi_{0,t}(\mathbf{y})|, \qquad \mathbf{t} \geq \mathbf{0}.$$

As before, the solution to this problem is immediate in the case of a shear flow. We shall see below that a complete solution can be found in the case of white noise velocity fields. But we do not know how to approach this problem in the general case of stationary homogeneous Gaussian fields with a Kolmogorov's spectrum.

1.4.3 Mass Transport

Let us consider the following realistic scenario. A ship is traveling along a specific trajectory at the surface of the ocean and buoys are dropped in the ocean at each of the times $s_1 < s_2 < \cdots < s_n$. Let us assume that the positions of the ship at these instants were $\mathbf{x_1}, \mathbf{x_2}, \cdots, \mathbf{x_n}$ and let us assume that each buoy is carried by the flow at the surface of the ocean as a passive tracer. When at a later time t a satellite takes an image of the population of drifters, the result is a sample from the (point) measure:

$$\nu_t(d\mathbf{y}) = \sum_{\mathbf{j}=\mathbf{1}}^{\mathbf{n}} \delta_{\varphi_{\mathbf{s_j},t}(\mathbf{x_j})}(d\mathbf{y})$$

equal to the sum of the unit Dirac point masses at the positions of the floats. This consideration leads to the following mathematical problem: let us fix a finite nonnegative measure ν_0 on $\mathbf{R}^d$ and let us think of it as the distribution of passive tracers at a fixed time. We then consider the pushed forward measure

$$\nu_t = \varphi_{0,t} \circ \nu_0$$

and view it as the distribution of tracers at time t resulting from the time evolutions of the (noninteracting) individual tracers distributed earlier (i.e., at time $t = 0$) according to the distribution ν_0. The family $\{\nu_t; t \geq 0\}$ is a measure valued process and its analysis is very difficult in general. But simple characteristics of ν_t, such as for example the center of mass:

$$c_t = \int x \nu_t(dx)$$

and the dispersion matrix

$$D_t = \int (x - c_t)(x - c_t)^t \, \nu_t(dx)$$

define stochastic processes with values in finite-dimensional spaces and their analyses are more reasonable. We review later some of the results obtained in [25] by Cinlar and Zirbel in the case of white noise velocity fields.

1.4.4 Curve Lengths

Let us consider a slightly less ambitious project, though still very useful in practice. Let us assume that a compact cloud of pollutant dust fell at the surface of the ocean at time $t = 0$. Let us assume that the initial cloud is bounded by a smooth curve which we denote by γ_0:

$$[0,1] \ni \alpha \hookrightarrow \gamma_0(\alpha) \in \mathbb{R}^d.$$

We assume that the dust particles are light enough to be passively transported by the velocity field at the surface of the ocean (while we assume at the same time that they are heavy enough so that they will not be blown away by the wind). The goal is to follow the time evolution of the cloud (and most importantly of its boundary), to try to understand the characteristics of this time evolution in order to be able to control the evolution of future disasters. This scenario is very simplistic, but the problems it raises are of the utmost environmental importance.

Let us denote by γ_t the curve at time t which started from γ_0 at time $t = 0$.

$$[0,1] \ni \alpha \hookrightarrow \gamma_t(\alpha) = \varphi_{0,t}(\gamma_0(\alpha)) \in \mathbb{R}^d.$$

One should think of γ_0 as enclosing a bounded domain $D \subset \mathbb{R}^d$ filled at time $t = 0$ with a set of passive tracer particles. Then, γ_t is the boundary of the set formed by the particles at time t.

The results of the numerical simulations reported in [15] (especially Figure 3 and Figure 2, reproduced here[2]) suggest that the length of γ_t increases in all cases but that this increase is more dramatic in the case of isotropic flows than in the case of shear flows. As we explained earlier, explicit computations are possible in the latter. Since:

---

[2]For technical reasons all figures are placed at the end of the chapter.

$$\ell(\gamma_t) = \int_0^1 \sqrt{1 + \left| \int_0^t \frac{\partial v(s, \gamma_0^1(\alpha))}{\partial x} ds \right|^2} \, d\alpha$$

and since $\{\partial v(t, \gamma_0^1(\alpha))/\partial x; \, ty \geq 0\}$ for each fixed $\alpha \in [0,1]$ is a stationary Gaussian process with a covariance which decays to 0 at infinity, we have:

$$\limsup_{t \to \infty} \frac{1}{t\sqrt{2\log t}} \left| \int_0^t \frac{\partial v(s, \gamma_0^1(\alpha))}{\partial x} ds \right| \leq \limsup_{t \to \infty} \frac{1}{\sqrt{2\log t}} \sup_{0 \leq s \leq t} \left| \frac{\partial v(s, \gamma_0^1(\alpha))}{\partial x} \right|$$

$$\leq \left. -\frac{\partial^2}{\partial t^2} \Gamma(t, 0) \right|_{t=0}$$

almost surely. This implies that:

$$\limsup_{t \to \infty} \frac{1}{t\sqrt{2\log t}} \ell(\gamma_t) < \infty \tag{1.22}$$

almost surely. The results of the numerical simulations indicate that, in the case of isotropic incompressible flows (see, for example, Figure 2) the large time behavior of the length $\ell(\gamma_t)$ of the curve γ_t could be exponential. The computation above shows that this is not the case for shear flows since (1.22) implies:

$$\lim_{t \to \infty} \frac{1}{t} \log \ell(\gamma_t) = 0.$$

In the general case, using Jensen's inequality and Fatou's lemma we get:

$$\liminf_{t \to \infty} \frac{1}{t} \log \ell(\gamma_t) = \liminf_{t \to \infty} \frac{1}{t} \log \int_0^1 \left\| \frac{d\gamma_t(\alpha)}{d\alpha} \right\| d\alpha$$

$$\geq \liminf_{t \to \infty} \int_0^1 \frac{1}{t} \log \left\| \frac{d\gamma_t(\alpha)}{d\alpha} \right\| d\alpha$$

$$\geq \int_0^1 \liminf_{t \to \infty} \frac{1}{t} \log \left\| \frac{d\gamma_t(\alpha)}{d\alpha} \right\| d\alpha.$$

Let us denote by ℓ_∞ the above right hand side (which we proved to be zero in the case of shear flows). We have:

$$\ell_\infty = \int_0^1 \liminf_{t \to \infty} \frac{1}{t} \log \left\| \nabla \varphi_{0,t}(\gamma_0(\alpha)) \frac{d\gamma_0(\alpha)}{d\alpha} \right\| d\alpha.$$

This motivates the analysis of the large time behavior of the norm of the Jacobian matrix $\nabla \varphi_{0,t}(\mathbf{x})$ which we take up now.

1.4.5   Jacobian Matrices and Lyapunov Exponents

As explained above, we need to understand the properties of the one-point motions as functions of their starting points. Hence we consider the gradient (i.e., derivative) of $\varphi_{s,t}(\mathbf{x})$ with respect to the initial position $\mathbf{x}$. We set:

$$Y_{s,t}(\mathbf{x}) = \nabla\varphi_{s,t}(\mathbf{x}).$$

As before, we concentrate on the qualitative properties of these objects and we leave the details of the proof of the existence as exercise. Minor changes to the arguments given in [45] can be used to do the job. In coordinate notation:

$$[Y_{s,t}(\mathbf{x})]_{j,j'} = \frac{\partial}{\partial x_j}\varphi^{j'}_{s,t}(\mathbf{x})$$

where we use the notation $\varphi^j_{s,t}(\mathbf{x})$ for the components of the element $\varphi_{s,t}(\mathbf{x})$ of $\mathbb{R}^d$. The analysis of the time behavior of the Jacobian matrix $Y_t(\mathbf{x}) = Y_{0,t}(\mathbf{x}) = \nabla\varphi_{0,t}(\mathbf{x})$ is going to show that $A(t,\mathbf{x}) = \nabla\vec{v}(t,\mathbf{x})$ is a first natural example of an observable which depends upon the velocity field and which we may want to compute along a Lagrangian trajectory. Indeed, if we recall the equation of motion:

$$\varphi_{0,t}(\mathbf{x}) = \mathbf{x} + \int_0^t \vec{v}(s, \varphi_{0,s}(\mathbf{x}))\ ds,$$

by taking gradients in both sides and using the chain rule one gets:

$$\nabla\varphi_{0,t}(\mathbf{x}) = I_d + \int_0^t \nabla\vec{v}(s, \varphi_{0,s}(\mathbf{x}))\nabla\varphi_{0,s}(\mathbf{x})\ ds, \qquad (1.23)$$

where we used the notation I_d for the $d \times d$ identity matrix. This shows that, for $\mathbf{x} \in \mathbb{R}^d$ fixed, the Jacobian matrix $Y_t = Y_t(\mathbf{x})$ is the solution of the $d \times d$ system of linear ordinary differential equations written in matrix form:

$$\begin{cases} d\mathbf{Y}_t &= d\mathbf{A}(t)\mathbf{Y}_t \\ \mathbf{Y}_0 &= I_d \end{cases} \qquad (1.24)$$

provided we set:

$$\mathbf{A}(t) = \nabla\vec{v}(t, \varphi_{0,t}(\mathbf{x})).$$

Since $\{\mathbf{A}(t) : t \geq 0\}$ is a Lagrangian observation, it is a stationary sequence of matrices and the subadditive ergodic theorem implies the existence for each $x \in \mathbb{R}^d$ of a measurable set Ω_x of full probability on which the limit:

$$\lambda_1(x) = \lim_{t\to\infty} \frac{1}{t}\log\|\nabla\varphi_{0,t}(\mathbf{x})\|$$

exists. This limit is nonrandom because of ergodicity. Moreover, because of the homogeneity assumption of the distribution of the velocity field, this limit did not depend upon x and we will set $\lambda_1(x) = \lambda_1$. This number is the (upper) Lyapunov exponent of the system since $\mathbf{Y}_t$ can be interpreted as a "product of random matrices". One of the most desirable properties of this Lyapunov exponent is its positivity. Such a result is known in the theory of products of random matrices as Furstenberg's theorem. It is known for some special products of independent matrices and for some other Markov chains of matrices. We shall show later that $\lambda_1 > 0$ can be proven in the case of white velocity fields by the direct use of stochastic calculus and in other cases by more sophisticated arguments.

The reader is referred to Chapter IV of [20] for details on the theory of products of random matrices and of their Lyapunov exponents.

1.4.6 Curve Lengths Revisited

Let us assume, for the purpose of the present discussion, that the upper Lyapunov exponent is positive, i.e., that $\lambda_1 > 0$. Modulo an extra technical condition (which can be checked in most of the cases in which one can prove the positivity of the upper Lyapunov exponent), one can use the Oceledec's ergodic theorem and one gets, for each $y \in \mathbf{R}^d$, the existence of a measurable set $\Omega_{x,y}$ of full probability on which:

$$\lim_{t \to \infty} \frac{1}{t} \log \|\nabla \varphi_{0,t}(x)y\| = \lambda_1.$$

Consequently the set:

$$\{(\omega, \alpha) \in \Omega \times [0,1]; \ \lim_{t \to \infty} \frac{1}{t} \log \|\nabla \varphi_{0,t}(\gamma_0(\alpha)) \frac{d\gamma_0(\alpha)}{d\alpha}\| = \lambda_1\}$$

is of full product measure, which proves (by Fubini's theorem) that we have, on a set of probability one that:

$$\lim_{t \to \infty} \frac{1}{t} \log \|\frac{\nabla \varphi_{0,t}(\gamma_0(\alpha)) d\gamma_0(\alpha)}{d\alpha}\| = \lambda_1$$

for almost every $\alpha \in [0,1]$. This proves that $\ell_\infty = \lambda_1$ and consequently that:

$$\liminf_{t \to \infty} \frac{1}{t} \log \ell(\gamma_t) \geq \lambda_1 > 0.$$

In fact, we believe that the limit actually exists and that it is equal to λ_1 but the above argument is not sufficient to prove this claim. Notice also that the above argument can be used to show in the case of shear flows that $\lambda_1 = 0$.

1.4.7 Intersection Number

We consider the problem of the estimation of the number of intersection points of the curve γ_t with a fixed nonrandom curve. For the purpose of the present discussion we restrict ourselves to the two-dimensional case $d = 2$. All of the results presented here can be restated (and proved) in the general context of general dimension d. We choose the deterministic curve L to be a line for the sake of definiteness, say, the horizontal line with equation $y = a$. We shall use the notation:

$$N_L(t) = \#\{\alpha \in [0, 1];\ \gamma_t(\alpha) \in L\}$$

for the number of intersection points of γ_t with L. We have:

$$
\begin{aligned}
N_L(t) &= \#\{\alpha \in [0, 1];\ \gamma_t^2(\alpha) = a\} \\
&= \int_0^1 \delta_a(\gamma_t^2(\alpha)) \left| \frac{d\gamma_t^2(\alpha)}{d\alpha} \right| d\alpha.
\end{aligned}
$$

Consequently, the expected number of intersections is given by the following formula à la Kac-Rice (see for example [26]):

$$\mathbf{E}\{N_L(t)\} = \int_0^1 \int_{I\!R} p_{t,\alpha}(a, u)|u|\, du\, d\alpha$$

where for each fixed $\alpha \in [0, 1]$ and $t \geq 0$, $p_{t,\alpha}(a, u)$ is the joint density of the couple $(\gamma_t^2(\alpha), d\gamma_t^2(\alpha)/d\alpha)$. By definition we have, for each fixed $\alpha \in [0, 1]$:

$$\gamma_t(\alpha) = \gamma_0(\alpha) + \int_0^t \vec{v}(s, \gamma_s(\alpha))\, ds.$$

So, by differentiating both sides and using the chain rule we get:

$$Z_t(\alpha) = \frac{d\gamma_0(\alpha)}{d\alpha} + \int_0^t \nabla \vec{v}(s, \gamma_s(\alpha)) Z_s(\alpha)\, ds$$

provided we set:

$$Z_t(\alpha) = \frac{d\gamma_t(\alpha)}{d\alpha} = \frac{d\varphi_{0,t}(\gamma_0(\alpha))}{d\alpha}.$$

Consequently, for each fixed $\alpha \in [0, 1]$, the vector $Z_t(\alpha)$ is a solution of the linear system:

$$
\begin{cases}
dZ_t(\alpha) &= A_t(\alpha) Z_t(\alpha)\, dt \\
Z_0(\alpha) &= d\gamma_0(\alpha)/d\alpha,
\end{cases}
$$

where the matrix $A_t(\alpha)$ is the Lagrangian observation of the Jacobian matrix $\nabla \varphi_{0,t}(\gamma_0(\alpha))$ along the trajectory starting from $\gamma_0(\alpha)$. Since the latter is a stationary dynamical system one has:

$$\lim_{t \to \infty} \frac{1}{t} \log |Z_t(\alpha)| = \lambda_1$$

where λ_1 is the (upper) Lyapunov exponent introduced earlier. Recall that it does not depend upon α because of the homogeneity in space.

We shall come back to these computations and complete the asymptotic analysis of the expected number of intersections when we can make more specific claims on the distribution of the density $p_{t,\alpha}$.

1.5   Numerical Simulations

In order to illustrate the results of the theory of Brownian flows reviewed in the previous section, and in order to visualize the transport properties of more general stochastic flows, we decided to use numerical simulations. Our first goal was to develop some intuition on statistical properties of the trajectories of the passive tracers and their velocities. Since the time series of their velocities (which we call Lagrangian velocities) are stationary, the numerical estimation of their spectral characteristics is natural. Since this analysis is done in practice for ocean drifters data, it seemed interesting to investigate, by large-scale numerical simulations, the spectral characteristics of the Lagrangian spectra (which should all be the same according to the theoretical results which we reviewed earlier) to the spectral characteristics of the Euler spectrum, and in particular to the parameters ϵ and z if we use the Kolmogorov spectrum for the simulations. See, for example, [22] for some preliminary results in this direction. In order to simulate large numbers of "drifters" at a minimal cost, we decided to implement the simulations on a massively parallel machine. But after a very short while, we realized that the tool we developed for statistical purposes was very versatile and could be used to illustrate and investigate new interesting mathematical results. The goal of this section is to present some of the salient features of the simulation package. See [15] for details.

The idea of the random simulation is based on the Markov property in time of the process $\{\vec{v}(t);\ t \geq 0\}$. We discretize the time into a sequence $(t_j)_{j=0,1,\dots}$ and we use the Markov property to construct the value $\vec{v}(t_j)$ from the last known value $\vec{v}(t_{j-1})$, ignoring the way the field "got there", i.e., ignoring the values of $\vec{v}(t_0)$, $\vec{v}(t_1)$, $\cdots$, $\vec{v}(t_{j-2})$. Once the velocity field is known, the positions of the passive tracers under investigation are updated according to the equation:

$$\mathbf{x}(t_j) = \mathbf{x}(t_{j-1}) + \Delta t\, \vec{v}(t_j, \mathbf{x}(t_{j-1})) \tag{1.25}$$

which is merely the discretized form of the equation of motion (1.1). Notice also that since each $\vec{v}(t)$ is itself an infinite-dimensional object, we shall need to use other finite-dimensional approximations to construct (or at least to approximate) the configurations $\vec{v}(t)$.

For the sake of simplicity we restrict ourselves to numerical simulations in the two-dimensional case $d = 2$. Moreover we concentrate on the incompressible case for which:

$$\mathrm{div}\{\vec{v}(t,\mathbf{x})\} = \partial_1 v_1(t,\mathbf{x}) + \partial_2 v_2(t,\mathbf{x}) \equiv 0.$$

This assumption guarantees the existence of a scalar function $\phi(t,\mathbf{x})$ satisfying:

$$v_1(t,\mathbf{x}) = -\partial_2\phi(t,\mathbf{x}) \qquad \text{and} \qquad v_2(t,\mathbf{x}) = \partial_1\phi(t,\mathbf{x}). \quad (1.26)$$

where we used the notations ∂_1 and ∂_2 for the partial derivatives $\partial/\partial x_1$ and $\partial/\partial x_2$, respectively. This function is called the *stream* function of the system. Formula (1.26) is extremely convenient because it reduces the computations on vector-valued functions to computations with real-valued functions and this is especially useful for simulation purposes. Because of the special form of the spectral density of $\{\vec{v}(t,\mathbf{x})\}$, and in particular because $E(\omega,\mathbf{k})$ vanishes near the origin, we can assume without any loss of generality that $\{\phi(t,\mathbf{x}); t \geq 0, \mathbf{x} \in \mathbb{R}^2\}$ is a stationary and homogeneous Gaussian field with covariance Γ_ϕ of the same form:

$$\Gamma_\phi(t,\mathbf{x}) = \int_{I\!R}\int_{I\!R} e^{i(\omega t + \mathbf{k}\cdot\mathbf{x})} \frac{|\mathbf{k}|^z}{|\mathbf{k}|^{2z} + \omega^2} \mathcal{E}(|\mathbf{k}|)\, d\omega d\mathbf{k}, \qquad t \geq 0, \mathbf{x} \in \mathbf{R}^2$$

for a function $\mathcal{E}(\nabla)$ as before, and using as before a partial Fourier transform in the variable ω we can argue that the random field $\phi(t,\mathbf{x})$ in the Fourier domain of the wave number $\mathbf{k}$ is a superposition of real-valued independent 0-U processes.

1.5.1 Discrete Approximations

This section is devoted to the choice of computable approximations of the random velocity fields. There are many possible ways to address this problem. From the point of view of the spectral theory of stationary and/or homogeneous random fields, the simplest approach is to approximate the spectral measure $\nu(d\omega, d\mathbf{k})$ by simple measures more amenable to random simulations on a computer. In particular, it is natural to approximate ν by a finite sum of point masses $\delta_{(\omega_i,\mathbf{k}_i)}$ in the frequency-wave number domain. This approach is especially useful because it makes it possible to generate the white noise directly without any concern with the possible dependence of the random variates. Nevertheless, we refrain from using this approach directly, especially in the time variable. Indeed, in order to save memory (and computing time) we want to use an adaptive approach. More precisely, we rely on an approximation scheme which uses the fact that $\vec{v}(t)$ is a time-homogeneous Markov process. For this reason we separate the problems of the discretization of the time and the space variables.

The discretization of the time is straightforward. We choose a time interval Δt and we sample the continuous time by restricting its values to:

$$t_0 = 0, \; t_1 = \Delta t, \; t_2 = 2\Delta t, \; \cdots, t_n = n\Delta t, \; \cdots\cdots$$

Because of the Markov property and because of our choice for the discretization of the time, at each instant t_j, one generates a sample of the field $\vec{v}(t_j)$ from the knowledge of the sample of the field $\vec{v}(t_{j-1})$ at the preceding "cycle".

The discretization of the space variable is much more involved. In the same way we replaced the continuous time t by a grid of discrete values, it is possible to replace the space variable $x \in \mathbf{R}^2$ by a discrete variable taking finitely many values in a lattice $h\mathbf{Z}^2$ for some mesh $h > 0$. This strategy has several drawbacks, especially from the point of view of Monte-Carlo simulation. Indeed, for each fixed time t there is a strong dependence between the values of $\vec{v}(t, \mathbf{x})$ and $\vec{v}(t, \mathbf{y})$ even when the points $\mathbf{x}$ and $\mathbf{y}$ are far apart. This will make the generation of the configuration of $\vec{v}(t_j)$ from $\vec{v}(t_{j-1})$ extremely costly in memory storage and in computing time: this approach is not reasonable for our application.

The crux of the generation method is the spectral representation of the stream function as the inverse Fourier transform of scalar O-U processes indexed by the wave numbers. We use a discretization procedure to replace such a continuous integral by a finite sum. In other words, we choose a finite subset, say $\mathcal{K}$, of the wave-number plane and we replace $\phi(t, \mathbf{x})$ by a finite sum of the form:

$$\phi(t, \mathbf{x}) = \sum_{\mathbf{k} \in \mathcal{K}} [\mathbf{a_k}(t) \cos(\mathbf{k} \cdot \mathbf{x}) + \mathbf{b_k}(t) \sin(\mathbf{k} \cdot \mathbf{x})] \tag{1.27}$$

where the real-valued processes $\{a_\mathbf{k}(t)\}_{t \geq 0}$ and $\{b_\mathbf{k}(t)\}_{t \geq 0}$ are independent stationary O-U processes. When needed, the velocity vector $\vec{v}(t, \mathbf{x})$ will be computed according to the formulas:

$$v_1(t, \mathbf{x}) = \sum_{\mathbf{k} \in \mathcal{K}} \mathbf{k_2} [\mathbf{a_k}(t) \sin(\mathbf{k} \cdot \mathbf{x}) - \mathbf{b_k}(t) \cos(\mathbf{k} \cdot \mathbf{x})] \tag{1.28}$$

and:

$$v_2(t, \mathbf{x}) = \sum_{\mathbf{k} \in \mathcal{K}} \mathbf{k_1} [-\mathbf{a_k}(t) \sin(\mathbf{k} \cdot \mathbf{x}) + \mathbf{b_k}(t) \cos(\mathbf{k} \cdot \mathbf{x})] \tag{1.29}$$

The choice of a trigonometric representation of $\phi(t, \mathbf{x})$ is made possible by the fact that its spectral density is spherically symmetric in the wave-number variable $\mathbf{k}$. This last property also suggests that we choose the set $\mathcal{K}$ to be approximately (if not exactly) spherically symmetric.

1.5.2   The Simulation Algorithm

Since discrete time O-U processes are nothing but order one auto-regressive processes, the simulation reduces to the generation of independent (over $\mathbf{k} \in \mathcal{K}$) sequences $\{a_{\mathbf{k}}(t_j)\}_{j=0,1,\dots}$ and $\{b_{\mathbf{k}}(t_j)\}_{j=0,1,\dots}$ according to the formulas:

$$a_{\mathbf{k}}(t_j) = \alpha_{\mathbf{k}} a_{\mathbf{k}}(t_{j-1}) + \sigma_{\mathbf{k}} \epsilon_{\mathbf{k}}^{(a)}(j) \tag{1.30}$$

$$b_{\mathbf{k}}(t_j) = \beta_{\mathbf{k}} b_{\mathbf{k}}(t_{j-1}) + \sigma_{\mathbf{k}} \epsilon_{\mathbf{k}}^{(b)}(j) \tag{1.31}$$

where the $\epsilon_{\mathbf{k}}^{(a)}(j)$ are independent (over the three parameters $\alpha = a, b$, $\mathbf{k} \in \mathcal{K}$ and $j = 0, 1, \cdots$) $N(0, 1)$ random variables and where the parameters of the model, i.e., the positive numbers $\alpha_{\mathbf{k}}$, $\beta_{\mathbf{k}}$, and $\sigma_{\mathbf{k}}$ are chosen as a function of the function $\mathcal{E}(\nabla)$. See [15] for details.

1.5.3   A First Implementation

We give a rather detailed presentation of one form of the simulation program which we developed to analyze the transport properties of the Gaussian velocity fields. We will briefly mention other forms of the program in Section 1.5.5. A full account of the various algorithms and of the simulation results can be found in [15].

Inputs

The spectral characteristics of the Eulerian spectrum, the number and the initial positions of the drifters are the inputs of the program. They can be supplied in the form of a text file or they can be generated by the program if the user so desires.

The Main Loop

At the beginning of each cycle j, the program generates, independently of all the preceding random generations, the $N(0, 1)$ independent random noise terms $\epsilon_{\mathbf{k}}^{(a)}(t_j)$ and $\epsilon_{\mathbf{k}}^{(b)}(t_j)$ for all the values of the wave vectors $\mathbf{k}$ in the selected set $\mathcal{K}$. Then, it computes the new values of the parameters $a_{\mathbf{k}}(t_j)$ and $b_{\mathbf{k}}(t_j)$ in terms of their previous values according to the Formulas (1.30) and (1.31).

The second step in the main loop is to update the positions of the drifters. This is done as follows. If a drifter is at the location $\mathbf{x}(t_{j-1})$

at time t_{j-1}, then the new position of this drifter is given by Formula (1.25) where the value of the velocity field is computed from the formulas (1.28) and (1.29).

Options

The program can operate in two modes.

• Eulerian mode: the limits of the surface of the ocean which is to be considered in the simulation are defined, boundary conditions are specified (they include killing, elastic reflection and periodic extension), and the region of interest is partitioned into $N = 4096$ subsets (typically rectangles). Each of the $N = 4096$ PEs of the MasPar is assigned to one of these regions.

• Lagrangian mode: in this mode the ocean has no limits! Each of the $N = 4096$ PEs of the MasPar is assigned a certain number of drifters, typically $[n/N]$ if n is the total number of drifters, and it follows the drifters wherever they go. In other words, at each cycle, each PE updates the positions of the drifters he has control of.

The Lagrangian mode has a certain number of advantages over the Eulerian mode.

◇ Each PE has the same computing load and the same memory constraints.

◇ The PE does not risk running out of memory.

Several figures show the time evolutions of a set of points, in particular, of curves (whether or not they are enclosing a region of interest). The numerical results (as well as the mathematical conjectures presented in the next section) indicate that points are drifting apart in the sense that the distance separating two given points is typically growing (presumably at an exponential rate). If the time evolution of a finite number of points is followed via the plots of their positions at each cycle, when the points form a smooth curve at the initial time, gaps appear rapidly inbetween separate segments of the curve, and after a relatively small number of cycles it is very difficult to visualize the set of points as a curve. We wanted to provide a remedy to this shortcoming of having the same number of drifters at each cycle.

• The program has an option which allows insertion of drifters inbetween two drifters whose separation reaches a preassigned threshold. The main drawback of the use of this option is that it gives output files with a non-rectangular format. But the main advantage is that the visual impressions left by the plots are very nice. For example, if one follows the time evolution of a set of drifters which form a (smooth) continuous curve at time $t = 0$, then at each subsequent time the plots of the positions of the points form a (smooth) continuous curve without artificial gaps.

1.5.4   Sample Outputs

Figure 1 shows the trajectories of four (real) drifters. These trajectories should be compared to some of the trajectories produced by the simulation program. We reproduced four of them in Figure 4 for the purpose of illustration.

There are obvious differences. We shall use them when we discuss possible alternatives to the Gaussian models. Some of the differences can be explained (presumably) by a poor choice of the set $\mathcal{K}$ of wave numbers used to generate the samples of the isotropic velocity field. But the lack of realism of the look of these trajectories is also due to the failure of this model to give an account of the effects of large eddies.

Figure 2 shows the time evolution of the point of a circle under a flow of a velocity field with an isotropic spectrum of the Kolmogorov type. Notice that the fact the curve remains continuous (even though the neighboring points do eventually separate) is not due to a postprocessing interpolation but by the fact that the simulation program inserts drifters when neighboring points drift apart by more than a preassigned threshold. It is interesting to notice that the particular choice of the discretization set $\mathcal{K}$ of wave vectors used in the simulation is such that the circular shape of the figure is preserved for a long time. Moreover, the asymptotic foliation phenomenon exhibited in [48] cannot be seen on these figures, either because the values of t_j of the time are not large enough or maybe because of the fact that the program does simulate a flow which is not exactly isotropic since (after all) the set $\mathcal{K}$ is finite.

Figure 3 shows the same type of time evolution (of the same circle as before), but in the case of a shear flow. The difference is obvious. Moreover, even though the length of the curve seems to grow to ∞ as before, the rate seems to be much slower!!

1.5.5   Other Forms and Uses of the Program

The simulation program was presented as a tool to generate a very large number of passive tracer trajectories at the same cost as with more traditional architectures. This corresponds to the original design of the program. The use of the first form of the program initiated new ideas of simulations and the need for new forms of the algorithm.

The program can now be used for the computation of discrete time approximations of the Lyapunov exponent λ_1. In fact, one can now compute the same approximations simultaneously for a large number of values of the spectral parameters in the same time it would take a serial computer to compute one of these values. Figure 5 shows the result of such a computation.

We are now in the process of changing the form of the simulation program to include a diffusion term and to be able to output the time evolution of quantities such as concentrations. In other words, we are trying to use it to evaluate numerical solutions of SPDEs.

1.6 The Case of White Velocity Fields

Throughout this section we consider the case of the white velocity fields introduced as the last special case in Subsection 1.2.2. In other words, we consider a mean zero Gaussian stationary and homogeneous velocity field $\{\vec{v}(t, \mathbf{x})\}$ with a covariance of the form:

$$\Gamma(t, \mathbf{x}) = \delta_0(t)\Gamma^{(0)}(\mathbf{x}) \tag{1.32}$$

where:

$$\Gamma^{(0)}(\mathbf{x}) = \int e^{i\mathbf{k}\cdot\mathbf{x}}\nu^{(0)}(d\mathbf{k}) \tag{1.33}$$

for some nonnegative definite matrix measure $\nu^{(0)}(d\mathbf{k})$. The isotropy condition can be rewritten as:

$$O\Gamma^{(0)}(\mathbf{x})O^* = \Gamma^{(0)}(O\mathbf{x}) \tag{1.34}$$

for all the matrices O in the orthogonal group $O(d)$. We shall monitor closely the consequences of the isotropy condition for the latter turns out to be a mathematical convenience. As explained earlier, this condition was introduced and analyzed in detail by workers in the mathematical theory of turbulence in fluid dynamics. See, for example, [53] and/or [61]. This isotropy condition forces the covariance matrix $\Gamma^{(0)}(\mathbf{x})$ and the spectral matrix measure $\nu^{(0)}(d\mathbf{k})$ to have special forms. More precisely, $\Gamma^{(0)}(\mathbf{x})$ can be written in the form:

$$\Gamma^{(0)}(\mathbf{x}) = \gamma_N^{(0)}(|\mathbf{x}|)\left[\delta_{j,\ell} - \frac{x_j x_\ell}{|\mathbf{x}|^2}\right]_{j,\ell} + \gamma_L^{(0)}(|\mathbf{x}|)\left[\frac{x_j x_\ell}{|\mathbf{x}|^2}\right]_{j,\ell} \tag{1.35}$$

for two functions $\gamma_N^{(0)}(r)$ and $\gamma_L^{(0)}(r)$ of the real variable r. The subscripts N and L stand for Normal and Longitudinal, respectively. As a consequence, the spectral matrix measure $\nu^{(0)}(d\mathbf{k})$ can be written in the form:

$$\nu^{(0)}(\mathbf{k}) = [\delta_{j,\ell} - \theta_j\theta_\ell]_{j,\ell}\,\sigma(d\theta)F_N(dr) + [\theta_j\theta_\ell]_{j,\ell}\,\sigma(d\theta)F_L(dr) \tag{1.36}$$

where $r = |\mathbf{k}|$, $\theta = \mathbf{k}/|\mathbf{k}|$, and $\sigma(d\theta)$ denotes the normalized surface measure on S^{d-1} and where $F_N(d\mathbf{k})$ and $F_L(d\mathbf{k})$ are positive measures on $\mathbb{R}_+$ with moments of all orders. We shall assume that these measures do not charge the origin. The constants:

$$A = \frac{1}{d} \int r^2 F_L(dr) \qquad \text{and} \qquad B = \frac{1}{d} \int r^2 F_N(dr)$$

will play a crucial role in the sequel. It is easy to show that:

$$\gamma_N^{(0)}(r) = \alpha - \beta r^2 + O(r^4)$$

and:

$$\gamma^{(0)}(r) = -\gamma r^2 + O(r^4)$$

for some positive constants α, β, and γ. The latter can be expressed in terms of the spectral characteristics of the covariance. Indeed, we have:

$$\alpha = \frac{1}{d} \int F_L(dr) + \frac{d-1}{d} \int F_N(dr).$$

and:

$$2\beta = \frac{1}{d+2}A + \frac{d+1}{d+2}B \qquad \text{and} \qquad \gamma = \frac{1}{d+2}(A - B).$$

These last two equations can be solved for A and B and we get:

$$A = 2\beta + (d+1)\gamma \qquad \text{and} \qquad B = 2\beta - \gamma.$$

As a last remark we give several equivalent forms to the physical condition of incompressibility.

$$
\begin{aligned}
\vec{v} \text{ is incompressible} \quad &\Longleftrightarrow \quad \text{div } \vec{v} \equiv 0 \\
&\Longleftrightarrow \quad \varphi_{s,t} \text{ preserves Lebesgue's measure} \\
&\Longleftrightarrow \quad A = 2\beta + (d+1)\gamma = 0 \\
&\Longleftrightarrow \quad F_L = 0.
\end{aligned}
$$

1.6.1 Equation of Motion

We now return to the equation of motion (1.1). Since the velocity field is a white noise in time (though a regular homogeneous field in the space variable), the equation of motion can only be written formally as:

$$d\mathbf{X_t} = \vec{v}(t, \mathbf{X_t})dt.$$

In order to make sense of such an equation, we rewrite it in an integral form:

$$\mathbf{X_t} = \mathbf{x_0} + \int_0^t \vec{\mathbf{w}}(\mathbf{dt}, \mathbf{X_t}) \tag{1.37}$$

where $\vec{\mathbf{w}}(t, \mathbf{x})$ is understood as the antiderivative of $\vec{\mathbf{v}}(t, \mathbf{x})$ in the sense that:

$$\vec{\mathbf{w}}(t, \mathbf{x}) = \int_0^t \vec{\mathbf{v}}(\mathbf{s}, \mathbf{x}) \mathbf{ds},$$

and the stochastic integral in (1.37) is to be understood as the limit of finite sums of the form:

$$\sum [\vec{\mathbf{w}}(t_{j+1}, \mathbf{X_{t_j}}) - \vec{\mathbf{w}}(t_j, \mathbf{X_{t_j}})].$$

$\{\vec{\mathbf{w}}(t, \mathbf{x}); \ t \geq 0, \ \mathbf{x} \in \mathbf{R^d}\}$ is a bona fide mean zero Gaussian field with covariance:

$$\mathbf{E}\{\vec{\mathbf{w}}(s, \mathbf{x}) \otimes \vec{\mathbf{w}}(t, \mathbf{y})\} = (\mathbf{s} \wedge \mathbf{t})\mathbf{\Gamma^{(0)}}(\mathbf{x} - \mathbf{y}).$$

It can be advantageously viewed as a $C(\mathbf{R}^d, \mathbf{R}^d)$-valued Brownian motion process $\{\vec{\mathbf{w}}(t); \ t \geq 0\}$. The existence and uniqueness theory of the stochastic equations involving this sort of integral is well understood. See, for example, [21] or [45]. In particular, the nature of the space covariance $\Gamma^{(0)}$ guarantees the existence of the stochastic flow $\{\varphi_{s,t} \ ; \ s, t \in \mathbf{R}\}$. The same remark as before applies because none of the derivatives of the velocity is bounded, and a modicum of care is needed to check that the results of [21] and [45] apply to the present situation. On the top of property (1.21) we have:

$$\varphi_{s,t} \text{ and } \varphi_{t,s} \text{ have the same distribution} \tag{1.38}$$

the latter depending only upon the difference $|t - s|$. Finally the white noise nature of the velocity field (or in other words the independence of the increments of the Brownian motion $\vec{\mathbf{w}}(t)$) implies that:

$$\varphi_{s,t} \text{ and } \varphi_{t,u} \text{ are independent whenever } s \leq t \leq u. \tag{1.39}$$

1.6.2 One-and Two-Point Motions and Mass Transport

For each fixed $s \in \mathbf{R}$ and $\mathbf{x} \in \mathbf{R}^d$ the $\mathbf{R}^d$-valued process $\{\varphi_{s,s+t}(x); \ t \geq 0\}$ describes the motion of a passive tracer particle which starts from the position $\mathbf{x}$ at time s. It is easy to see that this stochastic process is a process of Brownian motion in $\mathbf{R}^d$ with the covariance $\Gamma^{(0)}(0)$. Indeed this process has continuous sample paths and independent increments. One can also derive the same result by first remarking that it is an $\mathbf{R}^d$-valued martingale (as an integral with respect to the $C(\mathbf{R}^d, \mathbf{R}^d)$-valued martingale $\vec{\mathbf{w}}(t)$) and then computing the square bracket. In any case, the conclusion is that the one point motions generated by a white velocity field are Brownian motions!

In fact, if $\mathbf{x}$ and $\mathbf{y}$ are starting points in $\mathbf{R}^d$, then one has:

$$[\varphi_{0,\cdot}(\mathbf{x}), \varphi_{0,\cdot}(\mathbf{y})]_t = \int_0^t \mathbf{\Gamma}^{(0)}(\varphi_{0,\mathbf{s}}(\mathbf{x}) - \varphi_{0,\mathbf{s}}(\mathbf{y}))d\mathbf{s} \qquad (1.40)$$

which give the desired result when $\mathbf{x} = \mathbf{y}$. More precisely, one can check that the process $\{(\varphi_{0,t}(\mathbf{x}), \varphi_{0,t}(\mathbf{y}))\}_{t \geq 0}$ is a diffusion in $\mathbf{R}^d \times \mathbf{R}^d$ with infinitesimal generator L given by:

$$[Lf](\mathbf{x}, \mathbf{y}) = [\frac{1}{2}(\mathbf{\Delta}_{\mathbf{x}} + \mathbf{\Delta}_{\mathbf{y}})f + \sum_{j,l} \Gamma_{j,l}^{(0)}(\mathbf{x} - \mathbf{y})\frac{\partial^2 f}{\partial x_j \partial y_l}](\mathbf{x}, \mathbf{y}).$$

The coupling formula (1.40) makes it possible to determine properties of the separation process $\rho_t = |\varphi_{0,\cdot}(\mathbf{x}) - \varphi_{t,\cdot}(\mathbf{y})|$ in the case of white velocity fields. Indeed, it is easy to see that in this particular case ρ_t satisfies the (closed) equation:

$$d\rho_t = \sigma(\rho_t)dW_t + \eta(\rho_t)dt$$

for some one-dimensional Wiener process $\{W_t\}_{t \geq 0}$ and for some specific functions σ and η on the half axis $(0, \infty)$. See [47] and [10] for details. As a consequence, it is possible to show that:

$$\lim_{t \to \infty} \rho_t = +\infty \qquad (1.41)$$

almost surely if $d \geq 3$ while the process $\{\rho_t\}_{t \geq 0}$ is null recurrent and the limit (1.41) holds only in probability in the case $d = 2$. Notice that we are only recalling the results in the incompressible case. The reader can consult [47] and [10] for a discussion of the other cases.

When considering the time evolution ν_t of a mass distribution ν_0 under the flow (recall the meaning of the notations c_t and D_t for the center of mass and the dispersion matrix of ν_t), it is proven in [25] that the expected value of the entries of the matrix D_t are asymptotically linear in time while the variance of c_t is bounded when $d \geq 3$, and logarithmic in time when $d = 2$.

Again, these results concern only the incompressible case. The reader is referred to [25] for a discussion of the general case.

1.6.3 Jacobian Flows and Lyapunov Exponents

We momentarily fix $s = 0$ and the starting point $\mathbf{x}$. For each $t \geq 0$ we use the notation $\mathbf{Y}(t) = \mathbf{Y}_{s,s+t}(\mathbf{x})$ as before. We retrace, in the particular case of white velocity fields which we consider now, the various steps of the discussion of the Jacobian flow which we gave earlier. Taking (formally) the gradient in both sides of the defining equation:

$$\varphi_{s,t}(\mathbf{x}) = \mathbf{x} + \int_s^t \vec{\mathbf{w}}(d\mathbf{u}, \varphi_{s,u}(\mathbf{x})) \tag{1.42}$$

one gets:

$$\nabla\varphi_{s,t}(\mathbf{x}) = \mathbf{I_d} + \int_s^t \nabla\vec{\mathbf{w}}(d\mathbf{u}, \varphi_{s,u}(\mathbf{x}))\nabla\varphi_{s,u}(\mathbf{x}) \tag{1.43}$$

In other words we have:

$$\begin{cases} d\mathbf{Y}(t) &= d\mathbf{W}(t)\mathbf{Y}(t) \\ \mathbf{Y}(0) &= \mathbf{I_d} \end{cases} \tag{1.44}$$

provided we set:

$$\mathbf{W}(t) = \int_0^t \nabla\vec{\mathbf{w}}(d\mathbf{u}, \varphi_{s,u}(\mathbf{x})).$$

Notice that all the above stochastic integrals have to be understood in Ito's sense. The stochastic process $\{\mathbf{W}(t)\}$ has a simple interpretation when viewed as a process with values in the vector space $M(d) = \mathbf{R}^{d\times d}$ of $d \times d$ real matrices. Indeed, it has continuous sample paths and independent increments. In fact, it is a mean zero Gaussian process with covariance tensor:

$$\mathbb{E}\{\mathbf{W}(s) \otimes \mathbf{W}(t)\} = \mathbf{C}\, s \wedge t$$

for some positive definite tensor C. The entries of C are given by the covariances of the entries of the $d\times d$ Gaussian matrix $\mathbf{W}(1)$, more precisely:

$$C_{j,j'}^{\ell,\ell'} = \mathbb{E}\{\mathbf{W}_{j,j'}(1)\mathbf{W}_{\ell,\ell'}(1)\}.$$

It inherits the properties of the spatial covariance of the velocity field. In particular, the isotropy assumption forces C to be of the form:

$$C_{j,j'}^{\ell,\ell'} = a\delta_{j,\ell}\delta_{j',\ell'} + b\delta_{j,j'}\delta_{\ell,\ell'} + c\delta_{j,\ell'}\delta_{j',\ell}$$

for some real numbers a, b, and c, which satisfy:

$$\begin{cases} a + c > 0 \\ a - c > 0 \\ a + b + dc > 0 \end{cases}$$

These conditions are in fact equivalent to the existence of the isotropic Brownian motion on the space of matrices (see [47]). Moreover, under these conditions the incompressibility assumption becomes equivalent to the condition:

$$ad + b + c = 0$$

on the coefficients of the matrix C. The process $\{W(t)\}$ is called an *additive* Brownian motion. An $M(d)$-valued process is called a *multiplicative* Brownian motion in the space of matrices when it is the solution of Equation (1.24) when the latter is understood in the Stratonovich sense. In coordinate form, Equation (1.24) becomes the stochastic differential system:

$$dY_{j,j'}(t) = \sum_{\ell} dW_{j,\ell}(t) Y_{\ell,j'}(t), \qquad j, j' = 1, \cdots, d.$$

Using this form of the stochastic differential of $Y(t)$ and Ito's stochastic calculus, one gets:

$$
\begin{aligned}
d \log \|Y(t)\| &= \frac{1}{2\|Y(t)\|^2} d\|Y(t)\|^2 - \frac{1}{4\|Y(t)\|^2} d\left[\|Y(\cdot)\|^2, \|Y(\cdot)\|^2\right]_t \\
&= \sum_{j,m} \left(\sum_{l} Y_{j,1}(t) Y_{m,1}(t) \right) dW_{j,m}(t) + \frac{1}{2}(d-1)(a+c)\,dt \\
&= M_t + \frac{1}{2}(d-1)(a+c)\,dt.
\end{aligned}
$$

The martingale $\{M_t\}_{t \geq 0}$ is in fact a Brownian motion because its square bracket is proportional to t. Indeed a simple calculation gives:

$$[M, M]_t = a + c + d.$$

This implies that:

$$\lambda_1 = \frac{1}{2}(d-1)(a+c) > 0. \tag{1.45}$$

As before, the above computation was done in the incompressible case. The general case is given in [47] where the other Lyapunov exponents λ_j are also computed.

1.6.4 Intersection Number Revisited

Let us now come back to the discussion of the problem of the estimation of the number of intersection points of the curve γ_t with a fixed nonrandom straight line. We use the notations introduced earlier and we specialize the discussion to the Brownian flows analyzed in this section.

The random variables $\gamma_t^2(\alpha)$ and $d\gamma_t^2(\alpha)/d\alpha$ are always uncorrelated, in the case of the white noise velocity field (corresponding to the Brownian flows which we consider now), for each fixed $\mathbf{x} \in \mathbf{R}^d$ the process $\{\varphi_{0,t}(\mathbf{x})\}_{t\geq 0}$ is Gaussian. Consequently, for each fixed $t \geq 0$, $\varphi_{0,t}(\mathbf{x})$ and $\nabla\varphi_{0,t}(\mathbf{x})$ are jointly Gaussian and hence independent. This implies that for each $\alpha \in [0,1]$, the random variables $\gamma_t^2(\alpha)$ and $d\gamma_t^2(\alpha)/d\alpha$ are independent. Consequently:

$$p_{t,\alpha}(a,u) = p_{t,\alpha}(a)q_{t,\alpha}(u)$$

and:

$$
\begin{aligned}
\mathbf{E}\{N_L(t)\} &= \int_0^1 p_{t,\alpha}(a) \int_{I\!R} q_{t,\alpha}(u)|u|\, du\, d\alpha \\
&= \frac{1}{\sqrt{2\pi t}} \int_0^1 e^{-(\gamma_0(\alpha)-a)^2/2t} \mathbf{E}\{|\frac{d\gamma_t^2(\alpha)}{d\alpha)}|\} d\alpha\, d\alpha \\
&\approx e^{\lambda_1 t} \frac{1}{\sqrt{2\pi t}} \int_0^1 e^{-(\gamma_0(\alpha)-a)^2/2t} d\alpha
\end{aligned}
$$

which shows that the average number of intersections blows up exponentially when the upper Lyapunov λ_1 is strictly positive.

The joint normality property used above is presumably not true in general. Nevertheless, we expect that a central limit theorem type argument could be used to show asymptotic normality of this couple of random variables in the limit of large time $t \to \infty$, in this case we expect that the density factorization which we used above be approximately true and that the exponential nature of the growth of the expected number of intersections remain true in general.

1.7   Open Mathematical Problems

At the time the original version of these lecture notes was prepared, i.e., in November of 1994, most of the results discussed in the notes could only be proved in the case of Brownian flows (i.e., for velocity fields which are white in time). Despite some significant progress (see for example [16] through [18]) it is fair to say that there is still need for new methods of proof which can apply to the general case of Ornstein-Uhlenbeck velocity fields, and at least to those incompressible and isotropic fields with a Kolmogorov spectrum. We review briefly some of the main problems remaining open.

1.7.1   Analysis of the Lyapunov Exponents

The first problem to consider is of course the proof of the positivity of the upper Lyapunov exponent λ_1 in the general case of isotropic incompressible Ornstein-Uhlenbeck velocity fields with a Kolmogorov spectrum. The particular case of velocity fields with a spectrum with finitely many Fourier modes is considered in [16] but the general case is still resisting. Nevertheless, see [18].

The next question of interest is the dependence of λ_1 upon the parameters of the model. In the particular case of interest to us, the upper Lyapunov exponent is a mere function of the two parameters of the models, i.e., ϵ and z. It is interesting to understand what kind of a function is $(\epsilon, z) \hookrightarrow \lambda_1(\epsilon, z)$. The first results of the numerical simulations (see, for example, Figure 5) suggest some properties such as monotonicity and/or convexity, but our results are preliminary and it might be premature to formulate conjectures on the functional dependence of λ_1 upon the parameters of the model. Such results will have important applications if we find a practical way to estimate this Lyapunov exponent from the observations of a (relatively) small number of trajectories, say $\{\varphi_{0,t}(\mathbf{x_j}); \ \mathbf{t} = \mathbf{1}, \mathbf{1} \cdots, \mathbf{T}\}_{\mathbf{j}=\mathbf{1},\cdots,\mathbf{J}}$, of passive tracers. This is another of the difficult mathematical challenges of the subject!!

1.7.2   Fractal Dimensions

We considered the time evolution of a smooth curve $\{\gamma(s)\}_{0 \le s \le 1}$ in an incompressible flow. We saw that the curve remains smooth at all times but that its length grows (presumably exponentially in general, as suggested by [16] and [18]) without apparent bounds. The plots shown earlier (see Figure 6 for a blow-up) also suggest that the dimension of the subset of the plane formed by the points $\gamma_t(s)$ may, at least in an appropriate asymptotic regime, become strictly larger than 1. The numerical study [19] contains new evidence of an intermediate regime in which an effective dimension appears and some simulations are used to suggest a value close to 3/2 for this fractal dimension. Related problems have been considered in [5], where a slightly different problem involving the asymptotic dimension of interfaces was considered in a renormalization regime for time-independent shear flows. Another interesting challenge is the numerical computation of the singularity spectrum of γ_t in the multifractal formalism of [13].

1.7.3   An Alternative Model

Our numerical simulations of sample trajectories of the flow solving the equations of motion show clearly that some of the characteristics of the real-life drifter paths cannot be reproduced in the Gaussian model which we use, especially when the parameters of the simulation are not chosen appropriately. In particular, see the differences between Figure 1 and Figure 4. These shortcomings of the Gaussian models have been known and/or suspected for some time. The discrepancies have prompted researchers in the field to propose alternate models of velocity fields based on the existence of large long-lived structures trapping the tracers in smooth circular motions. These structures are called eddies. They can be made the main building blocks of the new models. A precise definition of such a model based on Poisson shot noise point processes can be found in [23] and [24].

1.8 Stochastic Flows and SPDEs

We end these lecture notes with a short discussion of some of the problems in stochastic partial differential equations (SPDEs for short) connected with our analysis of the transport properties of the Gaussian velocity fields. As explained in the introduction, this short section cannot do justice to the field of SPDEs. The first part of our discussion is modeled after [54].

As before, we consider a (possibly) time-dependent velocity field $\vec{v}(t, \mathbf{x})$. In the absence of diffusivity in our model, the Lagrangian trajectories are given by the solutions of the transport equation:

$$\varphi_{0,t}(\mathbf{x}) = \mathbf{x} + \int_0^t \vec{v}(\mathbf{s}, \varphi_{0,\mathbf{s}}(\mathbf{x})) \, d\mathbf{s}.$$

Let us consider the time evolution of the concentration $C(t, \mathbf{x})$ at time t and location $\mathbf{x}$ of some scalar passive tracer. In applications to physical oceanography, $C(t, \mathbf{x})$ could be the temperature, the salinity of the ocean, etc. This time evolution is given by the flow via the formula:

$$C(t, \mathbf{x}) = \mathbf{C_0}(\varphi_{0,t}^{-1}(\mathbf{x}))$$

where $C_0(\mathbf{x})$ is the initial concentration at time $t = 0$. It is the solution of the transport equation:

$$\frac{\partial C}{\partial t} = (\vec{v} \cdot \nabla)C.$$

More generally, one considers the transport of passive scalars by the velocity field $\vec{v}(t, \mathbf{x})$ in the presence of molecular diffusivity. The Lagrangian trajectories are obtained by solving the Ito stochastic differential equation:

$$dX_t = \vec{v}(t, X_t)dt + \sqrt{2\kappa} \, dW_t \tag{1.46}$$

where $\{W_t; t \geq 0\}$ is a d-dimensional standard Wiener process. The time evolutions of concentrations are now given by the solutions of the parabolic equation:

$$\frac{\partial C}{\partial t} = (\vec{\mathbf{v}} \cdot \nabla)C + \kappa \Delta C \tag{1.47}$$

and they can still be expressed in terms of the (stochastic) flow $\{\varphi_{s,t}\}$ generated by the stochastic differential Equation (1.46) via the formula:

$$C(t, \mathbf{x}) = \mathbf{E}^{\mathbf{W}}\{C_0(\varphi_{0,t}^{-1}(\mathbf{x}))\}$$

where the superscript ^{W} emphasizes the fact that the expectation is taken with respect to the Wiener process driving the stochastic differential Equation (1.46). The reader is referred to Kunita's book [45] for a detailed exposé of the standard facts used in this section.

In the classical theory of fluid flows the velocity field $\vec{\mathbf{v}}(t, \mathbf{x})$ is required to satisfy the equations derived from the physics of the problem. In most cases it appears as a solution of the Navier-Stokes equation. As explained in the Introduction, we are interested in the kinematic approach in which the velocity field is regarded as a random stationary and homogeneous field with a prescribed spectrum. As before, we shall assume that this field has mean zero and is Gaussian.

1.8.1   Moment Equations

Since the pioneering works of Bachelor and Taylor, one of the most important problems is to understand the properties of the moments

$$m_1(t, \mathbf{x}) = \mathbf{E}^{\mathbf{V}}\{\mathbf{C}(t, \mathbf{x})\}, \qquad
\begin{aligned}
\mathbf{m_p}(t, \mathbf{x}) &= \mathbf{E}^V\{C(t, \mathbf{x})^{\mathbf{p}}\}, \\
g_p(t, \mathbf{x}) &= \mathbf{E}^V\{|\nabla C(t, \mathbf{x})|^{\mathbf{p}}\},
\end{aligned}$$

of the concentrations. Notice that we now use the superscript ^{V} to indicate that the expectations are averages over the configurations of the velocity field. The time-honored method to analyze these (deterministic) functions of time and space is to show that they are solutions of some simple diffusion equations. Let us assume for the sake of illustration that the velocity field is white in time. In this case, the random partial differential Equation (1.47) can be interpreted as:

$$\begin{aligned}
dC(t, \mathbf{x}) &= (\vec{\mathbf{v}}(t, \mathbf{x})\mathbf{dt}) \cdot \mathbf{C}(t, \mathbf{x}) + \kappa \Delta \mathbf{C}(t, \mathbf{x})\mathbf{dt} \\
&= \vec{\mathbf{w}}(dt, \mathbf{x}) \cdot \mathbf{C}(t, \mathbf{x}) + \kappa \Delta \mathbf{C}(t, \mathbf{x})\mathbf{dt}
\end{aligned} \tag{1.48}$$

by using the notation $\vec{\mathbf{w}}$ for the antiderivative in time of $\vec{\mathbf{v}}$. Assuming that $\vec{\mathbf{v}}$ is a white noise in time is the same thing as assuming that $\vec{\mathbf{w}}$ is a Brownian motion in time. If Equation (1.48) is interpreted in the Ito sense, then the first moment is a solution of the heat equation:

$$\frac{\partial m_1(t, \mathbf{x})}{\partial t} = \kappa \Delta m_1(t, \mathbf{x}). \tag{1.49}$$

A correction term has to be included if the stochastic partial differential equation is understood in the sense of Stratonovich. Indeed, in this case the diffusion matrix κI of Equation (1.49) has to be replaced by a matrix of the form $(\kappa I + D)$, the extra term D being called the Taylor correction. See, for example, [14] for a rigorous justification by a diffusion approximation result. Similar equations can easily be derived for the moments of higher orders. Unfortunately the moments are not, in general, solutions of closed equations and a lot of energy has been devoted to the design of adhoc procedures intended to close the hierarchical structure of the moment equations. Most of these procedures are phenomenological in nature and as a consequence they are very artificial from a mathematical point of view. We shall review below several rigorous attempts to derive some of these equations. But first, we follow [54] and consider another toy model for the purpose of illustration. We assume that the velocity field changes in time but is constant over space. In other words we assume that:

$$\vec{v}(t, \mathbf{x}) \equiv \vec{v}(t)$$

for some $\mathbf{R}^d$-valued mean zero stationary Gaussian field $\{\vec{v}(t); \, t \geq 0\}$ with covariance:

$$\Gamma(t) = \mathbf{E}^V \{\vec{v}(t) \otimes \vec{v}(0)\}.$$

In this very particular case the Lagrangian flow and its inverse can be written explicitly. We get:

$$\varphi_{0,t}^{-1}(\mathbf{x}) = \mathbf{x} - \int_0^t \vec{v}(t - s) ds + \sqrt{2\kappa} \mathbf{W}_t$$

and the time evolution of the concentrations is given by the formula:

$$C(t, \mathbf{x}) = \int_{\mathbb{R}^d} \mathbf{C_0} \left(\mathbf{x} - \mathbf{y} - \int_0^t \vec{v}(s) ds \right) \frac{1}{\sqrt{4\pi t \kappa}^d} e^{-|y|^2/(4t\kappa)} \, d\mathbf{y}.$$

Rewriting the initial concentration C_0 in terms of its Fourier transform and using the notation:

$$\sigma^2(t) = \mathbf{E}^V \{| \int_0^t \vec{v}(s) ds|^2\},$$

one easily sees that the first moment $m_1(t, \mathbf{x})$ satisfies the parabolic equation:

$$\frac{\partial m_1(t, \mathbf{x})}{\partial t} = (\nabla \cdot D(t)\nabla)m_1(t, \mathbf{x}). \tag{1.50}$$

with:

$$D(t) = \kappa I + \frac{1}{2}\frac{d\sigma^2(t)}{dt} = \kappa I + \int_0^t \Gamma(s)ds. \tag{1.51}$$

The diffusion matrix $D = D(t)$ is called the *effective turbulent diffusivity tensor*. At the level of the first moment, the velocity field contributes to the overall diffusivity. We say that the latter has been enhanced.

The purpose of the following discussion is to identify regimes in which effective equations of the type:

$$\frac{\partial m_1(t, \mathbf{x})}{\partial t} = (\nabla \cdot D\nabla)m_1(t, \mathbf{x}), \tag{1.52}$$

can be derived rigorously. This is an important problem because the oceanographic literature (and more generally, the fluid mechanics literature) contains many publications in which a formula like (1.52) is stated as an assumption, and properties of the turbulent diffusion tensor D and of the physical system are derived from such an equation. As an example, we mention the *ANSATZ:*

$$\mathbb{E}^V\{(C(t, \mathbf{x}) - m_1(t, \mathbf{x}))\vec{v}(t, \mathbf{x}) \otimes \vec{v}(t, \mathbf{x})\} \equiv 0$$

which was used to derive a nonlinear integral equation relating the Eulerian spectral density matrix $E(\omega, \mathbf{k})$ to the Lagrangian spectral density. The existence of such an equation is of great practical importance, for drifter measurements can be used to estimate the Lagrange spectrum while the quantity of interest is the Euler spectrum. Unfortunately such an ansatz is yet to be fully justified at a mathematical level.

As explained in the introduction, we shall use the problems of physical oceanography to motivate and illustrate the assumptions made throughout. The multiscale nature of the physical phenomena imposes serious constraints on the scales and the correlation lengths of the mathematical models. Very little (as far as mathematical results are concerned) has been proved when all the scales are of the same order. Only when they are well separated is it possible to justify simplified mathematical models for which theorems can be proved.

Following [54] we introduce a *correlation radius* ℓ_v and a *correlation time* τ_E equal to the typical lifetime of an eddy. We use the notation τ_E to emphasize the Eulerian nature of this correlation time. We then define the *turnover time*

$$\tau_T = \frac{\ell_v}{\sigma_v}$$

where σ_v denotes the typical size of the velocity; more precisely $\sigma^2 = \mathbf{E}^V\{|\vec{v}(t,\mathbf{x})|^2\}$. Finally we define the molecular diffusion time τ_D by the formula:

$$\tau_D = \frac{\ell_0^2}{\kappa}$$

where ℓ_0 denotes a typical space scale and where κ denotes as usual the molecular diffusivity constant. For illustration purposes we discuss a few of the many ways in which these scales can separate.

- We can for example consider the case

$$t \ll \tau_D.$$

In this case the molecular diffusivity can be ignored and we can work with the mathematical model in which $\kappa = 0$. This is the case considered so far in these notes.

- If we consider the case

$$t \sim \tau_E \sim \tau_D \ll \tau_T$$

then the space variations become significant at such a large scale that it is reasonable to work with the mathematical model of a velocity field depending only on time, i.e., for which $\vec{v}(t,\mathbf{x}) = \vec{v}(\mathbf{t})$. The latter was considered earlier as a toy model.

- If we now consider the case

$$t \sim \tau_D \sim \tau_T \ll \tau_E$$

then the time variations of the velocity field can be neglected and it is reasonable to work with the mathematical model of a velocity field depending only on the space variable, i.e., for which $\vec{v}(t,\mathbf{x}) = \vec{v}(\mathbf{x})$. These time-independent velocity fields are called *stationary* in the non-probabilist literature. The flows which they generate are not necessarily easier to analyze, the difficulties being typical of the theory of random media. In any case, these models remain rich in their mathematical structure and numerous unsolved problems remain.

1.8.2 Renormalization Theories

If we consider the last scale separation regime considered above and if furthermore we assume that:

$$\tau_T \ll t \sim \tau_D \ll \tau_E,$$

then one possible way to analyze mathematically such a regime is to introduce a small parameter $\delta > 0$ and to consider, for example, velocity fields of the form:

$$\vec{v}_\delta(t, \mathbf{x}) = \frac{1}{\delta}\vec{v}(\frac{\mathbf{x}}{\delta})$$

This leads to the so-called *homogenization* regime in which the solution of the SPDE:

$$\frac{\partial C_\delta(t, \mathbf{x})}{\partial t} = \kappa\Delta C_\delta(t, \mathbf{x}) + \vec{v}_\delta(t, \mathbf{x}) \cdot \nabla C_\delta(t, \mathbf{x}) \qquad (1.53)$$

converges (say, in probability) to an effective concentration:

$$\lim_{\delta\searrow 0} C_\delta(t, \mathbf{x}) = \overline{c}(t, \mathbf{x})$$

which is a deterministic function which solves the diffusion equation:

$$\frac{\partial \overline{c}}{\partial t} = \nabla \cdot (\kappa I + \overline{D})\nabla\overline{c} \qquad (1.54)$$

The reader is referred to the classical works of [41] and [55] on the subject, and also to [5]. See below for more on homogenization.

- Diffusion Approximation

The introduction of a small parameter opens the door to all sorts of possible *renormalizations* of the SPDE (1.47). The simplest one concerns the *diffusion approximation* regime for which:

$$\vec{v}_\delta(t, \mathbf{x}) = \frac{1}{\delta}\vec{v}(\frac{t}{\delta^2}, \mathbf{x}). \qquad (1.55)$$

Notice that this regime corresponds to the scale separation:

$$\tau_E \ll \tau_T \ll 1.$$

In this case the rescaled velocity field $\vec{v}_\delta$ converges in distribution toward a velocity field which is white noise in time. In this case one expects that the concentration $C_\delta(t, \mathbf{x})$ converges in distribution (as a stochastic field parameterized by both time and space) toward the solution of the SPDE (1.48). This was proven in [14] using a technique first introduced in [12] in the analysis of diffusion approximation results for SPDEs.

Stirring and mixing processes are of great importance in physical oceanography. In the mathematical model, they correspond respectively to the increase and the decay of the gradients of the concentrations. This motivates the analysis of the quantities:

$$R_t(\mathbf{x}, \mathbf{y}) = \mathbb{E}^V\{\nabla C(t, \mathbf{x})^t \nabla C(t, \mathbf{y})\}$$

and:

$$g(t) = \mathbf{E}^V\{|\nabla C(t,\mathbf{x}) - \mathbf{m_1}(t,\mathbf{x})|^2\}.$$

In the diffusion approximation regime, i.e., for the solution of the SPDE (1.48), it is easy to derive closed equations for these quantities. Their time evolutions were completely analyzed in [52], where the physical interpretation of the results was given.

- Homogenization for Time-Dependent Velocity Fields

We now consider the scale separation given by:

$$\tau_T \sim \tau_E \ll t \sim \tau_D.$$

Modeling this scale separation with a small parameter, this case amounts to considering the velocity field of the form:

$$\vec{v}_\delta(t,\mathbf{x}) = \frac{1}{\delta}\vec{v}(\frac{t}{\delta^2}, \frac{\mathbf{x}}{\delta})$$

and analyzing the small δ limit for the solution of the SPDE (1.53) with initial condition in the right scale. This is the problem of homogenization for a time-dependent velocity field. One expects that under some conditions including:

$$\int |E(0,\mathbf{k})|\, \mathbf{dk} < \infty,$$

the solution of the SPDE (1.53) converges, say in probability, to an effective (deterministic) concentration which solves a diffusion equation of the type (1.54). Such a result has been proved rigorously in the case of shear flows in [4]. See below for more on the renormalization of shear flows. The formula:

$$\overline{D} = \int\int E(\omega,\mathbf{k})\frac{\mathbf{D_0 k \cdot k}}{(\mathbf{D_0 k \cdot k})^2 + \omega^2}\, \mathbf{d\omega dk}$$

with

$$D_0 = \kappa I + \pi \int E(0,\mathbf{k})\, \mathbf{dk}$$

was derived in [52] in the case of general Gaussian velocity fields, but unfortunately it seems that some of the arguments of [52] are not completely rigorous. An interesting feature of this formula is that it shows how the correction $\overline{D}$ to the molecular diffusivity κI appearing in the formula for the turbulent effective diffusivity D can depend upon the molecular diffusivity constant κ.

The proofs of homogenization results for time-dependent velocity fields are bound to be different than the proofs for time-independent ones. Several new homogenization results for time-dependent random velocity fields have been announced recently and preprints are circulating with such results for different classes of assumptions. We know of [40, 33] for fields under various time-correlation restrictions and, even more recently, [49] for general ergodic fields with smooth and bounded realizations, and finally [38] for PDEs with time-dependent periodic coefficients.

- Avellaneda-Majda Renormalization Theory of Shear Flows

The choice of the rescaling of the time by the factor δ^{-2} was justified by the uniform integrability of the spectral density near the origin and, as a consequence, the diffusive behavior $\mathbf{E}^V\{|\varphi_{0,t}(\mathbf{x})|^2\} \sim t$ of the flow. Avellaneda and Majda considered in [4] velocity fields like those given in Section 2.2.1, with an inertial range of the form $[r_0, r_1] = [\delta, 1]$ where δ, as above is the small parameter of the problem. Let us now denote by $C^{(\delta)}(t, \mathbf{x})$ the solution of the PDE:

$$\frac{\partial C^{(\delta)}(t, \mathbf{x})}{\partial t} = \kappa \Delta C^{(\delta)}(t, \mathbf{x}) + \vec{\mathbf{v}}(t, \mathbf{x}) \cdot \nabla \mathbf{C}^{(\delta)}(t, \mathbf{x})$$

with initial condition:

$$C^{(\delta)}(0, \mathbf{x}) = \mathbf{C}_0(\delta \mathbf{x})$$

for some smooth initial concentration C_0 with compact support. Avellaneda and Majda analyzed the existence and the behavior of the limit:

$$\overline{C}(t, \mathbf{x}) = \lim_{\delta \searrow 0} \mathbf{E}^{\mathbf{V}}\{\mathbf{C}^{(\delta)}(\frac{t}{\rho(\delta)^2}, \frac{\mathbf{x}}{\delta})\}$$

for appropriate functions $\rho(\delta)$. The results of their remarkable analysis is a finite partition of the (ϵ, z)-plane of spectral parameters corresponding to the different renormalization regimes. In each region of the partition, they found a function $\rho(\delta)$ for which they could prove the existence of $\overline{C}(t, \mathbf{x})$ as solution of an effective diffusion equation. We refer the interested reader to [4] for a complete analysis of the shear flows and to [18] for extension of this theory to the general Gaussian velocity fields with Kolmogorov spectra which we considered in these notes.

Acknowledgments: Our interest in the topic of these lectures originated in and was stimulated by a series of collaborative works with F. Cerou, S. Grishin, S.A. Molchanov, A. Wang, and L. Xin. I would like to thank them for letting me use material from joint works and works still in progress. Also, we benefited from the financial support of ONR N00014-91-1010 and NSF INT-9017002 during the preparation of the manuscript.

Bibliography

[1] R.J. Adler (1980): *Geometry of Random Fields*, John Wiley & Sons,, New York.

[2] A. Antoniadis and R. Carmona (1985): Infinite Dimensional Ornstein Uhlenbeck Processes, *Probab. Th. Rel. Fields*, **74**, 31-54.

[3] M. Avellaneda, F. Elliott, and A. Appelian (1993): Trapping, percolation and anomalous diffusion of particles in a @-dimensional random field (preprint).

[4] M. Avellaneda and A. Majda (1990): Mathematical models with exact renormalization for turbulent transport, *Commun. Math. Phys.*, **131**, 381-429.

[5] M. Avellaneda and A. Majda (1991): An integral representation and bounds on the effective diffusivity in passive advection by laminar and turbulent flows, *Commun. Math. Phys.*, **138**, 339-391.

[6] G.K. Batchelor (1982): The Theory of Homogeneous Turbulence, Cambridge University Press, London, U.K.

[7] P.H. Baxendale (1986): Asymptotic Behavior of Stochastic Flows of Diffeomorphisms: Two Case Studies, *Probab. Theor. Rel. Fields*, **73**, 51-85.

[8] P.H. Baxendale (1991): Statistical Equilibrium and Two-Point Motion for a Stochastic Flow of Diffeomorphisms, in *Spatial Stochastic Processes*, Eds. K.S. Alexander, and J.C. Watkins, pp 189-218, Birkhaüser, Boston.

[9] P.H. Baxendale (1992): Stability and Equilibrium of Stochastic Flows of Diffeomorphisms, in *Diffusion Processes and Related Problems in Analysis*, Vol. III Stochastic Flows, Eds. M.A. Pinsky and V. Wihstutz, pp 3-36, Birkhaüser, Boston.

[10] P.H. Baxendale and T.E. Harris (1986): Isotropic Stochastic Flows, *Ann. Probab.* **14**,1155-1179.

[11] A.F. Bennet (1987): A Lagrangian Analysis of Turbulent Diffusion. *Rev. Geophys.*, **25**,799-822.

[12] P. Bouc and E. Pardoux (1984): Asymptotic analysis of P.D.E.s with wide-band noise disturbances and expansion of the moments. *Stoch. Anal. Appl.*, **2**, 369-422.

[13] J.F. Buzy, E. Bacry, and A. Arneodo (1992): Multifractal formalism for fractal signals: the structure function approach versus the wavelet transform modulus maxima method. (Preprint.)

[14] R. Carmona and J. P. Fouque (1994): Diffusion-Approximation for the Advection-Diffusion of a Passive Scalar by a Space-Time Gaussian Velocity Field. *Proc. Intern. Conf. on SPDEs*, Ascona, June 1993. Birkhäuser, Basel.

[15] R. Carmona, S. Grishin, and S.A. Molchanov (1994): Massively Parallel Simulations of the Transport Properties of Gaussian Velocity Fields. to appear in *Mathematical Models for Oceanography*, Eds. R. Adler, P. Muller, and B. Rozovskii, Birkhaüser, Boston.

[16] R. Carmona, S. Grishin, S.A. Molchanov, and L. Xu (1996): Surface Stretching of Ornstein Uhlenbeck Velocity Fields, *Electr. Commun. Probab.* (to appear).

[17] R. Carmona and L. Xu (1996): Homogenization for Time Dependent 2-D Incompressible Gaussian Flows. (Submitted for publication.)

[18] R. Carmona and L. Xu (1996): Gaussian Velocity Fields with Kolmogorov Spectra. (In preparation.)

[19] R. Carmona and F. Cerou (1996): Transport Simulations with 2-D Incompressible OU Velocity Fields. (In preparation.)

[20] R. Carmona and J. Lacroix (1990): *Spectral Theory of Random Schrödinger Operators*, Birkhaüser, Boston.

[21] R. Carmona and D. Nualart (1990): *Nonlinear Stochastic Integrators Equations and Flows*, Stochastic Monographs, #6, Gordon & Breach, New York.

[22] R. Carmona and A. Wang (1994): Comparison Tests for the Spectra of Dependent Multivariate Time Series. (Preprint.)

[23] E. Cinlar (1994): On a Stochastic Velocity Field. (Preprint.)

[24] E. Cinlar (1994): Poisson Shot Noise Velocity Fields. (Preprint.)

[25] E. Cinlar and C.L. Zirbel (1994): Dispersion of Particle Systems in Brownian Flows, *Adv. Appl. Probab.*, **28**,53-74.

[26] H. Cramer and M.R. Leadbetter (1967): *Stationary and Related Stochastic Processes: Sample Function Properties and Their Applications*, John Wiley & Sons, New York.

[27] R.E. Davis (1982): On relating Eulerian and Lagrangian velocity statistics: single particle in homogeneous flows, *J. Fluid Mech.*, **114**, 1-26.

[28] R.E. Davis (1983): Oceanic property transport, Lagrangian particle statistics, and their prediction, *J. Mar. Res.*, **41**, 163-194.

[29] R.E. Davis (1991): Lagrangian ocean studies, *Ann. Rev. Fluid Mech.*, **23**, 43-64.

[30] D.A. Dawson (1972): Stochastic evolution equations. *Math. Biol. Sci.*, **15**, 287-316.

[31] D.A. Dawson (1975): Stochastic evolution equations and related measure processes, *J. Multivar. Anal.*, **5**, 1-52.

[32] A. Grorud and D. Talay (1995): Approximation of Upper Lyapunov Exponents of Bilinear Stochastic Differential Systems, *SIAM J. Numer. Anal.*, **28**(4), 1141-1164.

[33] A. Fangjiang and G. C. Papanicolaou (1996): Convection Enhanced Diffusion. (Preprint.)

[34] F. Elliott and A. Majda (1994): A wavelet Monte Carlo method for turbulent diffusion with many spatial scales. (Preprint.)

[35] F. Elliott, A. Majda, D. Horntrop, and R. McLaughlin (1994): Hierarchical Monte Carlo methods for fractal random fields. (Preprint.)

[36] U. Frisch (1995): Turbulence: the Legacy of A.N. Kolmogorov. Cambridge Univ. Press, London, UK.

[37] D. Geman and J. Horowitz (1975): Random shifts which preserve measure, *Proc. Am. Math. Soc.*, **49**, 143-150.

[38] J. Garnier (1994): Homogenization in Periodic and Time Dependent Potential. (Preprint.)

[39] R. Holley and D.W. Stroock (1978): Generalized Ornstein Uhlenbeck processes and infinite particle branching Brownian motions, *Publ. RIMS Kyoto Univ.*, **14**, 741-788.

[40] T. Komorowski and G. Papanicolaou (1986): Motion in a Random T-Dependent Gaussian Incompressible Flow. (Preprint.)

[41] S.M. Kozlov (1983): Reducibility of quasiperiodic operators and homogenization, *Trans. Moscow Math. Soc. #46*, 99-123.

[42] S.M. Kozlov (1985): The Method of Averaging and Walks in Inhomogeneous Environments, *Russ. Math. Surveys*, **40**, 73-145.

[43] R.M. Kraichnan (1959): The structure of isotropic turbulence at very high Reynolds numbers, *J. Fluids Mech.*, *#5*, 497-543.

[44] R.M. Kraichnan (1974): Convection of a passive scalar by a quasi uniform random straining field, *J. Fluids Mech.*, *#64*, 737-762.

[45] H. Kunita (1990): Stochastic Flows and Stochastic Differential Equations. Cambridge Univ. Press. London, UK.

[46] C. Landim, S. Olla, and H.T. Yau (1996): Convection-Diffusion Equation with Space-Time Ergodic Random Flow. (Preprint.)

[47] Y. Le Jan (1984): On isotropic Brownian motions, *Z. Wahrscheinlichkeitstheorie verw. Geb.*, **70**, 609-620.

[48] Y. Le Jan (1991): Asymptotic Properties of Isotropic Brownian Flows, in *Spatial Stochastic Processes*, Eds. K.S. Alexander, and J.C. Watkins, pp 219-232, Birkhaüser, Boston.

[49] C. Landim, S. Olla, and H.T. Yau (1966): Homogenization for general ergodic velocity fields. (Preprint.)

[50] J.L. Lumley (1962): The mathematical nature of the problem of relating Lagrangian and Eulerian statistical functions in turbulence, in *Mécanique de la Turbulence*, Coll. Intern. CNRS, Marseille, Ed. CNRS, Paris.

[51] S.A. Molchanov (1994): Lectures on Random Media, in St Flour Summer School in Probability, Lect. Notes in Math. Springer Verlag (to appear)

[52] S.A. Molchanov and L. Piterbarg (1992): Heat Propagation in Random Flows, *Russ. J. Math. Phys.#1*, 1-22.

[53] A.S. Monin and A.M. Yaglom (1971): *Statistical Fluid Mechanics: Mechanics of Turbulence*, MIT Press, Cambridge, Mass.

[54] L. Piterbarg (1994): Short Correlation Approximation in Models of Turbulent Diffusion. (Preprint.)

[55] G. Papanicolaou and S.R.S. Varadhan (1982): Boundary value problems with rapidly oscillating coefficients in random fields, *Coll. Math. Soc. Janos Bolay, #27*, Eds. J. Fritz, J.L. Lebowitz, and D. Szasz, 835-875.

[56] S.C. Port and C.J. Stone (1976): Random Measures and their Applications to Motion in an Incompressible Fluids, *J. Appl. Probab.*, **13**, 498-506.

[57] M. Röckner (1992): Dirichlet Forms on Infinite Dimensional State Spaces and Applications, in *Stochastic Analysis and Related Topics* Eds. H. Körezlioglu, and A.S. Üstunel, pp.131-186. Birkhaüser, Boston.

[58] D. Talay (1991): Lyapunov Exponents of Nonlinear Stochastic Differential Equations. *SIAM J. Appl. Math.* (to appear).

[59] G.I. Taylor (1953): Dispersion of Soluble Matter in Solvent flowing Slowly through a Tube, *Proc. R. Soc. A #219*, 186-203.

[60] J.B. Walsh (1981): Stochastic model of neural response, *Adv. Appl. Probab.* **13**, 231-281.

[61] A.M. Yaglom (1987): *Correlation Theory of Stationary and Related Random Functions*, Vol. I: Basic Results, Springer Verlag, New York.

[62] C.L. Zirbel (1993): Stochastic Flows: Dispersion of a Mass Distribution and Lagrangian Observations of a Random Field, Ph. D. thesis, Princeton University.

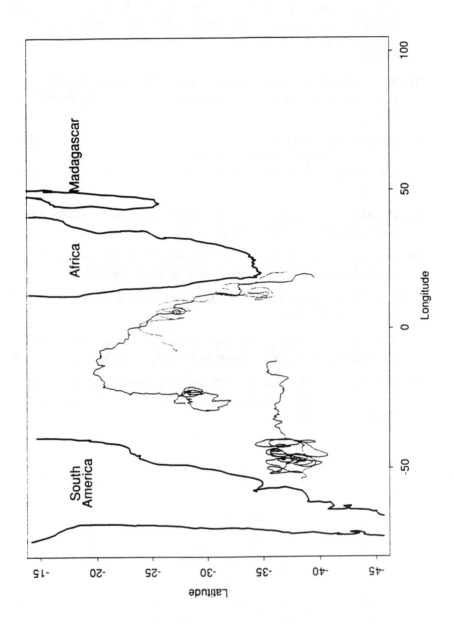

Figure 1: Plot of the daily positions measured by satellite of four drifters
Figures 1 and 2 also appear in publications [5] and [27].)

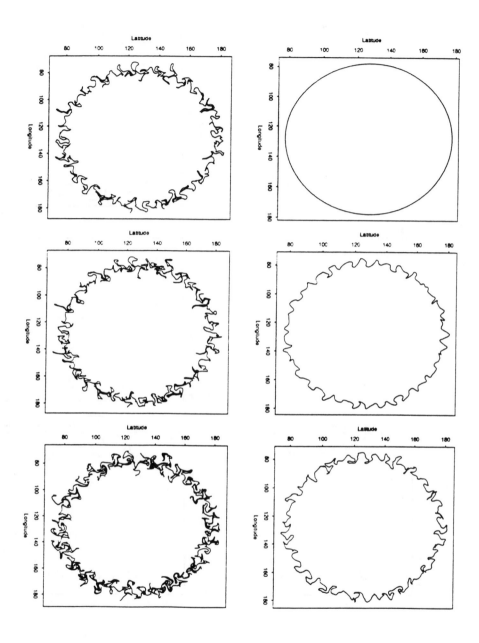

Figure 2: Time evolution of the boundary of a circle under the flow generated by an incompressible isotropic velocity field with a spectrum of the Kolmogorov's type.

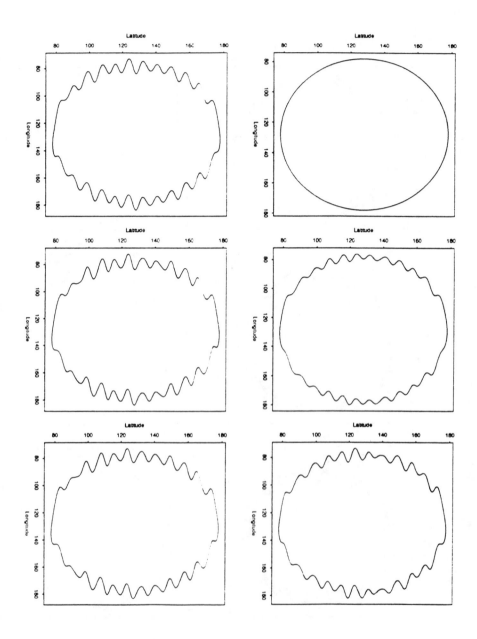

Figure 3: Time evolution of a circle under a shear flow.

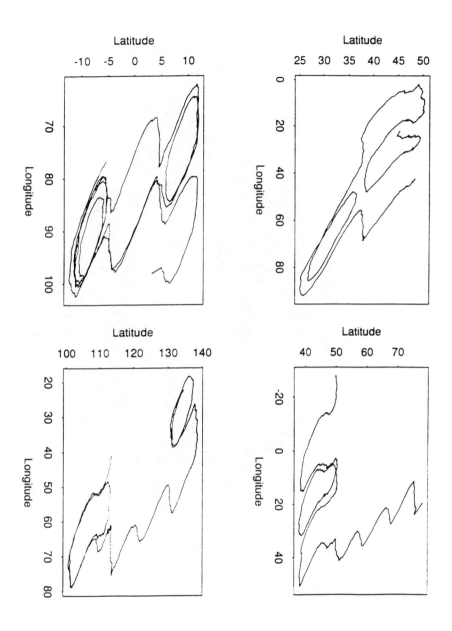

Figure 4: Plot of the trajectories of four simulated drifters.

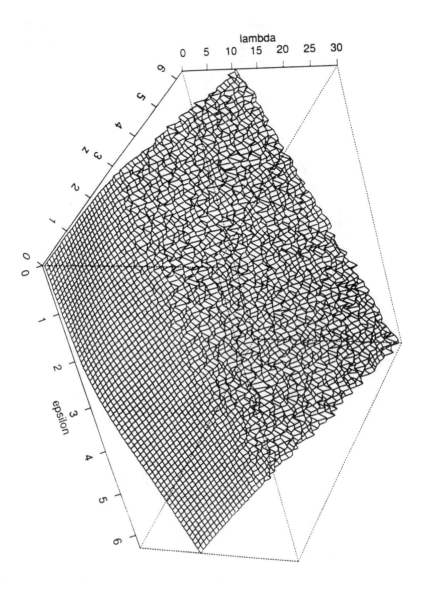

Figure 5: Example of a surface plot of the approximations of the upper Lyapunov exponents computed over a grid of spectral parameters (ϵ, z).

63

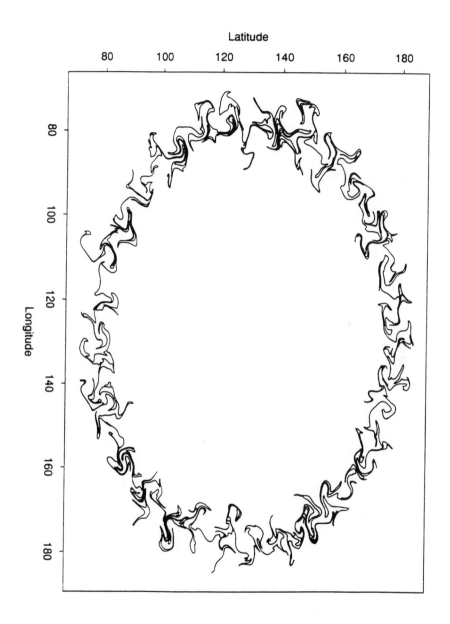

Figure 6: Blow up of the last of the six snapshots presented in Figure 2.

Chapter 2

Planar Stochastic Integration Relative to Quasimartingales

Michael L. Green

2.1 Introduction

Let $\{X_t, \mathcal{F}_t, t \in T \subset I\!\!R\}$ be a Brownian Motion (B.M.) process. If $f : T \longrightarrow I\!\!R$ is a bounded measurable function the integral $\int_T f(t) dX_t$ is undefined as a Lebesgue-Stieltjes integral, since X is of unbounded variation on every nonempty open subset of $I\!\!R$. This problem was overcome by N. Wiener who defined an integral $\int_T f(t) dX_t$ using a special type construction. Later this integral was generalized by Kolmogorov, Karhunen, Loève, and Cramér, among others, to the case when the integrator X is square integrable and with orthogonal increments or when the covariance of the increments is of (locally) bounded variation. This generalization is utilized for an analysis of stationary and related stochastic processes. On the other hand, if the integrand f is also a (bounded left continuous) stochastic function $f : T \times \Omega \longrightarrow \mathbb{R}$, and X is a B.M., then K. Itô has defined $\int_T f(t, \cdot) dX_t$. This integral plays a fundamental role in the study of stochastic differential equations, amongst its many other applications.

Since the B.M. X is also a square integrable martingale, it is natural to generalize this latter integral to more general classes of integrators. Here the resulting integral depended on a deep decomposition theorem for the submartingale $X^2 = \{X_t^2, t \in T\}$, a result called the Doob-Meyer decomposition, which was available only by the mid-1960s. With this, the martingale integral investigations, generalizing the work for Brownian motion, have taken a center stage in stochastic analysis, primarily in the hands of H. Kunita and S. Watanabe, and especially P.A.Meyer and associates, among others. In fact, Meyer went further and defined a very general class of processes, called semimartingales. These are processes X which can be written as $X = Y + A$ where Y is a (local) martingale and A is a process of (locally) bounded variation. Meyer extended the work on integration to these general integrators.

These two classes of stochastic integrals admit a unified treatment by using an extension of a boundedness principle formulated in 1956 by Bochner [2]. The original principle can be stated as follows. For $f = \sum_{i=0}^{n-1} a_i \chi_{(t_i, t_{i+1}]}$, $T = [a, b]$, $a = t_0 < t_1 < \cdots < t_n = b$ the following elementary integral is unambiguously defined:

$$\tau(f) = \int_a^b f(t) dX(t) = \sum_{i=0}^{n} a_i (X_{t_{i+1}} - X_{t+i}) \in L^p(P), \quad p > 0.$$

Suppose there exists an absolute constant C $(C_{p\rho} > 0)$ such that

$$\int_\Omega |\tau(f)|^p dP = E(|\tau(f)|^p) \le C \int_a^b |f(t)|^p dt dP. \qquad (2.1)$$

Then integrator X is called (by Bochner) an $L^{\rho,p}$-*bounded process.* Since $\tau : L^p(dt) \subset L^p(dtdP) \longrightarrow L^p(P)$ is a bounded linear operator on the simple functions, it follows that τ has a unique extension (by density) to all of $L^p(dt) \subset L^p(dtdP)$ into $L^p(P)$. The extended $\tau(f) = \int_a^b f dX, \quad f \in L^p(dt)$, is the stochastic integral. The above principle extends if $f : [a, b] \times \Omega \longrightarrow \mathbb{R}$ is a bounded stochastic function to a large class of functions f which are left continuous in t, for a.a.(ω), when (2.1) holds with $dtdP$ replaced by a σ-finite measure $\alpha : \mathcal{B}(T) \otimes \Sigma \longrightarrow \overline{\mathbb{R}}^+$. This simple modification actually unifies both the above classes of integrals for more general integrands f, and the integrators X can be more general than semimartingales. In the case when $\rho = p = 2$, the integrator process X is called $L^{2,2}$-bounded. This class is of particular interest because once the theory is established for $L^{2,2}$-bounded processes it can be extended to the more general $L^{\rho,p}$-classes using stopping time techniques. In the one-parameter case the optimal stochastic integrators are obtained via the extended Bochner principle. A unified treatment of all known stochastic integrals, in the form of $L^{2,2}$-bounded processes and its extensions to L^{ϕ_1,ϕ_2}-bounded processes, is formulated and detailed in Rao [14]. See especially Rao ([15], Chapter VI), for a longer discussion of the one-parameter theory.

Consider now a process $X = \{X_t, t \in T \subset \mathbb{R}^n\}, n > 1$. It is an n-parameter Brownian motion if it has independent increments, $E(X_t) = 0$, and $Var X_t = |t| = \prod_{i=1}^n |t_i|$. To be more precise, it is a Wiener-B.M., since there is another process called Lévy-B.M., both of which coincide for $n = 1$, but differ for $n > 1$. It is desirable to develop an integral calculus analogous to the one outlined above, so that eventually one can obtain the stochastic analogs of the Green and Stokes theorems after the necessary multiparameter calculus (and differential forms) are developed. Since the paths $t \rightarrow X_t(\omega)$ are generally of unbounded variation for a.a.(ω), the differential calculus must necessarily be defined in terms of integrals, in

contrast to the classical (deterministic) case. The fundamental paper, on
which almost all the subsequent developments are based, is due to Cairoli
and Walsh [4]. They considered for the most part the integrator process X
to be a Wiener-B.M. in the plane, or its extensions to certain martingales.
The first difficulty faced in multiparameter stochastic integration is the
lack of any reasonable linear order in $T \subseteq \mathbb{R}_+^2$. Using a coordinate-wise
partial ordering defined for $s = (s_1, s_2), \quad t = (t_1, t_2)$ in T where $s \prec t$
if and only if $s_i \leq t_i, \quad i = 1, 2$ (with $\prec\prec$ being the strict inequality),
they abstracted the necessary conditions from the one-parameter B.M. Let
$\{X_t, \mathcal{F}_t, t \in T\}$ be an adapted process on a probability space (Ω, Σ, P),
$\mathcal{F}_t \subset \Sigma$, and $\mathcal{F}_s^1 = \sigma(\cup_{s_2 \geq 0} \mathcal{F}_{s_1, s_2}), \mathcal{F}_s^2 = \sigma(\cup_{s_1 \geq 0} \mathcal{F}_{s_1, s_2})$. The conditions
on the filtration are the following.

(F_1): $s \prec t \Rightarrow \mathcal{F}_s \subset \mathcal{F}_t$,

(F_2): $\mathcal{F}_s$ is complete for each $s \in T$ relative to P,

(F_3): $\mathcal{F}_s = \cap_{t \succ s} \mathcal{F}_t$, and

(F_4): for each t, $\mathcal{F}_t^1$ and $\mathcal{F}_t^2$ are conditionally independent given $\mathcal{F}_t$.

The hypotheses $(F_1) - (F_3)$ are combined to say that $\{\mathcal{F}_t, t \in T\}$ is a
filtration satisfying the standard conditions, with $\mathcal{F}_3$ being the right (order)
continuity of the σ-algebras $\mathcal{F}_s \subset \Sigma$ with respect to the order $\prec$. $(\mathcal{F}_4)$ is
a technical condition used for many of the most crucial computations in
this work, and which is satisfied for the B.M. case. Condition (F_4) can also
be stated, using conditional expectations for each bounded Σ-measurable
$f : \Omega \longrightarrow \mathbb{R}$, equivalently as:

$$(F_4): \quad E\{f|\mathcal{F}_t\} = E\{E\{f|\mathcal{F}_t^1\}|\mathcal{F}_t^2\}, \quad a.e.[P], \quad t \in T.$$

Although $(F_1) - (F_3)$ remain the same if $T \subseteq \mathbb{R}_+^n$, $n \geq 2$, (F_4) needs a slight
change. Namely, if $\alpha \subset \beta \subset \{1, 2, \ldots, n\}$, and $t = (t_1, t_2, \ldots, t_n) \in T$, let
$\mathcal{F}_t^\alpha = \sigma(\cup_{s_i \geq 0, i \in \alpha} \mathcal{F}_s : s = (s_1, \ldots, s_n) \prec t)$ (replacing $\mathcal{F}_t^i$ above), then for

each bounded measurable $f : \Omega \longrightarrow I\!\!R$, and $\alpha \subset \beta$ as above, one has

$$(F_4') : \quad E\{f|\mathcal{F}_t^\beta\} = E\{E\{f|\mathcal{F}_t^\alpha\}|\mathcal{F}_t^{\beta-\alpha}\}, \quad a.e.[P].$$

If $n = 2$, so that $\beta = \{1, 2\}$, $\alpha = \{1\}$ or $= \{2\}^c$, (F_4') reduces to (F_4). Thus, to avoid the mostly notational complications, only the case $n = 2$ will be considered in this article.

A process $\{X_t, \mathcal{F}_t, t \in T \subset I\!\!R^2\} \subset L^1(P)$ is called a *planar martingale* if for each $s \prec t$, $E\{X_t|\mathcal{F}_s\} = X_s$ a.e.. Let $\mathcal{M}^2(z_0)$ be the space of square integrable planar martingales defined on the region $T = [(0, 0), z_0]$. To use the coordinate ordering effectively, however, both a weakening and a strengthening of the planar B.M. is needed. Let $(s, t]$ be the rectangle with s as the lower left-hand corner and t as the upper right-hand corner, and suppose $X((s, t])$ is the two-dimensional increment, i.e., $X((s, t]) = X_{t_1 t_2} - X_{t_1 s_2} - X_{s_1 t_2} + X_{s_1 s_2}$.

Definition 1. (1) X is a *weak martingale* whenever $E\{X(s, t]|\mathcal{F}_s\} = 0$ a.e.,
(2) X is an (adapted) *i-martingale* if X_t is $(\mathcal{F}_t$-) $\mathcal{F}_t^i$-measurable and
$\quad E\{X(s, t]|\mathcal{F}_s^i\} = 0$ a.e., $i = 1, 2$, and
(3) X is a *strong martingale* if $E\{X(s, t]|\mathcal{F}_s^1 \vee \mathcal{F}_s^2\} = 0$ where $\mathcal{F}_s^1 \vee \mathcal{F}_s^2 = \sigma(\mathcal{F}_s^1 \cup \mathcal{F}_s^2)$.

It can be verified that when X satisfies (F_4) above, X is a martingale if and only if X is an i-martingale for $i = 1, 2$. Thus laid out, the fundamental theory of these martingales and their integrals has been given by Cairoli and Walsh, who also obtained a stochastic analog of Green's theorem and then developed the theory of (real) holomorphic processes. In the case of one-parameter processes there are submartingales and supermartingales as well as quasimartingales; and a process containing all three called semimartingales (noted above) is available. Stochastic integration is well established for this latter class, and from an applicational point of view the semimartingales are more realistic than B.M., which is only an idealized "noise" process. Thus it is natural to seek the multidimensional versions of these processes to use in the theories of stochastic integration and differential equations. Because of the difficulties due to the lack of a strict linear ordering, the present work will use the coordinate-wise partial ordering defined earlier. This allows for sharper results than would a general partial ordering. In addition to the ordering $\prec$, another complementary ordering λ will be used in this work. Defined by $s \lambda t$ if and only if $s_1 \leq t_1$ and $s_2 \geq t_2$, and $\overset{\wedge}{\lambda}$ to denote the corresponding strict inequality. Some work on general partial orderings are found in the book by Métivier-Pellaumail [11] and articles by Ivanoff, Merzbach, Dozzi, and Schiopu-Kratina [10] and Dozzi and Ivanoff [7]. The theory is extended to manifolds, cf. Emry [8].

Corresponding to the semimartingale, the general integrator of a one-parameter process is the two-parameter quasimartingale. However, in the plane there are (at least) three different concepts of a quasimartingale, based on the Vitali, Arzelá, and Hardy definitions of a function of bounded variation. The stochastic versions of these have been defined and the basic properties (the inequalities and convergence results) presented in Brennan [3]. Using the Vitali concept, the corresponding integrals and some of their properties have been analyzed for a subclass of these martingales by Dozzi [6] in the n-dimensional case. Here are the definitions. As noted before, only the two-parameter case is considered and the parameter set T is taken as $[0, s_0] \times [0, t_0]$ in $\mathbb{R}^2_+$. If $z = (s, t) \in T$, and $0 < t_1 < \cdots < t_{n_t} \leq t_0$, $0 < s_1 < \cdots < t_{n_s} \leq s_0$, then the collection of points $g = \{(s_i, t_j) : 1 \leq i \leq n_s, 1 \leq j \leq n_t\}$ in T is called a *grid*, using Brennan's language. The evaluation of the increment will be noted by $\Delta_g X(z)$ where

$$\Delta X(s_i, t_j) = X\big((s_i, t_j), (s_{i+1}, t_{j+1})\big)$$

and the subscript g is used to indicate the grid over which summation is occurring. Let $\{X_z, \mathcal{F}_z\}_{z \in T}$ be an adapted integrable process on (Ω, Σ, P).

Definition 2. X is termed a

(i) *V-quasimartingale* if

$$K_X = \sup E \left\{ \left(\sum_{z \in g} |E \left(\Delta_g X(z) | \mathcal{F}_z \right)| \right) : g \text{ is a grid on } T \right\} < \infty,$$

and K_X the variation constant;

(ii) *A-quasimartingale* if

$$K_X = \sup E \left\{ \sum_{i=1}^{n-1} \left| E \left\{ (X(z_{i+1}) - X(z_i) | \mathcal{F}_{z_i} \right\} \right| \right.$$

$$\left. : (0,0) \prec z_1 \prec \cdots \prec z_n \prec (s_0, t_0) \right\} < \infty;$$

(iii) *H-quasimartingale* if it is a V-quasimartingale and the boundary processes $\{X(s, t_0), \mathcal{F}_{s,t_0} : 0 \le s \le s_0\}$ and $\{X(s_0, t), \mathcal{F}_{s_0,t} : 0 \le t \le t_0\}$ are one-parameter quasimartingales.

These are based (respectively) on Vitali, Arzelá, and Hardy definitions of bounded variation of a function of two variables. It can be verified that a V-quasimartingale X, with variation $K_X = 0$, is a weak martingale as in Definition 1. Moreover, one can show that an adapted process is an H-process if and only if it is both a V-quasimartingale and an A-quasimartingale (Brennan [3]). The processes considered by Brennan contain all the processes used in the two-parameter stochastic integrals analyzed in the literature. For this reason the $L^{2,2}$-bounded processes of this sort are considered in this article. In Sections 6 and 7 of this article a Doob inequality for a class of V-quasimartingales will be needed. This is stated here for reference.

Theorem 3 (Brennan-Doob Inequality).   *Let $\{X_t, \mathcal{F}_t\}_{t \in T}$ be a separable process with $D = \{(\frac{i}{2^n}, \frac{j}{2^n}) : 0 \le i \le 2^n, 0 \le j \le 2^n\}$ as the set of separability. Let $\Phi(x) = x \log^+ x$ for $p = 1$, and $\Phi(x) = |x|^p$ for $p > 1$.*

Suppose X satisfies the following conditions:

(I) $\Phi_V(X) = \sup\left\{E\left\{\Phi\left(\sum_{t\in g}|E\{\Delta_g X \mid \mathcal{F}_t\}|\right)\right\} : g \text{ is a grid on } T\right\} < \infty$

(II) $\Phi_j(X) = \sup\left\{E\left\{\Phi\left(\sum_{i=0}^{n-1}|E\{X_{t_{i+1}} - X_{t_i} \mid \mathcal{F}_{t_i}\}|\right)\right\} : t_i \prec t_{i+1}, \right.$

$\left. i = 0,...,n-1, \text{ with } t_i \in \Gamma_j \right\} < \infty, \text{ for } j = 1, 2.$

Then there exist constants a_p and b_p depending on the variations Φ_V, Φ_1, and Φ_2 only, such that

(a) $\lambda P\left\{\sup_{t\in T}|X_t| > \lambda\right\} \le a_1 + b_1 E\{\Phi(X(s_0, t_0))\}$, $p = 1$,

(b) $E\left\{\sup_t \Phi(X_t)\right\} \le a_p + b_p E\{\Phi(X(s_0, t_0))\}$, $p > 1$.

(See Brennan [3] for Part (a) and Green [9] for Part (b).)

A decomposition given by Brennan [3] will be needed for further work and thus the following definition is required. The decomposition holds for those V-quasimartingales which are of class D'.

Definition 4. A process $\{X_z, \mathcal{F}_z\}_{z\in T}$ is said to be of *class D'* if for all $\epsilon > 0$ there exists a $\delta_\epsilon > 0$ such that if g is a grid on T and $\{B_z \in \mathcal{F}_z, z \in g\}$ is any collection of sets satisfying $P(\cup_{z\in g}B_z) < \delta$ then

$$\sum_{z\in g}|\int_{B_z} \Delta_g X_z dP| < \epsilon.$$

The processes of bounded variation in the plane are now defined.

Definition 5. $\{A_z, \mathcal{F}_z\}_{z\in T}$ is a process of bounded variation if

(1) $A(z) = 0$ a.e. for z on the coordinate axes of T,

(2)   The map $z \mapsto A(z, \omega)$ is right continuous for a.a. ω,

(3) $\|A\| = \sup\left\{\sum_{z\in g}|\Delta_g A(z)| : g \subset T \text{ for } g \text{ a grid}\right\} \in L^1(P)$.

With these definitions, a classification of the decomposable V-quasimartingales is possible.

Theorem 6 (Brennan). *Let $\{Y_z, \mathcal{F}_z\}_{z \in T}$ be an L^1 right continuous V-quasimartingale. Then $Y_z = X_z + A_z$ where $\{X_z, \mathcal{F}_z\}_{z \in T}$ is a weak martingale and $\{A_z, \mathcal{F}_z\}_{z \in T}$ is a process of bounded variation if and only if Y is of class D'.*

In this article the V-quasimartingales will be assumed to be of class D' without mention. The V-quasimartingales can be further decomposed using a classification of weak martingales due to Wong and Zakai [16].

Theorem 7 (Wong and Zakai). $\{X_z, \mathcal{F}_z\}_{z \in T}$ *is a weak martingale on T if and only if X is expressible as $X_z = X_z^{(1)} + X_z^{(2)}$ where $X_z^{(i)}$ is an adapted i-martingale for $i = 1, 2$, and the filtration $\{\mathcal{F}_z\}_{z \in T}$ satisfies the (F4) condition.*

Remark 8. (1) The decomposition of the above proposition is not unique. However, if $X = X^{(1)} + X^{(2)}$ and $X = Y^{(1)} + Y^{(2)}$ then $X^{(1)} - Y^{(1)}$ and $X^{(2)} - Y^{(2)}$ are both 1-martingales and 2-martingales, hence by (F4) are martingales.

(2) If the σ-algebras $\{\mathcal{F}_{s0}\}_{s \in [0, s_0]}$ and $\{\mathcal{F}_{0t}\}_{t \in [0, t_0]}$ are trivial and $X_{(0,0)} = 0$ then $X_{0,t}^{(1)} = X_{s,0}^{(1)} = X_{0,t}^{(2)} = X_{s,0}^{(2)} = 0$.

Thus a V-quasimartingale X of class D' can be decomposed into

$$X_z = X_z^{(1)} + X_z^{(2)} + A_z, \quad z \in T$$

where $\{X_z^i, \mathcal{F}_z\}_{z \in T}$ is an adapted i-martingale ($i = 1, 2$) and $\{A_z, \mathcal{F}_z\}_{z \in T}$ is a predictable process of bounded variation. This decomposition will be used in the sections to follow.

The generalized Bochner boundedness principle for the multi-parameter case is an obvious extension of (2.1), and can be stated as:

$$E\left\{|\tau(f)|^2\right\} \le C \int_{T \times \Omega} |f|^2 \, d\alpha(z, \omega) \tag{2.2}$$

where $T \subseteq \mathbb{R}_+^n$, $\tau(f)$ is the integral of a simple function f relative to X, $C > 0$ an absolute constant, and $\alpha : \mathcal{B}(T) \otimes \Sigma \longrightarrow \overline{\mathbb{R}}^+$ is a σ-finite measure.

Then the additive operator τ on simple functions f on σ-(sub) algebras of $\mathcal{B}(T) \otimes \Sigma$, admits a unique extension onto the appropriate closure of these functions in $L^2(T \times \Omega, \mathcal{B}(T) \otimes \Sigma, \alpha)$, giving the desired stochastic integral. This immediately provides us with the fundamental dominated convergence property. By analogy to the properties of the one-parameter case, it appears that $L^{2,2}$-boundedness gives essentially the optimal class of stochastic integrators (and integrals) to possess the dominated convergence property (see Rao [14]). This property is suggestive, but has not been established here. Thus, the problem reduces to finding the dominating measure α for X, which initially involves verifying that the known stochastic integrators do satisfy this boundedness principle. Section 2 is devoted to various constructions of stochastic surface integrals which then are utilized in defining the line integrals and their properties in Section 3. Also, the mixed integral "$\partial_i X dF$", the product integral "$dX dX$", and a stochastic Fubini theorem are established in Sections 4 and 5. For the work here an additional technical condition, called "cross-term" domination, is introduced and employed. It is automatic in the case of martingale integrators used by Cairoli and Walsh, as well as Dozzi, but plays a key role in the general $L^{2,2}$-bounded case. Finally, Sections 6 and 7 are devoted to a stochastic Green theorem for the integration theory developed in the preceding sections.

2.2   The Bochner Bounding Measure for V-Quasimartingales

The two-parameter integration theory for strong martingales and planar martingales studied by Cairoli and Walsh [4] is extended to the more inclusive class of processes satisfying the Bochner boundedness principle. In order to construct the Bochner bounding measure it is necessary to consider the products of multiparameter processes. Two processes, $\{M_z, \mathcal{F}_z\}_{z \in T}$ and

$\{N_z, \mathcal{F}_z\}_{z \in T}$, are said to be *orthogonal*, denoted $M \perp N$, if their product is a weak martingale. That is, for $z \prec w$ and a rectangle $D = (z, w]$, $E\{MN(D)|\mathcal{F}_z\} = 0$. For $M \in \mathcal{M}^2(z_0)$ the following proposition due to Cairoli and Walsh [4] gives the properties of products of planar martingales.

Proposition 9. *Let $M, N \in \mathcal{M}^2(z_0)$. Then*

(a) $E\{MN(D)|\mathcal{F}_z\} = E\{M(D)N(D)|\mathcal{F}_z\}$ *for each rectangle*

$D = (z, z'] \subset T$,

(b) MN *is a weak martingale if and only if* $E\{M(D)N(D)|\mathcal{F}_z\} = 0$

for each rectangle $D = (z, z'] \subset T$.

The generalized Bochner boundedness principle applies for all $M \in \mathcal{M}^2(z_0)$ as these processes are $L^{2,2}$-bounded processes.

Theorem 10. *Let $M \in \mathcal{M}^2(z_0)$. Then for every predictable simple function ϕ, $E\left\{\left(\int_T \phi dM\right)^2\right\} = \int_{\Omega \times T} \phi^2 d\alpha$ where $d\alpha = d\langle M \rangle_{st} \otimes dP$.*

Proof. Let $\phi = \sum_{i=1}^n \alpha_i \chi_{(s_i, t_i]}$ where $\alpha_i \in \mathcal{F}_{s_i}$ and $(s_i, t_i] \cap (s_j, t_j] = \emptyset$, $i, j = 1, 2, \ldots, n$. Then

$$E\left\{\left(\int_T \phi dM\right)^2\right\} = E\left\{\left(\sum_{i=1}^n \alpha_i M(s_i, t_i]\right)^2\right\}$$

$$= E\left\{\sum_{i=1}^n \alpha_i^2 M(s_i, t_i]^2\right\}$$

$$+ \sum_{1 \leq i,j \leq n} E\{\alpha_i \alpha_j M(s_i, t_i] M(s_j, t_j]\}.$$

$$(2.3)$$

Since the rectangles $(s_i, t_i]$ and $(s_j, t_j]$ are disjoint, each pair of rectangles can be separated by either a vertical or horizontal line. Suppose the rectangles lie on either side of a vertical line with $(s_i, t_i]$ to the left of $(s_j, t_j]$.

Then α_i, α_j, and $M(s_i, t_i]$ are $\mathcal{F}^1_{s_j}$-measurable and the summands on the far right of (2.3) satisfy

$$E\left\{\alpha_i\alpha_j M(s_i, t_i]M(s_j, t_j]\right\} = E\left\{\alpha_i\alpha_j M(s_i, t_j]E\left\{M(s_j, t_j]|\mathcal{F}^1_{s_j}\right\}\right\}$$
$$= 0.$$

Applying Proposition 9, the Bochner bounding measure can be constructed.

$$E\left\{\sum_{i+1}^n \alpha_i M(s_i, t_i]^2\right\} = E\left\{\sum_{i=1}^n \alpha_i M^2(s_i, t_i]\right\}$$
$$= E\left\{\sum_{i=1}^n \alpha_i^2 \langle M\rangle(s_i, t_i]\right\}$$
$$= E\left\{\int_T \phi^2 d\langle M\rangle\right\}$$
$$= \int_{\Omega\times\Sigma} \phi^2 d\alpha.$$

$\square$

So for planar martingales, and hence for strong martingales, the generalized Bochner boundedness principle holds. Let $\{X_z, \mathcal{F}_z\}_{z\in T}$ be a separable V-quasimartingale with decomposition $X = X^{(1)} + X^{(2)} + A$ where $\{X^{(i)}_z, \mathcal{F}_z\}_{z\in T}$, $i = 1, 2$, are adapted i-martingales and $\{A_z, \mathcal{F}_z\}_{z\in T}$ is a predictable process of bounded variation. An analog of Proposition 9 will be needed for i-martingales in the construction of the Bochner bounding measure for V-quasimartingales. It is given here for the more general weak martingales though the usefulness, for this article at least, is solely as a stepping stone to the special case for i-martingales. It is assumed as always, that the processes are zero on the axes.

Proposition 11. *Let X and Y be weak martingales with decompositions $X = X^{(1)} + X^{(2)}$ and $Y = Y^{(1)} + Y^{(2)}$ where $X^{(i)}$ and $Y^{(i)}$, $i = 1, 2$, are* *adapted i-martingales. Then for* $D = (z, w]$ *it follows that:*

(1) $E\{XY(D)|\mathcal{F}_z\} = E\{X(D)Y(D)|\mathcal{F}_z\}$

$$+ E\left\{Y_{z_1 w_2} X^{(2)}(D)|\mathcal{F}_z\right\} + E\left\{X_{z_1 w_2} Y^{(2)}(D)|\mathcal{F}_z\right\}$$

$$+ E\left\{X_{w_1 z_2} Y^{(1)}(D)|\mathcal{F}_z\right\} + E\left\{Y_{z_1 w_2} X^{(2)}(D)|\mathcal{F}_z\right\}$$

$$+ E\left\{\left(X^{(2)}_{w_1 z_2} - X^{(2)}_{z_1 z_2}\right)\left(Y^{(1)}_{z_1 w_2} - Y^{(1)}_{z_1 z_2}\right)|\mathcal{F}_z\right\}$$

$$+ E\left\{\left(X^{(1)}_{z_1 w_2} - X^{(1)}_{z_1 z_2}\right)\left(Y^{(2)}_{w_1 z_2} - Y^{(2)}_{z_1 z_2}\right)|\mathcal{F}_z\right\}.$$

(2) *If X and Y are 1-martingales then*

$$E\{XY(D)|\mathcal{F}_z\} = E\{X(D)Y(D)|\mathcal{F}_z\}$$

$$+ E\{X_{w_1 z_2} Y(D)|\mathcal{F}_z\} + E\{Y_{w_1 z_2} X(D)|\mathcal{F}_z\}.$$

(3) *If X and Y are 2-martingales then*

$$E\{XY(D)|\mathcal{F}_z\} = E\{X(D)Y(D)|\mathcal{F}_z\}$$

$$+ E\{X_{z_1 w_2} Y(D)|\mathcal{F}_z\} + E\{Y_{z_1 w_2} X(D)|\mathcal{F}_z\}.$$

(4) *If X is a 1-martingale and Y is a 2-martingale then*

$$E\{XY(D)|\mathcal{F}_z\} = E\{X(D)Y(D)|\mathcal{F}_z\} + E\{X_{z_1 w_2} Y(D)|\mathcal{F}_z\}$$

$$+ E\{Y_{w_1 z_2} X(D)|\mathcal{F}_z\} + E\left\{\left(X^{(1)}_{z_1 w_2} - X^{(1)}_{z_1 z_2}\right)\left(Y^{(2)}_{w_1 z_2} - Y^{(2)}_{z_1 z_2}\right)|\mathcal{F}_z\right\}.$$

(5) *If X = Y are 1-martingales (respectively, 2-martingales) then*

$$E\{X^2(D)|\mathcal{F}_z\} = E\{X(D)^2|\mathcal{F}_z\} + 2E\{X_{w_1 z_2} X(D)|\mathcal{F}_z\}.$$

Proof. The claims (2) through (5) follow from (1) by making clever choices for the decompositions, for example $X = X^{(1)} + 0$ for 1-martingales. By doing some algebra it can be shown that for the division of the rectangle $(\bar{0}, w]$ into the subrectangles $D = (z, w]$, $A = (\bar{0}, z]$, B, and C the following equations hold:

$$XY(D) = X(D)Y(D) + X(A)Y(D) + X(D)Y(A)$$

$$+ X(D)Y(B) + X(B)Y(D) + X(C)Y(D)$$

$$+ X(D)Y(C) + X(C)Y(B) + X(B)Y(C).$$

Since X and Y are zero on the axes it follows that

$$E\{X(A)Y(D)|\mathcal{F}_z\} = E\{X_{z_1z_2}Y(D)|\mathcal{F}_z\}$$

$$= X_{z_1z_2}E\{Y(D)|\mathcal{F}_z\} = 0.$$

Similarly, it can be seen that $E\{X(D)Y(A)|\mathcal{F}_z\} = 0$. Considering the other terms,

$$E\{X(D)Y(B)|\mathcal{F}_z\} = E\{(Y_{z_1w_2} - Y_{z_1z_2})X(D)|\mathcal{F}_z\}$$

$$= E\{Y_{z_1w_2}X(D)|\mathcal{F}_z\} - Y_{z_1z_2}E\{X(D)|\mathcal{F}_z\}$$

$$= E\{Y_{z_1w_2}X(D)|\mathcal{F}_z\}.$$

$$E\{X(B)Y(D)|\mathcal{F}_z\} = E\{X_{z_1w_2}Y(D)|\mathcal{F}_z\}.$$

$$E\{X(C)Y(D)|\mathcal{F}_z\} = E\{(X_{w_1z_2} - X_{z_1z_2})Y(D)|\mathcal{F}_z\}$$

$$= E\{X_{w_1z_2}Y(D)|\mathcal{F}_z\} - X_{z_1z_2}E\{Y(D)|\mathcal{F}_z\}$$

$$= E\{X_{w_1z_2}Y(D)|\mathcal{F}_z\}.$$

$$E\{X(D)Y(C)|\mathcal{F}_z\} = E\{Y_{w_1z_2}X(D)|\mathcal{F}_z\}.$$

$$E\{X(C)Y(B)|\mathcal{F}_z\} = E\{(X_{w_1z_2} - X_{z_1z_2})(Y_{z_1w_2} - Y_{z_1z_2})|\mathcal{F}_z\}.$$

$$E\{X(B)Y(C)|\mathcal{F}_z\} = E\{(X_{z_1w_2} - X_{z_1z_2})(Y_{w_1z_2} - Y_{z_1z_2})|\mathcal{F}_z\}.$$

Applying the decomposition further reduction can be done.

$$E\{X(D)Y(B)|\mathcal{F}_z\} = E\left\{Y_{z_1w_2}\left(X^{(1)}(D) + X^{(2)}(D)\right)|\mathcal{F}_z\right\}$$

$$= E\left\{Y_{z_1w_2}X^{(1)}(D)|\mathcal{F}_z\right\} + E\left\{Y_{z_1w_2}X^{(2)}(D)|\mathcal{F}_z\right\}$$

$$= E\left\{Y_{z_1w_2}E\left\{X^{(1)}(D)|\mathcal{F}_z^1\right\}|\mathcal{F}_z\right\}$$

$$+ E\left\{Y_{z_1w_2}X^{(2)}(D)|\mathcal{F}_z\right\}$$

$$= E\left\{Y_{z_1w_2}X^{(2)}(D)|\mathcal{F}_z\right\}.$$

$$E\{X(B)Y(D)|\mathcal{F}_z\} = E\left\{X_{z_1w_2}Y^{(2)}(D)|\mathcal{F}_z\right\}.$$

$$E\{X(C)Y(D)|\mathcal{F}_z\} = E\left\{X_{w_1z_2}\left(Y^{(1)}(D) + Y^{(2)}(D)\right)|\mathcal{F}_z\right\}$$

$$= E\left\{X_{w_1z_2}Y^{(1)}(D)|\mathcal{F}_z\right\}$$

$$+ E\left\{ E\left\{ X_{w_1 z_2} Y^{(2)}(D)|\mathcal{F}_z^2 \right\} |\mathcal{F}_z \right\}$$

$$= E\left\{ X_{w_1 z_2} Y^{(1)}(D)|\mathcal{F}_z \right\}.$$

$$E\left\{ X(D)Y(C)|\mathcal{F}_z \right\} = E\left\{ Y_{w_1 z_2} X^{(1)}(D)|\mathcal{F}_z \right\}.$$

$$E\left\{ X(C)Y(B)|\mathcal{F}_z \right\} = E\left\{ (X_{w_1 z_2} - X_{z_1 z_2})(Y_{z_1 w_2} - Y_{z_1 z_2}) X(D)|\mathcal{F}_z \right\}$$

$$= E\left\{ \left(Y^{(1)}_{z_1 w_2} - Y^{(1)}_{z_1 z_2} \right) E\left\{ \left(X^{(1)}_{w_1 z_2} - X^{(1)}_{z_1 z_2} \right) |\mathcal{F}_z^1 \right\} \mathcal{F}_z \right\}$$

$$+ E\left\{ \left(Y^{(2)}_{z_1 w_2} - Y^{(2)}_{z_1 z_2} \right) E\left\{ \left(X^{(1)}_{w_1 z_2} - X^{(1)}_{z_1 z_2} \right) |\mathcal{F}_z^1 \right\} \mathcal{F}_z \right\}$$

$$+ E\left\{ \left(X^{(2)}_{w_1 z_2} - X^{(2)}_{z_1 z_2} \right) \left(Y^{(1)}_{z_1 w_2} - Y^{(1)}_{z_1 z_2} \right) |\mathcal{F}_z \right\}$$

$$+ E\left\{ \left(X^{(2)}_{w_1 z_2} - X^{(2)}_{z_1 z_2} \right) E\left\{ \left(Y^{(2)}_{z_1 w_2} - Y^{(2)}_{z_1 z_2} \right) |\mathcal{F}_z^2 \right\} \mathcal{F}_z \right\}$$

$$= E\left\{ \left(X^{(2)}_{w_1 z_2} - X^{(2)}_{z_1 z_2} \right) \left(Y^{(1)}_{z_1 w_2} - Y^{(1)}_{z_1 z_2} \right) |\mathcal{F}_z \right\}.$$

$$E\left\{ X(B)Y(C)|\mathcal{F}_z \right\} = E\left\{ \left(X^{(1)}_{z_1 w_2} - X^{(1)}_{z_1 z_2} \right) \left(Y^{(1)}_{w_1 z_2} - Y^{(1)}_{z_1 z_2} \right) |\mathcal{F}_z \right\}.$$

The sum of the terms is (1) of the proposition.   □

Remark 12. The equation above could be further reduced; however, the resulting equations would throw no further light on the problem. Relation (5) of Proposition 11 will be used in the following work.

The consideration of existence of the bounding measure for V-quasimartingales leads naturally to the consideration of weak martingales, given Brennan's decomposition. The weak martingale, however, is too general for applications in the stochastic calculus of multiparameter processes. To illustrate this difficulty Bakry [1] constructs a two-parameter increasing predictable process $\{A_t\}_{t\in[0,t_0]}$ on a filtered probability space $(\Omega, \Sigma, \mathcal{F}_s, P, s \in T)$, having total variation equal to 1 a.s.(ω), such that for a process $X_{st} = E\{A_t|\mathcal{F}_s^1\}$ there exists a sequence of simple $\phi_n \to 0$ for which the integral $\int_T \phi_n dX_{st}$ does not converge in probability, hence is not a 0-stochastic integrator. So if one is to construct a Bochner bounding measure it is to be expected that some additional conditions will be required, unlike the one-parameter case. This need for more conditions is not surprising when the mathematics moves from one-parameter to two-parameter. One need only recall the difference in the definitions of continuity for functions of one and two variables. The condition used here is "cross-term" domination.

Definition 13. A process $\{X_z, \mathcal{F}_z\}_{z \in T}$ is called a *cross-term dominated process* if for every simple predictable function $\phi = \sum_{i=1}^{n} \alpha_i X_{(s_i, t_i]}$, X satisfies the relation

$$\left| E\left\{ \sum_{1 \leq i,j \leq n} \alpha_i \alpha_j X(s_i, t_i] X(s_j, t_j] \right\} \right| \leq CE\left\{ \sum_{i=1}^{n} \alpha_i^2 X(s_i, t_i]^2 \right\} \qquad (2.4)$$

for some absolute constant $C > 0$.

Since for martingales and strong martingales the left-hand side of (2.4) is zero, this concept applies and includes the work of Cairoli and Walsh. Suppose the process $\{X_z, \mathcal{F}_z\}_{z \in T}$ satisfies the following *conditional independence* property: for disjoint rectangles $(s, t]$ and $(z, w] \subset T$

$$E\left\{ X(s, t] X(z, w] | \mathcal{F}_{s \vee z} \right\} = E\left\{ X(s, t] | \mathcal{F}_{s \vee z} \right\} E\left\{ X(z, w] | \mathcal{F}_{s \vee z} \right\}.$$

Then in (2.4) it follows that

$$E\{\alpha_i \alpha_j X(s_i, t_i] X(s_j, t_j]\}$$

$$= E\left\{ \alpha_i \alpha_j E\left\{ X(s_i, t_i] X(s_j, t_j] | \mathcal{F}_{s_i \vee s_j} \right\} \right\}$$

$$= E\left\{ \alpha_i \alpha_j E\left\{ X(s_i, t_i] | \mathcal{F}_{s_i \vee s_j} \right\} E\left\{ X(s_j, t_j] | \mathcal{F}_{s_i \vee s_j} \right\} \right\}.$$

This leads to the next assertion.

Corollary 14. *Suppose X is a 1-martingale (or a 2-martingale) which obeys the conditional independence condition. Then X has the cross-term domination property.*

Proof. For disjoint rectangles $(s_i, t_i] \cap (s_j, t_j] = \emptyset$ there exists either a horizontal or a vertical line separating the two. Consequently, either $\mathcal{F}_{s_i \vee s_j} \subset$

$\mathcal{F}_{s_i}^1$ or $\mathcal{F}_{s_i \vee s_j} \subset \mathcal{F}_{s_j}^1$, so

$$E\left\{\alpha_i \alpha_j X(s_i, t_i] X(s_j, t_j]\right\}$$

$$= E\left\{\alpha_i \alpha_j E\left\{X(s_i, t_i]|\mathcal{F}_{s_i \vee s_j}\right\} E\left\{X(s_j, t_j]|\mathcal{F}_{s_i \vee s_j}\right\}\right\}$$

$$= E\left\{\alpha_i \alpha_j E\left\{E\left\{X(s_i, t_i]|\mathcal{F}_{s_i}^1\right\}|\mathcal{F}_{s_i \vee s_j}\right\} E\left\{X(s_j, t_j]|\mathcal{F}_{s_i \vee s_j}\right\}\right\}$$

$$= 0.$$

Hence X has the cross-term domination property. $\qquad \square$

The generalized Bochner boundedness principle will be proven for V-quasi-martingales X with the decomposition $X = X^{(1)} + X^{(2)} + A$, as mentioned before, where the adapted i-martingales $\{X_z^{(i)}, \mathcal{F}_z\}_{z \in T}$ satisfy the cross-term domination condition. First the general Bochner boundedness principle is shown to hold for predictable processes of bounded variation (as given in Definition 5), and then for the cross-term dominated adapted i-martingales. These will be used later in the general case.

Proposition 15. *Let* $\{A_z, \mathcal{F}_z\}_{z \in T}$ *be a process of bounded variation. Then* $\{|A|_z, \mathcal{F}_z\}_{z \in T}$ *is an increasing process with respect to the partial ordering* $\prec$, *and* $\sup_{z \in t} |A|_z \leq \|A\|$, *a.e.*

Proof. Since A is of bounded variation, A can be decomposed into positive and negative parts, $A = A^+ - A^-$ where A^+ and A^- are both positive increasing processes. Then the total variation satisfies $|A|_z = A_z^+ + A_z^-$ for all $z \in T$. Thus for $z \prec z'$ and denoting the rectangle $(\overline{0}, z]$ by R_z

$$|A|_z(\omega) = \sup_{g \subset R_z} \sum_{y \in g} |\Delta_g A(y)|(\omega)$$

$$\leq \sup_{g \subset R_z} \sum_{y \in g} |\Delta_g A(y)|(\omega)$$

$$+ \sup_{g \subset (R_{z'}/R_z)} \sum_{y \in g} |\Delta_g A(y)|(\omega)$$

$$\leq \sup_{g \subset R_{z'}} \sum_{y \in g} |\Delta_g A(y)|(\omega)$$

$$= |A|_{z'}(\omega).$$

Also,

$$\sup_{z \in R_{z_0}} |A|_z = \sup_{z \in R_{z_0}} \left(\sup_{g \subset R_z} \sum_{y \in g} |\Delta_g A(y)|(\omega) \right)$$

$$= \sup_{g \subset R_{z_0}} \sum_{y \in g} |\Delta_g A(y)|(\omega)$$

$$= \|A\|.$$

□

Using this property for processes of bounded variation the Bochner bounding measure is now constructed.

Theorem 16. *Let* $\{A_z, \mathcal{F}_z\}_{z \in T}$ *be a process of bounded variation, then for all simple predictable* $\phi = \sum_{i=1}^n \alpha_i \chi_{(s_i, t_i]}$

$$E\{ \left(\int_T \phi dA \right)^2 \} \leq \int_{\Omega \times T} \phi^2 d\mu_A \qquad (2.5)$$

where $d\mu_A = \|A\| d|A|_z \otimes dP$.

Proof. Computing:

$$E\left\{ \left(\int_T \phi dA \right)^2 \right\} \leq E\left\{ \left(\int_T |\phi| d|A| \right)^2 \right\}$$

$$\leq E\left\{ \sup_{z \in t} |A|_z \int_T |\phi|^2 d|A| \right\} \quad \text{(by Jensen's Inequality)}$$

$$= E\left\{ \int_T |\phi|^2 \sup_{z \in t} |A|_z d|A| \right\}$$

$$= \int_{\Omega \times T} \phi^2 (\|A\| d|A|_z) \otimes dP$$

$$= \int_{\Omega \times T} \phi^2 d\mu_A.$$

□

For the i-martingales the construction is similar for both $i = 1$ and $i = 2$, requiring simply a change of indices. Thus the Bochner bounding measure is given here for only the 1-martingales.

Theorem 17. *Let $\{X_z, \mathcal{F}_z\}_{z \in T}$ be an adapted 1-martingale which is cross-term dominated. Then for every simple function $\phi = \sum_{i=1}^{n} \alpha_i \chi_{(s_i, t_i]}$, $\alpha_i \in \mathcal{F}_{s_i}$, $(s_i, t_i] \cap (s_j, t_j] = \emptyset$, $i, j = 1, 2, \ldots, n$, and for some constant $C > 0$*

$$E\left\{\left(\int_T \phi \, dX\right)^2\right\} \leq C \int_{\Omega \times T} \phi^2 \, d\mu \qquad (2.6)$$

where $d\mu = \mu_2(d\omega_2, \omega_1)\nu(d\omega_1)$ is a non-Cartesian product measure with ν as any probability measure on $[0, t_0]$, and μ_2 the Doleans-Dade measure generated by X^2 for fixed second coordinates. (The analogous theorem for 2-martingales holds with obvious adjustments.)

Proof. There is a natural question which arises. Since the left side of (2.6) does not depend on ν, the infimum over all probability measures on $\mathcal{B}[0, t_0]$ can be considered with the hope of finding a single ν_0 giving the infimum. The existence of such a measure ν_0 is an interesting problem, not considered in the present work, requiring the investigation and application of a deep theorem (of metric geometry) due to Grothendieck. Thus only (2.6) is considered. By the definition of the integral,

$$E\left\{\left(\int_T \phi \, dX\right)^2\right\} = E\left\{\left(\sum_{i=1}^{n} \alpha_i X(s_i, t_i]\right)^2\right\}$$

$$= \sum_{i=1}^{n} E\left\{\alpha_i^2 X(s_i, t_i]^2\right\}$$

$$+ 2 \sum_{1 \leq i, j \leq n} E\left\{\alpha_i \alpha_j X(s_i, t_i] X(s_j, t_j]\right\}$$

$$\leq (K + 1) \sum_{i=1}^{n} E\left\{\alpha_i^2 X(s_i, t_i]^2\right\}, \qquad (2.7)$$

since X is cross-term dominated. By Proposition 11(5) for 1-martingales the following equation holds,

$$E\left\{\alpha_i^2 X^2(s_i, t_i]\right\} = E\left\{\alpha_i^2 X(s_i, t_i]^2\right\} + 2E\left\{\alpha_i^2 X_{t_i^1, s_i^2} X(s_i, t_i]\right\}.$$

So by substitution,

$$\sum_{i=1}^{n} E\left\{\alpha_i^2 X(s_i, t_i]^2\right\} = \sum_{i=1}^{n} E\left\{\alpha_i^2 X^2(s_i, t_i]\right\} - \sum_{i=1}^{n} 2E\left\{\alpha_i^2 X_{t_i^1, s_i^2} X(s_i, t_i]\right\}.$$

It suffices to consider one term due to the linearity of the expectation. Thus consider the equation for a rectangle $(s, t]$,

$$E\left\{\alpha^2 X(s, t]^2\right\} = E\left\{\alpha^2 X^2(s, t]\right\} - 2E\left\{\alpha^2 X_{t_1, s_2} X(s, t]\right\}. \qquad (2.8)$$

The second summand on the right side of (2.8) satisfies the relation

$$-2E\left\{\alpha^2 X_{t_1 s_2} X(s, t]\right\} = -2E\left\{\alpha^2 X_{t_1 s_2}\left(X_{t_1 t_2} - X_{t_1 s_2} - X_{s_1 t_2} + X_{s_1 s_2}\right)\right\}$$

$$= -2E\left\{\alpha^2 \left(X_{t_1 s_2} X_{t_1 t_2} - X_{s_1 s_2} X_{s_1 t_2}\right)\right\}$$

$$+ 2E\left\{\alpha^2 \left(X_{t_1 s_2}^2 - X_{s_1 s_2}^2\right)\right\}.$$

The above relation used the property

$$E\left\{X_{t_1 s_2} X_{s_1 t_2}\right\} = E\left\{E\left\{X_{t_1 s_2} X_{s_1 t_2} | \mathcal{F}_{s_1 t_2}^1\right\}\right\}$$

$$= E\left\{X_{s_1 t_2} E\left\{X_{t_1 s_2} | \mathcal{F}_{s_1 t_2}^1\right\}\right\} = E\left\{X_{s_1 t_2} X_{s_1 s_2}\right\},$$

since X is a 1-martingale, and

$$E\left\{X_{t_1 s_2} X_{s_1 s_2}\right\} = E\left\{X_{s_1 s_2} E\left\{X_{t_1 s_2} | \mathcal{F}_{s_1 s_2}^1\right\}\right\} = E\left\{X_{s_1 s_2}^2\right\}.$$

With the identity $(ab - cd) = (a - c)(b - d) + c(b - d) + d(a - c)$,

$$-2E\left\{\alpha^2 \left(X_{t_1 s_2} X_{t_1 t_2} - X_{s_1 s_2} X_{s_1 t_2}\right)\right\} + 2E\left\{\alpha^2 \left(X_{t_1 s_2}^2 - X_{s_1 s_2}^2\right)\right\}$$

$$= -2E\left\{\alpha^2 \left[\left(X_{t_1 s_2} - X_{s_1 s_2}\right)\left(X_{t_1 t_2} - X_{s_1 t_2}\right)\right.\right.$$

$$+ X_{s_1 s_2} \left(X_{t_1 t_2} - X_{s_1 t_2} \right)$$

$$\left. \left. + X_{s_1 t_2} \left(X_{t_1 s_2} - X_{s_1 s_2} \right) \right] \right\}$$

$$+ 2E \left\{ \alpha^2 \left(X_{t_1 s_2}^2 - X_{s_1 s_2}^2 \right) \right\}$$

$$= -2E \left\{ \alpha^2 \left(X_{t_1 s_2} - X_{s_1 s_2} \right) \left(X_{t_1 t_2} - X_{s_1 t_2} \right) \right\}$$

$$+ 2E \left\{ \alpha^2 \left(X_{t_1 s_2}^2 - X_{s_1 s_2}^2 \right) \right\}$$

$$+ E \left\{ \alpha^2 X_{s_1 t_2} E \left\{ X_{t_1 t_2} - X_{s_1 t_2} | \mathcal{F}_{s_1 t_2}^1 \right\} \right\}$$

$$+ E \left\{ \alpha^2 X_{s_1 t_2} E \left\{ X_{t_1 s_2} - X_{s_1 s_2} | \mathcal{F}_{s_1 t_2}^1 \right\} \right\} \tag{A}$$

$$= -2E \left\{ \alpha^2 \left(X_{t_1 s_2} - X_{s_1 s_2} \right) \left(X_{t_1 t_2} - X_{s_1 t_2} \right) \right\}$$

$$+ 2E \left\{ \alpha^2 \left(X_{t_1 s_2}^2 - X_{s_1 s_2}^2 \right) \right\}, \tag{B}$$

since the last two terms of (A) are zero. Also

$$(B) \le 2E \left\{ \alpha^2 | \left(X_{t_1 s_2} - X_{s_1 s_2} \right) \left(X_{t_1 t_2} - X_{s_1 t_2} \right) | \right\}$$

$$+ 2E \left\{ \alpha^2 \left(X_{t_1 s_2}^2 - X_{s_1 s_2}^2 \right) \right\} \qquad \text{(Jensen's Inequality)}$$

$$\le E \left\{ \alpha^2 \left(X_{t_1 s_2} - X_{s_1 s_2} \right)^2 \right\} + E \left\{ \alpha^2 \left(X_{t_1 t_2} - X_{s_1 t_2} \right)^2 \right\}$$

$$+ 2E \left\{ \alpha^2 \left(X_{t_1 s_2}^2 - X_{s_1 s_2}^2 \right) \right\} \quad (|xy| \le \tfrac{1}{2}(x^2 + y^2))$$

$$= E \left\{ \alpha^2 \left(X_{t_1 s_2}^2 - 2 X_{t_1 s_2} X_{s_1 s_2} + X_{s_1 s_2}^2 \right) \right\}$$

$$+ E \left\{ \alpha^2 \left(X_{t_1 t_2}^2 - 2 X_{t_1 t_2} X_{s_1 t_2} + X_{s_1 t_2}^2 \right) \right\}$$

$$+ 2E \left\{ \alpha^2 \left(X_{t_1 s_2}^2 - X_{s_1 s_2}^2 \right) \right\},$$

since $E \left\{ X_{t_1 s_2} X_{s_1 s_2} \right\} = E \left\{ X_{s_1 s_2}^2 \right\}$ and $E \left\{ X_{t_1 t_2} X_{s_1 t_2} \right\} = E \left\{ X_{s_1 t_2}^2 \right\}$,

$$= 3E \left\{ \alpha^2 \left(X_{t_1 s_2}^2 - X_{s_1 s_2}^2 \right) \right\} + E \left\{ \alpha^2 \left(X_{t_1 t_2}^2 - X_{s_1 t_2}^2 \right) \right\}. \tag{2.9}$$

The first term of (2.8) is now considered,

$$E\left\{\alpha^2 X^2(s,t]\right\} = E\left\{\alpha^2 \left(X_{t_1 t_2}^2 - X_{t_1 s_2}^2 - X_{s_1 t_2}^2 + X_{s_1 s_2}^2\right)\right\}$$

$$= E\left\{\alpha^2 \left(X_{t_1 t_2}^2 - X_{s_1 t_2}^2\right)\right\} - E\left\{\alpha^2 \left(X_{t_1 s_2}^2 - X_{s_1 s_2}^2\right)\right\}.$$

$$(2.10)$$

Substituting (2.9) and (2.10) into (2.8) it follows that

$$E\left\{\alpha^2 X(s,t]^2\right\} \leq 2E\left\{\alpha^2 \left(X_{t_1 t_2}^2 - X_{s_1 t_2}^2\right)\right\} + 2E\left\{\alpha^2 \left(X_{t_1 s_2}^2 - X_{s_1 s_2}^2\right)\right\},$$

and returning to the sum to get:

$$\sum_{1=1}^{n} E\left\{\alpha_i^2 X(s_i,t_i]^2\right\} \leq 2\sum_{i=1}^{n} E\left\{\alpha_i^2 \left(X_{t_i^1 t_i^2}^2 - X_{s_i^2 t_i^2}^2\right)\right\}$$

$$+ 2\sum_{i=1}^{n} E\left\{\alpha_i^2 \left(X_{t_i^1 s_i^2}^2 - X_{s_i^1 s_i^2}^2\right)\right\}.$$

$$(2.11)$$

The process $\{X_{st}^2, \mathcal{F}_{st}\}_{(s,t)\in T}$ is for fixed t a one-parameter submartingale, sometimes called a *1-submartingale* in the literature. By a theorem due to Doleans-Dade (see Rao [12], page 383) there exists a unique signed measure (positive in the case being considered since M^2 is a 1-submartingale) μ_{s_2} such that for each $s_2 \in [0, t_0]$

$$\mu_{s_2}\left((s_1, s_2), (t_1, s_2)\right] \times B) = \begin{cases} \int_B \left(X_{t_1 s_2}^2 - X_{s_1 s_2}^2\right) dP & , B \in \mathcal{F}_{s_1 s_2}^1, t_1 < s_0, \\ \int_B X_{s_0 s_2}^2 dP & , \quad t_1 = s_0. \end{cases}$$

$$(2.12)$$

A digression will be made here before continuing with the proof. A process $X = \{X_s, \mathcal{F}_s\}_{s\geq 0}$ is called a $(*)$-process if for any $0 \leq a < b \leq \infty$ and any partition $a \leq s_1 < s_2 < \cdots < s_{n-1} \leq b$, X satisfies

$$E\left\{\sum_{i=1}^{n} \left|E\left\{X_{s_{i+1}} - X_{s_i}|\mathcal{F}_{s_i}\right\}\right|\right\} \leq K_X^b < \infty,$$

where K_X^b depends only on $\{a, b\}$ and the process X, but not on the partition. If the supremum is taken over the partitions, one returns to the definition of a quasimartingale. Associated to any process $\{X_s, \mathcal{F}_s\}_{s \geq 0}$ is the additive set function $\mu_\alpha^x : \mathcal{S} \to I\!\!R$ defined by:

$$\mu_\alpha^x \left((s, s'] \times A\right) = \begin{cases} \int_A (X_s - X_{s'}) \, dP & , \quad A \in \mathcal{F}_s, \quad s' < \alpha \\ \int_A X_\alpha dP & , \text{ for } s = \alpha, \end{cases}$$

where $\mathcal{S} = \{(s, s'] \times A : 0 \leq s < s', A \in \mathcal{F}_s\}$. Set $\mathcal{P} = \sigma(\mathcal{S})$, the *predictable* σ-algebra generated by the predictable semi-ring $\mathcal{S}$. The present case is restricted to an interval of $I\!\!R$, say $[0, \alpha]$, and the ring $\mathcal{S}_\alpha$ and σ-algebra $\mathcal{P}_\alpha$ are similarly defined. It is clear that for $\alpha < \alpha'$, $\mathcal{P}_\alpha \subset \mathcal{P}_{\alpha'}$. The following result due to Doleans-Dade [5] will be used.

Theorem 18 (Doleans-Dade). *Let $\{\mathcal{F}_t, t \geq 0\}$ be a right continuous increasing family of σ-subalgebras from (Ω, Σ, P) and let $\{X_t, \mathcal{F}_t\}_{t \geq 0}$ be an adapted right continuous process in $L^1(P)$. Suppose X is a $(*)$-process and μ_α^x is the associated additive set function for $\alpha \geq 0$. Then there exists a unique signed measure $\mu^x : \quad \mathcal{P} \to I\!\!R$ such that $\mu^x|_{\mathcal{S}_\alpha} = \mu_\alpha$, $\alpha \geq 0$ if and only if X is of class (DL).*

The submartingale $\{X_{st}^2\}_{s \in [0, s_0]}$ is being considered which by hypothesis is a right continuous process. Since the processes considered here are in $L^2(P)$, the process $\{X_{st}^2\}_{s \in [0, s_0]}$ is a $(*)$-process being in $L^1(P)$ and a submartingale, so the signed measure is well defined on the predictable σ-algebra. That X^2 is of class (DL) is also proven in Rao ([12], Theorem 5, page 387) where it is shown that a $(*)$-process is a semimartingale if and only if it is of class (DL). Since a submartingale is a (local) semimartingale it is of class (DL), in fact it is of class (D), since it is restricted to a bounded interval. Finally (Rao [12], Corollary 4, page 387) a $(*)$-process of class (DL) which is right continuous is a positive supermartingale if and only if the measure μ^x is positive and supported by $I\!\!R \times \Omega$. Thus with the definition (2.12) $\{-X_{st}^2, \mathcal{F}_{st}^1\}_{s\in[0,s_0]}$ is a negative supermartingale and thus a negative measure μ^x, hence $-\mu^x$ is a positive measure. It should be remarked that in (2.12) it is the difference $(X_{s't}^2 - X_{st}^2)$ vs. the difference $(X_{st}^2 - X_{s't}^2)$ in the definition of the additive function, so carrying the minus sign over to the other side of the equality of (2.12) gives the original definition. Thus the generated measure is positive and bounded since the $L^2(P)$ integrable process $\{X_z, \mathcal{F}_z\}_{z\in T}$ is defined on a bounded interval $[0, s_0]$.

Proof of theorem (continued).

Since $\{X_{st}^2\}_{(s,t)\in T}$ is a 1-martingale with respect to both $\{\mathcal{F}_{st}\}_{(s,t)\in T}$ and $\{\mathcal{F}_{st}^1\}_{(s,t)\in T} = \{\mathcal{F}_{s\cdot}^1\}_{s\in[0,s_0]}$, there is a choice about which filtration to use in (2.12). The positive bounded measure μ_{s_2} is defined on the σ-algebras:

(1) $\mathcal{P}_{s_2}$ generated by sets of the form $((s_1, s_2), (t_1, s_2)] \times B$ for $B \in \mathcal{F}_{s_1 s_2}$ where $0 \le s_1 < t_1 \le s_0$, that is with respect to the filtration $\{\mathcal{F}_{st}\}_{(s,t)\in T}$.

(2) $\Sigma_{s_2}^1$ generated by sets of the form $((s_1, s_2), (t_1, s_2)] \times B$ where $B \in \mathcal{F}_{s_1\cdot}^1$. where $0 \le s_1 < t_1 \le s_0$, that is with respect to the filtration $\{\mathcal{F}_{s\cdot}^1\}_{s\in[0,s_0]}$.

The first filtration has the property $\mathcal{P}_s \subset \mathcal{P}_t$ for $0 \le p < q \le t_0$. Since $((s_1, s_2), (t_1, s_2)] \times B = ((s_1, t_2) \times B)_{s_2}$ the inclusion follows from the inclusion of $\mathcal{F}_{s_1 p} \subset \mathcal{F}_{s_1 q}$. This is a disadvantage in defining a measure on $\Omega \times T$, so the filtration $\Sigma_{s_2}^1$ is used. This second filtration $\Sigma_{s_2}^1$ has the advantage that Σ_p^1 corresponds to Σ_q^1 for all $p, q \in [0, t_0]$. For if π_1 is the projection defined by

$$\pi_1(s, t, \omega) = (s, \omega) \text{ where } \pi_1 : [0, s_0] \times [0, t_0] \times \Omega \to [0, s_0] \times \Omega$$

then

$$\pi_1\left(\Sigma_p^1\right) = \pi_1\left(\Sigma_q^1\right) \stackrel{\text{def}}{=} \Sigma^1$$

for all $p, q \in [0, t_0]$, because $\mathcal{F}^1_{s_1 p} = \mathcal{F}^1_{s_1 q}$ in contrast to the previous case $\mathcal{F}_{s_1 p} \subseteq \mathcal{F}_{s_1 q}$. Now define for $A = (s_1, t_1] \times B$, $B \in \mathcal{F}^1_{s_1 \cdot}$, $0 \le s_1 < t_1 \le s_0$ the function $\mu_2 : [0, t_0] \times \Sigma^1 \to I\!\!R^+$ by

$$\mu_2(t, A) = \mu_2(t, (s_1, t_1] \times B)$$

$$\stackrel{\text{def}}{=} \mu((s_1, t), (t_1, t)] \times B)$$

$$= \begin{cases} \int_\Omega \chi_B \left(X^2_{t_1 t} - X^2_{s_1 t} \right) dP & , \quad t_1 < s_0 \\ \int_\Omega \chi_B X^2_{s_0 t} dP & , \quad t_1 = s_0. \end{cases}$$

By the Doleans-Dade Theorem, using the same symbol for the extension, it follows that

$$\mu_2(s, \cdot) : \Sigma^1 \to I\!\!R$$

is a positive measure with respect to $\mathcal{B}([0, t_0])$, the Borel σ-algebra on $[0, t_0]$. Now let $\nu : [0, t_0] \to [0, 1]$ be a probability measure with $([0, t_0], \mathcal{B}([0, t_0]), \nu)$ being the associated probability space. By the Tonelli Theorem for non-Cartesian products (see Rao [13]), given ν and μ_2 are finite measures, it follows that for all $\sigma(\mathcal{S}) = \mathcal{B}([0, t_0]) \otimes \Sigma^1$ measurable functions ($\mathcal{S}$ is the algebra of sets $A \times B$ where $A \in \mathcal{B}([0, t_0])$ and $B \in \Sigma^1$)

$$\int_{[0,t_0]} \left[\int_{[0,s_0] \times \Omega} f(\omega_1, \omega_2) \mu_2 \, (d\omega_2, \omega_1) \right] \nu(d\omega_1) = \int_{T \times \Omega} f(\omega_1, \omega_2) d\mu$$

where μ is the Caratheodory measure generated by $(\alpha, \mathcal{S})$ for α defined by

$$\alpha(A \times B) = \int_B \mu_2(\omega, A) \nu(d\omega), \quad A \times B \in \mathcal{S}.$$

Thus for the simple functions of the form

$$f = \sum_{i=1}^{n} \alpha_i^2 \chi_{(s^i, t^i]}$$

$$= \sum_{i=1}^{n} \alpha_i^2 \chi_{((s^i_1, s^i_2), (t^i_1, t^i_2)]}$$

$$= \sum_{i=1}^{n} a_i^2 \chi_{B_i} \chi_{((s^i_1, s^i_2), (t^i_1, t^i_2)]}$$

where $B \in \mathcal{F}_{s_1 s_2} \subseteq \mathcal{F}_{s_1 s_2}^1$. (Recall that for every $\mathcal{F}_{s_1 s_2}$-measurable function α there exists a sequence of simple functions of the form $\phi_n = \sum_{i=1}^{m_n} a_i \chi_{B_i}$, $a_i \in \mathbb{R}$, $B_i \in \mathcal{F}_{s_1 s_2}$, such that $\phi_n \to \alpha$ by the structure theorem)

$$\int_{[0,t_0]} \left(\int_{[0,s_0] \times \Omega} \sum_{i=1}^{n} a_i^2 \chi_{B_i \times (s^i, t^i]} \mu_2(d\omega_2, \omega_1) \right) \nu(d\omega_1) = \int_{\Omega \times T} \phi^2 d\mu.$$

Using the notation $\chi_B = \chi(B)$ it follows that

$$E\left\{a^2 \chi_B \left(X_{t_1 s_2}^2 - X_{s_1 s_2}^2\right)\right\} = \int_{\Omega} a^2 \chi(B) \left(X_{t_1 s_2}^2 - X_{s_1 s_2}^2\right) dP$$

$$= a^2 \mu_2(s_2, (s_1, t_1] \times B)$$

$$= \int_{[0,s_0] \times \Omega} a^2 \chi\left(((s_1, s_2), (t_1, s_2)] \times B\right) d\mu_2$$

(where s_2 is fixed)

$$\leq \int_{[0,s_0] \times \Omega} a^2 \chi\left(((s_1, s_2 - \epsilon), (t_1, t_2)] \times B\right)$$

$$\cdot \chi\left(((s_1, s_2), (t_1, s_2)] \times B\right) d\mu_2,$$

since $((s_1, s_2 - \epsilon), (t_1, t_2)] \times B \supseteq ((s_1, s_2), (t_1, s_2)]$ and μ_2 is defined along the horizontal segment determined by s_2, for all $\epsilon > 0$. Now integrating over $[0, t_0]$ by ν,

$$\int_{[0,t_0]} \left(\int_{[0,s_0] \times \Omega} a^2 \chi\left(((s_1, s_2 - \epsilon), (t_1, t_2)] \times B\right)\right.$$

$$\left. \cdot \chi\left(((s_1, s_2), (t_1, s_2)] \times B\right) d\mu_2 \right) d\nu$$

$$= \int_{T \times \Omega} a^2 \chi\left(((s_1, s_2 - \epsilon), (t_1, t_2)] \times B\right)$$

$$\cdot \chi\left(((s_1, s_2), (t_1, s_2)] \times B\right) d\mu$$

(Tonelli's Theorem)

$$\leq \int_{T \times \Omega} a^2 \chi\left(((s_1, s_2 - \epsilon), (t_1, t_2)] \times B\right) d\mu.$$

Summing over $i = 1, \ldots, n$, for the simple function $\phi = \sum_{i=1}^{n} a_i \chi_{B_i \times (s^i, t^i]}$

$$E\left\{ \sum_{i=1}^{n} a_i^2 \chi \left(B_i\right) \left(X_{s_1^i t_2^i}^2 - X_{s_1^i s_2^i}^2 \right) \right\}$$

$$\leq \int_{T \times \Omega} \left(\sum_{i=1}^{n} a_i \chi \left(B_i \times ((s_1^i, s_2^i - \epsilon), (t_1, t_2)] \right) \right)^2 d\mu.$$

(2.13)

Also,

$$E\left\{ a^2 \chi_B \left(X_{t_1 t_2}^2 - X_{s_1 t_2}^2 \right) \right\}$$

$$= \int_{\Omega} a^2 \chi_B \left(X_{t_1 t_2}^2 - X_{s_1 t_2}^2 \right) dP$$

$$= a^2 \mu_2 \left((t_2, (s_1, t_1] \times B) \right)$$

$$= \int_{[0, s_0] \times \Omega} a^2 \chi \left(((s_1, t_2), (t_1, t_2)] \times B \right) \cdot \chi \left(((s_1, s_2), (t_1, t_2)] \times B \right) d\mu_2.$$

Integrating by ν over $[0, t_0]$,

$$\int_{[0, t_0]} \left(\int_{[0, s_0] \times \Omega} \chi \left(((s_1, t_2), (t_1, t_2)] \times B \right) \right.$$

$$\left. \cdot \chi \left(((s_1, s_2), (t_1, t_2)] \times B \right) d\mu_2 \right) d\nu$$

$$= \int_{T \times \Omega} a^2 \chi \left(((s_1, t_2), (t_1, t_2)] \times B \right)$$

$$\cdot \chi \left(((s_1, s_2), (t_1, t_2)] \times B \right) d\mu$$

$$\leq \int_{T \times \Omega} a^2 \chi \left(((s_1, s_2), (t_1, t_2)] \times B \right) d\mu.$$

(2.14)

Thus substituting the Equations (2.11), (2.13), and (2.14) into (2.6), with the simple function $\phi = \sum_{i=1}^{n} a_i \chi \left(B_i \times (s_i, t_i] \right)$

$$E\left\{ \left(\int_T \phi \, dX \right)^2 \right\}$$

$$\leq (K+1) \sum_{i=1}^{n} E\left\{ a_i \chi_{B_i} X(s_i, t_i]^2 \right\} \qquad \text{(by (2.11))}$$

$$\leq 2(K+1) \sum_{i=1}^{n} \left(E\left\{ a_i \chi_{B_i} \left(X_{t_1^i t_2^i}^2 - X_{s_1^i t_2^i}^2 \right) \right\} \right.$$

$$+ E\left\{a_i \chi_{B_i}\left(X^2_{t^i_1 s^i_2} - X^2_{s^i_1 s^i_2}\right)\right\}\right) \qquad \text{(by (2.13))}$$

$$\leq 2(K+1)\left(\int_{T \times \Omega}\left(\sum_{i=1}^{n} a_i \chi_{B_i} \chi\left(((s^i_1, s^i_2 - \epsilon), (t_1, t_2)]\right)\right)^2 d\mu\right.$$

$$\left. + \int_{T \times \Omega}\left(\sum_{i=1}^{n} a_i \chi_{B_i} \chi\left((s^i_1, s^i_2), (t^i_1, t^i_2)]\right)\right)^2 d\mu\right). \quad (A)$$

The integration with ν does not change the value of the left-hand side of (A), since the left-hand side of (A) is independent of t. By the dominated convergence theorem, taking the limit as $\epsilon \to 0$, (through a sequence)

$$(A) \longrightarrow 2(K+1)\left(\int_{T \times \Omega} \phi^2 d\mu + \int_{T \times \Omega} \phi^2 d\mu\right) = 4(K+1)\int_{T \times \Omega} \phi^2 d\mu.$$

$$(2.15)$$

□

Now the generalized Bochner boundedness principle will be shown to hold for a class of V-quasimartingales.

Theorem 19. *Suppose $\{X_{st}, \mathcal{F}_{st}\}_{(s,t) \in T}$ is a square integrable V-quasimartingale whose decomposition into the sum of predictable process of bounded variation, adapted 1-martingale, and adapted 2-martingale*

$$X = X^{(1)} + X^{(2)} + A$$

has the property that the i-martingales are cross-term dominated. Then X satisfies the Bochner $L^{2,2}$-boundedness principle with bounding measure

$$d\beta = 3d\mu_1 + 3d\mu_2 + 3\|A\|d|A| \otimes dP$$

where $d\mu_i$ is the bounding measure for the $X^{(i)}$ process as defined in Theorem 17.

Proof. Using Theorem 5 and Theorem 17 the calculation is routine.

$$E\left\{\left(\int_T \phi dX\right)^2\right\} = E\left\{\left(\int_T \phi d\left(X^{(1)} + X^{(2)} + A\right)\right)^2\right\}$$

$$\leq 3E\left\{\left(\int_T \phi dX^{(1)}\right)^2\right\} + 3E\left\{\left(\int_T \phi dX^{(2)}\right)^2\right\}$$

$$+ 3E\left\{\left(\int_T \phi dA\right)^2\right\}$$

$$\leq 3\int_{T\times\Omega} \phi^2 d\mu_1 + 3\int_{T\times\Omega} \phi^2 d\mu_2$$

$$+ 3\int_{T\times\Omega} \phi^2 \left(\|A\|d|A|\right) \otimes dP$$

$$= \int_{T\times\Omega} \phi^2 \left(3d\mu_1 + 3d\mu_2 + 3\|A\|d|A| \otimes dP\right).$$

$\square$

As seen in the proof of Theorem 17, the problem of constructing Bochner bounding measures for other processes lies solely in the problem of dominating the cross-terms

$$\sum_{1\leq i,j\leq n} E\left\{\alpha_i \alpha_j X(z_i, w_i] X(z_j, w_j]\right\}$$

by some other manageable sum. The other term in (11), namely

$$\sum_{i=1}^n E\left\{\alpha_i^2 X(z_i, w_i]^2\right\}$$

can be dominated for any 1-martingale X, and consequently for any weak martingale.

2.3 Line Integrals

The work of the last section on the stochastic surface integral leads naturally to a stochastic line integral. This section is devoted to the line integral with respect to $L^{2,2}$-bounded processes. The approach that Cairoli and Walsh developed for Brownian Motion and strong martingales is adapted to this purpose using the processes $X_i^{\Gamma(z)}$ of the integral sum of X along the path Γ up to $z \in \Gamma$. The original case (used by Cairoli and Walsh [4]) for martingales used a process $X_i^{\Gamma}(z)$ defined to be the integral sum over all of Γ conditioned back to z by $\mathcal{F}_z^i$ which coincides with $X_i^{\Gamma(z)}$ for the class of planar martingales. These two processes are compared and studied for general $L^{2,2}$-bounded processes. Construction of the bounding measures for X can be considered in two ways. The process X can be thought of having a bounding measure α on $T \times \Omega$ which is restricted to the paths $\Gamma \subset T$. On the other hand, the process X could first be restricted to the path Γ and a bounding measure constructed for this restricted process. The later proves to be quite involved, whereas the former is well suited for any $L^{2,2}$-bounded process. Finally some convergence theorems for the thus developed general line integrals are considered. In the following suppose $\{X_z, \mathcal{F}_z\}_{z \in T}$ is a right continuous $L^2(P)$ process such that for all simple ϕ there exists a σ-finite measure $\alpha : \mathcal{B}(T) \otimes \Sigma \to \mathbb{R}$ such that

$$E\left\{\left(\int_T \phi dX\right)^2\right\} \le C \int_{T \times \Omega} \phi^2 d\alpha. \tag{2.16}$$

Then X is an $L^{2,2}$-bounded process and the integral $\int_T \phi dX$ is well defined for the closure of ϕ under the $L^2(d\alpha)$ metric. To begin, the question of the existence of the line integrals with respect to general $L^{2,2}$-bounded processes will be considered. Then the special properties of $X_i^{\Gamma(z)}$ and $X_i^{\Gamma}(z)$ as well as their relation to the Bochner boundedness principle together with the previous work will be discussed. It will be assumed throughout that the constant C in (2.16) is equal to 1 by absorbing it into $\alpha(\cdot)$ if necessary. It will be shown that the natural development for the line integral is through $X_i^{\Gamma(z)}$ for general $L^{2,2}$-bounded processes.

One can follow the direct approach if the path Γ is increasing and X is an A-quasimartingale. Let the path Γ have the parameterization

$$\Gamma = \{z : z = \gamma(\sigma), 0 \le \sigma \le 1\}.$$

Then the increasing path Γ is linearly ordered with respect to $\prec$, i.e., if $z = \gamma(s)$, $w = \gamma(t) \in \Gamma$ such that $s \le t$ then $z \prec w$. Since an A-quasimartingale satisfies

$$\sup E \left\{ \sum_{i=1}^{n-1} \left| E\left\{ X_{t_{i+1}} - X_{t_i} | \mathcal{F}_{t_i} \right\} \right| : \right.$$

$$\left. (0,0) \prec t_1 \prec \cdots \prec t_n \prec (s_0, t_0) \in \Gamma \right\} < \infty,$$

where the index is linear, the process $\{X_z, \mathcal{F}_z\}_{z \in \Gamma}$ is a one-parameter quasi-martingale and has the generalized Doob-Meyer decomposition

$$X_{\gamma(\sigma)} = M_{\gamma(\sigma)} + A_{\gamma(\sigma)}$$

where $\{M_{\gamma(\sigma)}, \mathcal{F}_{\gamma(\sigma)}\}$ is a martingale and $\{A_{\gamma(\sigma)}, \mathcal{F}_{\gamma(\sigma)}\}$ is a predictable process of bounded variation. For predictable ϕ let

$$\int_T \phi dX = \int_0^1 \phi(\gamma(\sigma)) \, dX_{\gamma(\sigma)}$$

$$= \int_0^1 \phi(\gamma(\sigma)) \, dM_{\gamma(\sigma)} + \int_0^1 \phi(\gamma(\sigma)) \, dA_{\gamma(\sigma)}$$

which is unambiguously defined by the one-parameter theory. This X is an $L^{2,2}$-bounded process along the increasing paths $\Gamma \subset T$ where for simple ϕ

$$E \left\{ \left(\int_\Gamma \phi dX \right)^2 \right\} = E \left\{ \left(\int_0^1 \phi(\gamma(\sigma)) \, dX_{\gamma(\sigma)} \right)^2 \right\}$$

$$\le \int_{[0,1] \times \Omega} \phi^2 \left(d\langle M \rangle_{\gamma(\sigma)} \otimes dP + |A|_{\gamma(1)} d|A|_{\gamma(\sigma)} \otimes dP \right)$$

using the bound for quasimartingales of one parameter (see Rao[14]). So the theory of line integration can be directly developed for A-quasimartingales on increasing paths Γ. However, it will be necessary to integrate over circles and other more complicated curves in T, and the direct approach then fails because of the lack of a single linear ordering.

Denote by Γ a continuous path in $T = [0, s_0] \times [0, t_0]$ with the parameterization as above for some continuous function $\gamma : [0, 1] \to T$. Denote by $\widehat{\Gamma}$ the opposite orientation of Γ with parameterization

$$\widehat{\Gamma} = \{z : z = \widehat{\gamma}(\sigma) = \gamma(1 - \sigma), 0 \leq \sigma \leq 1\}.$$

There are four types of paths which are as follows:

Type I   if for $\sigma \leq \sigma'$ one has $\gamma(\sigma) \prec \gamma(\sigma')$,

Type II   if for $\sigma \leq \sigma'$ then $\gamma(\sigma) \curlywedge \gamma(\sigma')$,

Type I'   if $\widehat{\Gamma}$ is of type I, and,

Type II'   if $\widehat{\Gamma}$ is of type II.

Collectively these path types are called *pure types*. Given the filtration $\{\mathcal{F}_z\}_{z \in T}$ the following properties hold when considering pure paths with respect to the generated filtrations $\{\mathcal{F}_z^1\}_{z \in T}$ and $\{\mathcal{F}_z^2\}_{z \in T}$. Since $\{\mathcal{F}_z^i\}_{z \in T}$, $i = 1, 2$ are linearly ordered, i.e., $\{\mathcal{F}_z^1\}_{z \in T} = \{\mathcal{F}_{st_0}^1\}_{s \in [0, s_0]}$ and $\{\mathcal{F}_z^2\}_{z \in T} = \{\mathcal{F}_{s_0t}^2\}_{t \in [0, t_0]}$, a process X when restricted to a curve Γ of type I or II is a one-parameter linearly ordered process under the ordering $\prec$. Similarly, if Γ is of type I or II' then $\{X_z, \mathcal{F}_z^2\}_{z \in \Gamma}$ is linearly ordered under $\curlywedge$. In both cases the filtrations are right continuous relative to the respective order relations. Also the vertical and horizontal curves are each of two types: vertical of type I and II, and horizontal of type I and II'. As usual it is assumed that $X = 0$ on the $s-$ and $t-$ axes. Given a curve Γ with initial point $\gamma(0) = z_1$ and final point $\gamma(1) = z_f$, the following sets for $z \in \Gamma$ will be needed:

$$D^1(z) = \overline{D}^1(z) - V_{z_1}$$

where V_z is the vertical line segment joining $z \in \Gamma$ to the s-axis and $\overline{D}^1(z)$

is the closure of the region bounded by V_{z_1}, V_{z_0}, and Γ. Likewise,

$$D^2(z) = \overline{D}^2(z) - H_{z_1}$$

where H_z is the horizontal line segment joining $z \in \Gamma$ to the t-axis. Using these regions, the line integrals can be developed as the sum of the integrals over $D^1(z)$ and $D^2(z)$ as z varies over Γ. The $L^2(P)$ limits and approximations by a subclass of pure curves called polygonal curves will be used to achieve this objective. These curves are defined now.

Definition 20. Γ is said to be a *polygonal curve* if it is a path of consecutive horizontal and vertical segments. Denote the vertices of a polygonal curve by $(a_1, b_1) \prec (a_1, b_2) \prec \cdots \prec (a_n, b_n)$, or as $(a_1, b_1) \prec (a_2, b_1) \prec \cdots \prec (a_n, b_n)$ depending on whether the first segment is horizontal or vertical. Assume the first segment is horizontal for convenience.

In the work below the following convergence for polygonal curves will be used. Let Γ be a fixed path of pure type and suppose $\{\Gamma^n\}_{n=1}^{\infty}$ is a sequence of polygonal curves of the same type, non-intersecting, such that Γ and $\{\Gamma^n\}_{n=1}^{\infty}$ have the same initial point z_1 and final point z_f. Define

$$\text{Area}(\Gamma^n \cup \Gamma)^{\circ} \longrightarrow \emptyset$$

to mean the open area enclosed by $\Gamma^n \cup \Gamma$ tends to the empty set in the sense that for each $z \in \Gamma$ there exists a vertical or horizontal line upon which $\tilde{z}$ and z lie such that

$$|X_{\tilde{z}}^{\Gamma^n} - X_z^{\Gamma}| \longrightarrow 0 \text{ as } n \longrightarrow \infty$$

where X^{Γ} denotes the restriction of X to Γ with $\tilde{z} \in \Gamma^n$ for $n = 1, 2, \ldots$, $z \in \Gamma$, is the point of intersection determined by $z \in \Gamma$ and the horizontal or vertical line through z. This is similar to the Lévy convergence on the space of distribution functions which is known to be a metric on that space.

For polygonal curves Γ of type I or II when $i = 1$, and for polygonal curves of type I or II' when $i = 2$, set

$$X_i^\Gamma(z_f) = X(D^i(z_f)) = \begin{cases} \sum_{j=1}^n \left(X_{(a_{j+1},b_j)} - X_{(a_j,b_j)} \right), & i = 1, \\ \sum_{j=1}^n \left(X_{(a_j,b_{j+1})} - X_{(a_j,b_j)} \right), & i = 2, \end{cases} \tag{2.17}$$

where the polygonal curves Γ are with initial point $z_1 = (a_1, b_1)$ and final point $z_f = (a_{n+1}, b_n)$. It will be necessary to add the integral sum over these polygonal curves up to a point $z \in \Gamma$. This leads to the next definition.

Definition 21. Let $X_i^{\Gamma(z)}$ denote the sum of X along a polygonal curve Γ from z_1 to $z \in \Gamma$, i.e.,

$$X_i^{\Gamma(z)} = \begin{cases} \sum_{\text{up to } z} \left(X_{(a_{j+1},b_j)} - X_{(a_j,b_j)} \right), & i = 1, \\ \sum_{\text{up to } z} \left(X_{(a_j,b_{j+1})} - X_{(a_j,b_j)} \right), & i = 2, \end{cases} \tag{2.18}$$

where the last point is $(a_j, b_j) = z$.

For general curves the process will be defined using L^2-limits of polygonal curves. Cairoli and Walsh use the following definition to develop the line integral for planar martingales $M \in \mathcal{M}^2(z_0)$. The generalization is natural.

Definition 22. For $z \in \Gamma$ and X an $L^{2,2}$-bounded process set

$$X_i^\Gamma(z) = E \left\{ X \left(D^i(z_f) \right) | \mathcal{F}_z^i \right\}, \text{ for } i = 1, 2, \tag{2.19}$$

where $X(D^i(z_f))$ is defined by (2.17).

For planar martingales $M \in \mathcal{M}^2(z_0)$,

$$M_i^\Gamma(z) = E \left\{ M \left(D^i(z_f) \right) | \mathcal{F}_z^i \right\}$$
$$= M \left(D^i(z) \right) = M_i^{\Gamma(z)} \quad i = 1, 2, \quad z \in \Gamma. \tag{2.20}$$

Hence Definitions 21 and 22 coincide for $M \in \mathcal{M}^2(z_0)$. For general $L^{2,2}$-bounded processes, however, the equality in (2.20) does not hold. For example, if an $L^{2,2}$-bounded 1-martingale X is considered then

$$
\begin{aligned}
X_i^\Gamma(z) &= E\left\{X\left(D_{z_f}^i\right) | \mathcal{F}_z^i\right\} \\
&= \begin{cases}
X\left(D^1(z)\right) = X_1^{\Gamma(z)} & , \ i = 1, \quad (2.21) \\
X\left(D^2(z)\right) + E\left\{X\left(D^2(z_f) - D^2(z)\right) | \mathcal{F}_z^2\right\} & , \ i = 2.
\end{cases}
\end{aligned}
$$

So the theory is more involved for general $L^{2,2}$-bounded processes. At the final point of the path, z_f, $X_i^\Gamma(z_f) = X_i^{\Gamma(z_f)}$ for $i = 1,2$ and consequently the right-hand summand of (2.21) for the case $i = 2$, may be looked at as $X_2^{\Gamma(z)} +$ (correction term). Consider first the process $\{X_i^\Gamma(z), \mathcal{F}_z^i\}$ as given in Definition 22 which includes the work of Cairoli and Walsh [4]. The existence of the $L^{2,2}$-bounding measure α for the process X will be needed to show the polygonal paths $X_i^{\Gamma_n}$, $n = 1, 2, \cdots$, will approximate X_i^Γ ($i = 1, 2$) for pure paths Γ. All the statements concerning $X_i^\Gamma(z_f)$ apply for $X^{\Gamma(z_f)}$ as well, and vice versa. For all other points $z \in \Gamma$ this is not true, unless X is a planar martingale.

Proposition 23. *Let Γ and Γ' be polygonal curves of type I or II (respectively, curves of type I or II' with $i = 2$) both having the same initial point z_1 and final point z_f. Suppose for $i = 1$ that Γ' lies entirely above Γ ($i = 2$, to the right). Denote by A the open area enclosed by Γ and Γ', i.e., $A = \text{Area}(\Gamma \cup \Gamma')^\circ$. Then, for $\{X_z, \mathcal{F}_z\}_{z \in T}$ an $L^{2,2}$-bounded process having Bochner bounding measure $\alpha : \Sigma \otimes \mathcal{B}(T) \to \mathbb{R}^+$,*

$$
E\left\{\left(X_i^{\Gamma'}(z_f) - X_i^\Gamma(z_f)\right)^2\right\} \leq \int_{\Omega \times T} \chi_{\overline{A} - \Gamma} \ d\alpha.
$$

Proof. This proposition will be proved for the case $i = 1$, the case $i = 2$ being similar. Denote the vertices of Γ' by $(a_1, b_1) \prec (a_1, b_2) \prec \cdots \prec (a_{n-1}, b_n) \prec (a_n, b_n)$ and of Γ by $(c_1, d_1) \prec (c_2, d_1) \prec \cdots \prec (c_m, d_{m-1}) \prec (c_m, d_m)$ where by hypothesis $z_1 = (a_1, b_1) = (c_1, d_1)$ and $z_f = (a_n, b_n) =$

(c_m, d_m). Then

$$X\left(D_1^{\Gamma'}(z_f)\right) = \sum_{j=1}^{n} \left(X_{(a_{j+1}, b_{j+1})} - X_{(a_j, b_{j+1})}\right)$$

and

$$X\left(D_1^{\Gamma}(z_f)\right) = \sum_{k=1}^{m} \left(X_{(c_{k+1}, d_k)} - X_{(c_k, d_k)}\right).$$

Since $X = 0$ on the axes the above equations can be written as follows:

$$X\left(D_1^{\Gamma'}(z_f)\right) = \sum_{j=1}^{n} X(A_j), \qquad \text{for } A_j = ((a_j, 0), (a_{j+1}, b_{j+1})],$$

and

$$X\left(D_1^{\Gamma}(z_f)\right) = \sum_{k=1}^{m} X(B_k), \qquad \text{for } B_k = ((c_k, 0), (c_{k+1}, d_k)].$$

Then using the vertices of both Γ and Γ' the region can be partitioned as

$$\bigcup_{j=1}^{n} A_j = \bigcup_{j=1}^{n} \bigcup_{i=1}^{q_i} A_{ji}$$

and

$$\bigcup_{k=1}^{m} A_k = \bigcup_{k=1}^{m} \bigcup_{l=1}^{p_k} B_{kl}$$

by slicing with vertical lines from each corner (horizontal for $i = 2$). Then

$$X\left(D_1^{\Gamma'}(z_f)\right) - X\left(D_1^{\Gamma}(z_f)\right) = \sum_{j=1}^{n} X(A_j) - \sum_{k=1}^{m} X(B_k)$$

$$= \sum_{j=1}^{n} \sum_{i=1}^{q_i} X(A_{ji}) - \sum_{k=1}^{m} \sum_{l=1}^{p_k} X(B_{kl}).$$

Now for each B_{kl} there exists some A_{jl} which contains it, in fact there is a one-to-one correspondence between these sets determined by the vertical lines from the corners of Γ and Γ'.

$$X\left(D_1^{\Gamma'}(z_f)\right) - X\left(D_1^{\Gamma}(z_f)\right) = \sum_{w=1}^{n} X(C_w), \quad C_w = A_{ji} - B_{kl}, \quad B_{kl} \subseteq A_{ji}.$$

Thus

$$E\left\{\left(X_i^{\Gamma'}(z_f) - X_i^{\Gamma}(z_f)\right)^2\right\} = E\left\{\left(X\left(D_1^{\Gamma'}(z_f)\right) - X\left(D_1^{\Gamma}(z_f)\right)\right)^2\right\}$$

$$= E\left\{\left(\sum_{w=1}^{n} X(C_w)\right)^2\right\}$$

$$= E\left\{\left(\int_T \chi_{\cup_{w=1}^n C_w} dX\right)^2\right\}$$

$$= E\left\{\left(\int_T \chi_{\overline{A}-\Gamma} dX\right)^2\right\}$$

$$\leq \int_{T\times\Omega} \chi_{\overline{A}-\Gamma} d\alpha,$$

since $\chi_{\overline{A}-\Gamma}$ is predictable and the Bochner boundedness principle applies.

□

The next corollary is immediate.

Corollary 24. *If X does not charge Γ, then for Γ and Γ' as in the above proposition*

$$E\left\{\left(X_i^{\Gamma'}(z_f) - X_i^{\Gamma}(z_f)\right)^2\right\} \leq \int_{\Omega\times T} \chi_A d\alpha.$$

The previous inequalities were for X along the whole of Γ. If the process is restricted up to a point $z \in \Gamma$ Definition 21 will be needed. The next proposition is false for Definition 22, unless $X \in \mathcal{M}^2(z_0)$ or $z = z_f$ as mentioned before. $[T \times \Omega, \Omega \times T$ denote the same space.]

Proposition 25. *Let $\{X_z, \mathcal{F}_z\}_{z\in T}$ be an $L^{2,2}$-bounded process with Γ a curve of type I or II for $i = 1$, and of type I or II' for $i = 2$ (not necessarily a polygonal curve), then there exists a sequence $\{\Gamma^n\}_{n=1}^{\infty}$ of polygonal curves of respective type such that*

$$E\left\{\left(X_i^{\Gamma(z)} - X_i^{\Gamma^n(z')}\right)^2\right\} \leq \int_{T\times\Omega} \chi_{\overline{A}_z^n - \Gamma} d\alpha$$

where $A_z^n = Area\,(\Gamma(z) \cup \Gamma^n(z'))^\circ$, i.e., A_z^n is the open area up to the vertical line V_z between Γ^n and Γ, and $z' = \inf\{z : z \in V_z \cap \Gamma^n\}$. Consequently,

$$X_i^{\Gamma^n(z)} \overset{L^2}{\to} X_i^{\Gamma(z)}.$$

Also, if X does not charge Γ then

$$E\left\{\left(X_i^{\Gamma(z)} - X_i^{\Gamma^n(z')}\right)^2\right\} \leq \int_{T\times\Omega} \chi_{A_z^n}\, d\alpha.$$

Proof. The proposition is shown for $z = z_f$, the arbitrary case follows by attaching for $z \in \Gamma$ the vertical line V_z to Γ at z and then proceeding as before. So, let Γ be a continuous curve of type I (or II as before) and $i = 1$. Set $\{\Gamma^n\}_{n=1}^\infty$ to be a sequence of polygonal curves having the same initial and final points laying above Γ such that

$$\lim_{n\to\infty} Area\,(\Gamma \cup \Gamma^n)^\circ = \emptyset.$$

Since $A_n = Area\,(\Gamma \cup \Gamma^n)^\circ$ is an open set $\chi_{A_n \times \Omega}$ is predictable. On Γ for $m = 1, 2, \ldots$ let π_m be the collection of points $\pi_m = \left\{\gamma\left(\frac{k}{2^n}\right) : 0 \leq k \leq 2^n\right\}$ where γ is the parameterization of Γ. Construct the polygonal path Φ_m by passing successive horizontal and vertical line segments joined at the ends with the points of π_m, i.e., $\gamma(\frac{k}{2^n})$, $k = 1, 2, \ldots 2^n$. Then $\Phi_m \to \Gamma$ as $n \to \infty$ in the sense mentioned at the beginning of the section. Denote the open area between $\Gamma^n \cup \Phi_m$ by $B_{mn} = Area(\Gamma^n \cup \Phi_m)^\circ$ which satisfies the limit

$$\lim_{m\to\infty} B_{mn} = Area\,(A_n \cup \Gamma).$$

The regions satisfy the containment

$$D^1_{\Phi_m}(z_f) \subseteq D^1_{\Gamma^n}(z_f) \text{ for } m, n = 1, 2, \ldots.$$

By Proposition 23, setting $B_{mn} = \cup_{k=1}^{p} C_k^{mn}$ and $\Gamma_1^{\Gamma^n(z_f)} = X_1^{\Gamma^n}(z_f)$,

$$E\left\{\left(X_1^{\Gamma^n}(z_f) - X_1^{\Phi_m}(z_f)\right)^2\right\} = E\left\{\left(X\left(D_1^{\Gamma^n}(z_f)\right) - X\left(D_1^{\Phi_m}(z_f)\right)\right)^2\right\}$$

$$= E\left\{\left(\sum_{k=1}^{p} X\left(C_k^{mn}\right)\right)^2\right\}$$

$$= E\left\{\left(\int_T \chi_{\cup_{k=1}^p C_k^{mn}} dX\right)^2\right\}$$

$$= E\left\{\left(\int_T \chi\left(D_1^{\Gamma^n}(z_f) - D_1^{\Phi_m}(z_f)\right) dX\right)^2\right\}$$

$$= E\left\{\left(\int_T \chi(B_{mn}) dX\right)^2\right\}$$

$$\leq \int_{T \times \Omega} \chi(B_{mn}) d\alpha,$$

by the boundedness principle. Also,

$$E\left\{\left(\int_T \chi(B_{mn}) dX - \int_T \chi(A_n \cup \Gamma) dX\right)^2\right\}$$

$$= E\left\{\left(\int_T (\chi(B_{mn}) - \chi(A_n \cup \Gamma)) dX\right)^2\right\}$$

$$\leq \int_{T \times \Omega} (\chi(B_{mn}) - \chi(A_n \cup \Gamma))^2 d\alpha \longrightarrow 0$$

as $m \longrightarrow \infty$ by monotone convergence. Thus

$$\lim_{m \to \infty} E\left\{\left(X_1^{\Gamma^n}(z_f) - X_1^{\Phi_m}(z_f)\right)^2\right\} = \lim_{m \to \infty} E\left\{\left(\int_T \chi(B_{mn}) dX\right)^2\right\}$$

$$= E\left\{\left(\int_T \chi(A_n \cup \Gamma) dX\right)^2\right\}$$

$$= E\left\{\left(X_1^{\Gamma^n}(z_f) - X_1^{\Gamma}(z_f)\right)^2\right\}.$$

Hence

$$E\left\{\left(X_1^{\Gamma^n}(z_f) - X_1^{\Gamma}(z_f)\right)^2\right\} = \lim_{m \to \infty} E\left\{\left(X_1^{\Gamma^n}(z_f) - X_1^{\phi_m}(z_f)\right)^2\right\}$$

$$= \lim_{m \to \infty} E\left\{\left(\int_T \chi(B_{mn}) dX\right)^2\right\}$$

$$\leq \liminf_{m \to \infty} \int_{T \times \Omega} \chi(B_{mn}) \, d\alpha$$

$$= \int_{T \times \Omega} \chi(\overline{A}_n - \Gamma) \, d\alpha \text{ by Fatou's Lemma.}$$

(2.22)

Taking the limit as $n \to \infty$ on both sides of (2.22)

$$X_1^{\Gamma^n}(z_f) \xrightarrow{L^2} X_1^{\Gamma}(z_f).$$

For the region determined by $\Gamma \cup \Gamma^n$ and the vertical line V_z, the above argument holds by replacing z_f for $z \in \Gamma$. Thus

$$X_1^{\Gamma^n(z)} \xrightarrow{L^2} X_1^{\Gamma(z)}.$$

$\square$

Thus for $L^{2,2}$-bounded processes $\{X_z, \mathcal{F}_z\}_{z \in T}$ defined along pure paths Γ the restricted process $\{X^{\Gamma(z)}, \mathcal{F}_z\}_{z \in \Gamma}$ can be approximated by a sequence of processes restricted to polygonal paths $\{X^{\Gamma^n(z)}, \mathcal{F}_z\}_{z \in \Gamma^n}$, $n = 1, 2, \cdots$, in $L^2(P)$. In Definition 22 the process $\{X_i^{\Gamma}(z), \mathcal{F}_z^i\}_{z \in \Gamma}$, $i = 1, 2$, is defined using conditional expectation and as a consequence is a martingale along Γ. The continuity of this martingale is discussed below and is shown to depend on the continuity of the filtration.

Proposition 26. *Let $\{X_z, \mathcal{F}_z\}_{z \in T}$ be a right continuous $L^{2,2}$-bounded process. Then $\{X_i^{\Gamma}(z), \mathcal{F}_z^i\}_{z \in \Gamma}$, $i = 1, 2$, is a one-parameter right continuous martingale and the process $\{X_i^{\Gamma}(z), \mathcal{F}_z^i\}_{z \in T}$, $i = 1, 2$, will be continuous if X is continuous and if the filtration has no jump discontinuities. As usual for $i = 1$, it is assumed that Γ is of type I or II, and for $i = 2$, Γ is type I or II'.*

Proof. The result will be proven for $i = 1$. The martingale property follows immediately from the definition $X_1^{\Gamma}(z) = E\{X(D_{z_f}^1) | \mathcal{F}_z^1\}$, since the conditional expectation operator satisfies $E\{E\{Y | \mathcal{F}_z\} | \mathcal{F}_{z'}\} = E\{Y | \mathcal{F}_{z'}\}$ for

$\mathcal{F}_{z'} \subseteq \mathcal{F}_z$. In Proposition 25 it was shown that for Γ there exists a sequence of polygonal curves $\Gamma^n \to \Gamma$ such that $X_1^{\Gamma^n}(z_f) \overset{L^2}{\to} X_1^{\Gamma}(z_f)$. Thus a subsequence can be selected, denoted by the same indices, such that

$$E\left\{\left(X_1^{\Gamma^n}(z_f) - X_1^{\Gamma}(z_f)\right)^2\right\} < \frac{1}{2^n}$$

for $n = 1, 2, \ldots$. Let the points $z_1 = (s_1, t_1)$ and $z_f = (s_f, t_f)$, then for $s_1 \leq s \leq s_f$ define

$$N_n(s) = E\left\{X_1^{\Gamma^n}(z_f)|\mathcal{F}_{s0}^1\right\}, \text{ and}$$

$$N(s) = E\left\{X_1^{\Gamma}(z_f)|\mathcal{F}_{s0}^1\right\}.$$

Clearly, for each $n = 1, 2, \ldots$, $\{N_n(s), \mathcal{F}_{s0}^1\}_{s_1 \leq s \leq s_f}$ is a martingale. Also

$$E\left\{(N_n(s) - N(s))^2\right\} = E\left\{\left(E\left\{X_1^{\Gamma^n}(z_f) - X_1^{\Gamma}(z_f)|\mathcal{F}_{s0}^1\right\}\right)^2\right\}$$

$$\leq E\left\{\left(X_1^{\Gamma^n}(z_f) - X_1^{\Gamma}(z_f)\right)^2\right\}$$

$$< \frac{1}{2^n}.$$

Hence by the submartingale maximal inequality,

$$P\left\{\max_{1 \leq k \leq n}|N_k(s) - N(s)|^2 > \epsilon^2\right\} \leq \frac{1}{\epsilon^2}E\left\{|N_n(s) - N(s)|^2\right\}$$

$$\leq \frac{1}{\epsilon^2 2^n}.$$

So $N_n(s) \to N(s)$, a.e. ω, uniformly in s. If X_{st} is continuous in the first parameter it follows that

$$N_n(s) = E\left\{X_1^{\Gamma^n}(z_f)|\mathcal{F}_{s0}^1\right\}$$

$$= E\left\{X\left(D_1^{\Gamma^n}(z_f)\right)|\mathcal{F}_{s0}^1\right\}$$

$$= X_1^{\Gamma^n}\left((s, \inf\{t : (s, t) \in \Gamma^n\})\right),$$

and

$$\lim_{t \to s+} N_n(t) = \lim_{t \to s+} E\left\{ X_1^{\Gamma^n}(z_f) | \mathcal{F}_{t0}^1 \right\}$$

$$= E\left\{ X_1^{\Gamma^n}(z_f) | \cap_{t>s} \mathcal{F}_{t0}^1 \right\}$$

(by the right continuity of the filtration)

$$= N_n(s).$$

However, for the left continuity

$$\lim_{t \to s-} N_n(t) = \lim_{t \to s-} E\left\{ X_1^{\Gamma^n}(z_f) | \mathcal{F}_{t0}^1 \right\}$$

$$= E\left\{ X_1^{\Gamma^n}(z_f) | \cup_{t<s} \mathcal{F}_{t0}^1 \right\}$$

$$= E\left\{ X\left(D_1^{\Gamma^n}(z_f) \right) | \mathcal{F}_{s0}^1 \right\} = N_n(s)$$

if and only if the filtration has no jump discontinuities. So $\{X_i^{\Gamma}(z), \mathcal{F}_z^i\}_{z \in \Gamma}$ is right continuous martingale being continuous only if the filtration has no jump discontinuities. □

Using the martingale property of the process $\{X_i^{\Gamma}(z), \mathcal{F}_z^i\}_{z \in \Gamma}$, $i = 1, 2$, a Bochner bounding measure can be constructed.

Proposition 27. *Let* $\{X_z, \mathcal{F}_z\}_{z \in T}$ *be an* $L^{2,2}$*-bounded process and* Γ *a curve of type I or II for* $i = 1$, *type I or II' for* $i = 2$. *Then* $\{X_i^{\Gamma}(z), \mathcal{F}_z^i\}_{z \in \Gamma}$, $i = 1, 2$, *is an* $L^{2,2}$*-bounded process with bounding measure* $\alpha = \langle X_i^{\Gamma} \rangle_z \otimes P$.

Proof. By Proposition 26, $\{X_i^{\Gamma}(z), \mathcal{F}_z^i\}_{z \in \Gamma}$ is a right continuous martingale. Then for simple ϕ and the fact that the $z \in \Gamma$ are linearly ordered with respect to $\prec$ for $i = 1$, and λ for $i = 2$, it is immediate that

$$E\{ (\int_T \phi dX_i^{\Gamma}(z))^2 \} \le \int_{\Gamma \times \Omega} \phi^2 d\langle X_i^{\Gamma} \rangle_z \otimes dP$$

from the one-parameter theory where $\{(X_i^{\Gamma})^2 - \langle X_i^{\Gamma} \rangle_z, \mathcal{F}_z^i\}$, $z \in \Gamma$, $i = 1, 2$ is a martingale. □

Thus for the case that X is a martingale (and that $X_i^\Gamma(z) = X_i^{\Gamma(z)}$ for martingales) the stochastic line integral is well defined with respect to dX_i^Γ, $i = 1, 2$. In general, $X_i^{\Gamma(z)}$ and $X_i^\Gamma(z)$ are related as follows:

$$X_i^\Gamma(z) = E\left\{ X\left(D_i^\Gamma\right) | \mathcal{F}_z^i \right\}$$

$$= E\left\{ X\left(D_i^\Gamma(z)\right) + X\left(D_i^\Gamma(z_f) - D_i^\Gamma(z)\right) | \mathcal{F}_z^i \right\}$$

$$= X\left(D_i^\Gamma(z)\right) + E\left\{ X\left(D_i^\Gamma(z_f) - D_i^\Gamma(z)\right) | \mathcal{F}_z^i \right\}$$

$$= X_i^{\Gamma(z)} + E\left\{ X\left(D_i^\Gamma(z_f) - D_i^\Gamma(z)\right) | \mathcal{F}_z^i \right\}, \qquad (2.23)$$

for $i = 1, 2$. The second term on the right is zero for planar martingales. The goal is to define the line integral along a general curve Γ. Let X^Γ be the restriction of X to the curve Γ. Then for polygonal curves Γ with initial point z_1 and final point z_f,

$$X_z^\Gamma - X_{z_1}^\Gamma = X\left(D_1^\Gamma(z)\right) + X\left(D_2^\Gamma(z)\right)$$

$$= X_1^{\Gamma(z)} + X_2^{\Gamma(z)}$$

$$= X_1^\Gamma(z) + X_2^\Gamma(z) - E\left\{ X\left(D_1^\Gamma(z_f) - D_1^\Gamma(z)\right) | \mathcal{F}_z^1 \right\}$$

$$- E\left\{ X\left(D_2^\Gamma(z_f) - D_2^\Gamma(z)\right) | \mathcal{F}_z^2 \right\}, \text{ by } (2.23) .$$

So in the general case a more natural development is via $X_i^{\Gamma(z)}$, since

$$X_z^\Gamma - X_{z_1}^\Gamma = X_1^{\Gamma(z)} + X_2^{\Gamma(z)}.$$

Unlike the case for the process $\{X_i^\Gamma(z), \mathcal{F}_z^i\}_{z \in \Gamma}$, the process $\{X_i^{\Gamma(z)}, \mathcal{F}_z^i\}_{z \in \Gamma}$, $i = 1, 2$, of Definition 21 is not a martingale. It is true that if X is an i-martingale that $\{X_i^{\Gamma(z)}, \mathcal{F}_z^i\}_{z \in \Gamma}$ is an $\mathcal{F}_z^i$ martingale, but $\{X_i^{\Gamma(z)}, \mathcal{F}_z^{\tilde{i}}\}_{z \in \Gamma}$ is not a martingale where $\tilde{i} = \{1, 2\} \setminus \{i\}$, $i = 1, 2$. The bounding measure for a line integral with respect to a $L^{2,2}$-bounded process X is given by a direct calculation for the integral with respect to $X^\Gamma = X_1^\Gamma + X_2^\Gamma$.

Theorem 28. *Let* $\{X_z, \mathcal{F}_z\}_{z \in T}$ *be an* $L^{2,2}$-*bounded process having the Bochner* *bounding measure α, and let Γ be a curve of pure type I. Then $X_1^{\Gamma(z)} + X_2^{\Gamma(z)}$ is $L^{2,2}$-bounded with respect to α.*

Proof. It suffices to consider simple functions ϕ,

$$E\left\{\left(\int_T \phi\left(dX_1^{\Gamma(z)} + dX_z^{\Gamma(z)}\right)\right)^2\right\} = E\left\{\left(\int_T \phi dX_z^\Gamma\right)^2\right\}$$

$$= E\left\{(X_\Gamma \phi dX)^2\right\}$$

(the restriction to Γ)

$$\leq \int_{T \times \Omega} (X_\Gamma \phi)^2 \, d\alpha$$

$$= \int_{\Gamma \times \Omega} \phi^2 \, d\alpha.$$

□

This theorem was restricted to curves of type I since $X_1^{\Gamma(z)}$ and $X_2^{\Gamma(z)}$ are simultaneously defined for such curves. It is necessary to find the bounding measure for $\{X_i^{\Gamma(z)}, \mathcal{F}_z^i\}_{z \in \Gamma}$, $i = 1, 2$, as well, if the development of the line integral is to be completed.

Theorem 29. *Let $\{X_z, \mathcal{F}_z\}_{z \in T}$ be an $L^{2,2}$-bounded process. Then for $i = 1$ with polygonal curve Γ of type I or II $\{X_1^{\Gamma(z)}, \mathcal{F}_z^1\}_{z \in \Gamma}$ is $L^{2,2}$-bounded, and for $i = 2$ with polygonal curve Γ of type I or II' $\{X_2^{\Gamma(z)}, \mathcal{F}_z^2\}_{z \in \Gamma}$ is $L^{2,2}$-bounded.*

Proof. The proof is for the case $i = 1$, the other being similar. Let $\phi = \sum_{i=1}^n \alpha_i X_{(s_i, t_i]}$ be an $\mathcal{F}_z$-predictable simple function. It will be assumed that the rectangles $(s_i, t_i]$ intersect the curve Γ at the points s_i' and t_i' taking the infimum if necessary. Then $X_1^{\Gamma(z)}(s_i, t_i] = X(t_i') - X(s_i')$ for $i = 1, 2, \ldots, n$ where for polygonal curves

$$X(t_i') = \sum_{\text{up to } t_i'} \left(X_{a_{j+1}, b_j} - X_{a_j, b_j}\right),$$

and with

$$A_j^{s_i'} = \begin{cases} X_{a_{j+1},b_j} - X_{a_j,b_j} & , \text{if } t_i' \geq (a_{j+1},b_j) \\ X_{t_i'} - X_{a_j,b_j} & , \text{if } (a_j,b_j) < t_i' \leq (a_{j+1},b_j), \end{cases}$$

it follows that

$$X(t_i') - X(s_i') = \sum_{\text{up to } t_i'} \left(X_{a_{j+1},b_j} - X_{a_j,b_j} \right) - \sum_{\text{up to } s_i'} \left(X_{a_{j+1},b_j} - X_{a_j,b_j} \right)$$

$$= \sum_{\text{up to } t_i'} X(A^{t_i'}) - \sum_{\text{up to } s_i'} X\left(A^{s_i'} \right)$$

$$= \sum_{j=1}^{m(t_i')} X\left(A_j^{t_i'} \right) - \sum_{j=1}^{m(s_i')} X\left(A_j^{s_i'} \right), \tag{say}$$

where from the increment containing s_i' up to the increment containing t_i' the rectangles $A^{t_i'}$ and $A^{s_i'}$ are identical. The $m(s_i')$ is defined to mean the sum over the first m rectangles with the last rectangle being sliced at s_i'. So one can let

$$\sum_{j=1}^{m(t_i')} X(A_j^{t_i'}) - \sum_{j=1}^{m(s_i')} X(A_j^{s_i'}) = \sum_{j=m(s_i')}^{m(t_i')} X(A_j^{t_i' \backslash s_i'})$$

where the set $A_j^{t_i' \backslash s_i'}$ is defined as follows

$$A_j^{t_i' \backslash s_i'} = \begin{cases} A_{m(s_i')}^{t_i'} - A_{m(s_i)}^{t_i'} & , \ j = m(s_i') \\ A_j^{t_i'} & , \text{for } j \neq m(s_i'). \end{cases}$$

Then

$$E\left\{ \left(\int_\Gamma \phi dX_1^{\Gamma(z)} \right)^2 \right\} = E\left\{ \left(\chi_\Gamma \sum_{i=1}^n \alpha_i X_1^{\Gamma(z)}(s_i,t_i] \right)^2 \right\}$$

$$= E\left\{ \left(\chi_\Gamma \sum_{i=1}^n \alpha_i \left(X(t_i') - X(s_i') \right) \right)^2 \right\}$$

$$= E\left\{ \left(\chi_\Gamma \sum_{i=1}^n \alpha_i \left(\sum_{j=m(s_i')}^{m(t_i')} X\left(A_j^{t_i' \backslash s_i'} \right) \right) \right)^2 \right\}$$

$$= E\left\{\left(\int_T \sum_{i=1}^n \alpha_i \chi_\Gamma \sum_{j=m(s_i')}^{m(t_i')} \chi_{A_j^{t_i'\backslash s_i'}} dX\right)^2\right\}$$

$$= E\left\{\left(\int_T \chi_\Gamma \sum_{i=1}^n \alpha_i \chi_{(s_i,t_i]} dX\right)^2\right\}$$

$$\leq \int_{T\times\Omega}\left(\chi_\Gamma \sum_{i=1}^n \alpha_i \chi_{(s_i,t_i]}\right)^2 d\alpha$$

$$= \int_{\Gamma\times\Omega}\phi^2 d\alpha.$$

$\square$

For general curves Γ let Γ^n be a sequence of polygonal curves converging to Γ. Then by Proposition 25,

$$X_i^{\Gamma^n(z)} \xrightarrow{L^2} X_i^{\Gamma(z)}$$

for all $z \in \Gamma$, $i = 1, 2$. Thus for simple ϕ,

$$E\left\{\left(\int_T \phi dX_1^{\Gamma^n(z)} - \int_T \phi dX_1^{\Gamma(z)}\right)^2\right\} = E\left\{\left(\int_T \phi d\left(X_1^{\Gamma^n(z)} - X_1^{\Gamma(z)}\right)\right)^2\right\}$$

$$\to 0 \text{ as } n \to \infty.$$

So

$$E\left\{\left(\int_T \phi dX_1^{\Gamma(z)}\right)^2\right\} = \lim_{n\to\infty} E\left\{\left(\int_T \phi dX_1^{\Gamma^n(z)}\right)^2\right\}$$

$$\leq \lim_{n\to\infty}\int_{\Gamma^n\times\Omega}\phi^2 d\alpha$$

$$= \int_{\Gamma\times\Omega}\phi^2 d\alpha.$$

Hence the line integrals are well defined with respect to $\{X_i^{\Gamma(z)}, \mathcal{F}_z^i\}_{z\in\Gamma}$ for all $\phi \in L^2(\alpha)$. The definition will now be completed for the line integral. For Γ of type I or II and $\phi = \{\phi_z, \mathcal{F}_z^1\}_{z\in\Gamma}$, a predictable function, the line integral with respect to X_1^Γ is denoted as

$$\int_\Gamma \phi dX_1^\Gamma, \text{ or as } \int_\Gamma \phi\partial_1 X, \quad z \in \Gamma.$$

For Γ of type I or II' and $\phi = \{\phi_z, \mathcal{F}_z^2\}_{z \in \Gamma}$ a predictable function, the line integral with respect to X_2^Γ is denoted as

$$\int_\Gamma \phi \, dX_2^\Gamma \text{ or as } \int_\Gamma \phi \partial_2 X, \quad z \in \Gamma.$$

If Γ is of type I' or II' then

$$\int_\Gamma \phi \partial_1 X = -\int_{\widehat{\Gamma}} \phi \partial_1 X,$$

or for Γ of type I' or II then

$$\int_\Gamma \phi \partial_2 X = -\int_{\widehat{\Gamma}} \phi \partial_2 X,$$

where $\widehat{\Gamma}$ is the opposite orientation of Γ. Finally, if Γ is of pure type,

$$\int_\Gamma \phi \partial X = \int_\Gamma \phi \partial_1 X + \int_\Gamma \phi \partial_2 X.$$

This definition can be extended to piecewise pure curves by writing the integral as a sum of pure types. Thus integration over the boundary of circles can be performed using this definition. Also, if $\{X_z, \mathcal{F}_z\}_{z \in T}$ is an $L^{2,2}$-bounded A-quasimartingale and Γ is increasing (of type I) then

$$\begin{aligned}
\int_\Gamma \phi \partial X &\stackrel{\text{def}}{=} \int_\Gamma \phi \left(\partial_1 X + \partial_2 X \right) \\
&\stackrel{\text{def}}{=} \int_\Gamma \phi \left(dX_1^{\Gamma(z)} + dX_2^{\Gamma(z)} \right) \\
&= \int_\Gamma \phi \, dX^\Gamma
\end{aligned}$$

since $X_z^\Gamma - X_z^{\Gamma(z_1)} = X_1^{\Gamma(z)} + X_2^{\Gamma(z)}$. So the definition of the line integral is complete and the development for A-quasimartingales over increasing paths can be continued. Thus the work of Cairoli and Walsh can be extended to this (sub)class of $L^{2,2}$-bounded processes. In the case of an A-quasimartingale the Bochner bound along increasing paths was developed using the one-parameter theory. The development is limited due to the lack of a single linear order on the pure type paths. The approach above utilizes the Bochner bounding measure of the process defined on the whole two-dimensional parameter set T. Some convergence theorems for line integrals are now presented which will be useful in further investigations.

Proposition 30. *Suppose* $\{X_z, \mathcal{F}_z\}_{z \in T}$ *is an* $L^{2,2}$*-bounded process with a bounding measure* α*, and suppose* $\{\phi_z, \mathcal{F}_z\}_{z \in \Gamma}$ *and* $\{\phi_z^n, \mathcal{F}_z^n\}_{z \in \Gamma}$*,* $n = 1, 2, \ldots$ *are predictable processes. If there exists a* ψ *such that* $|\phi_n| \leq \psi$ *a.e.* (ω) *with*

$$\int_{T \times \Omega} \psi^2 \, d\alpha < \infty,$$

and $\phi^n \longrightarrow \phi$ *a.e., then*

$$\int_{\Gamma^n} \phi^n \partial X \xrightarrow{L^2} \int_{\Gamma} \phi \partial X$$

where $\{\Gamma^n\}_{n=1}^{\infty}$ *and* Γ *are pure curves of the same type.*

Proof. By hypothesis $\chi_{\Gamma^n} \phi^n \longrightarrow \chi_\Gamma \phi$ a.e., where for each $n = 1, 2, \ldots$ $\chi_{\Gamma^n} \phi^n$ is predictable with $|\chi_{\Gamma^n} \phi^n| \leq |\phi^n| \leq \psi$. Hence

$$\lim_{n \to \infty} \int_T \chi_{\Gamma^n} \phi^n dX = \int_T \chi_\Gamma \phi dX$$

in L^2 by dominated convergence and the Bochner boundedness principle. In fact the following is true:

$$\int_T \chi_{\Gamma^n} \phi^n dX = \int_{\Gamma^n} \phi^n \partial X$$

$$= \int_{\Gamma^n} \phi^n \partial_1 X + \int_{\Gamma^n} \phi^n \partial_2 X$$

$$\longrightarrow \int_\Gamma \phi \partial_1 X + \int_\Gamma \phi \partial_2 X$$

$$= \int_\Gamma \phi \partial X.$$

□

This result does not say what convergence properties the integrators $\partial_1 X$ and $\partial_2 X$ have individually. For this a few additional hypotheses will be needed. Let Γ^n and Γ, $n = 1, 2, \ldots$ be curves of the same type, either I or II, all having the same initial point $z_1 = (s_1, t_1)$ and final point $z_f = (s_f, t_f)$. Suppose that $\Gamma^n \to \Gamma$ as $n \to \infty$. For $s \geq 0$ and any curve Λ let

$$\nu_\Lambda(s) = (s, \inf\{t : (s, t) \in \Lambda\}).$$

Then for an $L^{2,2}$-bounded process X relative to $\{\mathcal{F}_s^1\}_{s \in [0, s_0]} = \{\mathcal{F}_z^1\}_{z \in \Gamma}$ set

$$Y(s) = X_1^\Gamma(\nu_\Gamma(s)), \quad s_1 \leq s \leq s_f \text{ and}$$

$$Y_n(s) = X_1^{\Gamma^n}(\nu_{\Gamma^n}(s)), \quad n = 1, 2, \ldots, s_1 \leq s \leq s_f.$$

Proposition 31. *Let $\{X_z, \mathcal{F}_z\}_{z \in T}$ be an $L^{2,2}$-bounded process, $\{\phi_z, \mathcal{F}_z\}_{z \in \Gamma}$ and $\{\phi_z^n, \mathcal{F}_z^n\}_{z \in \Gamma}$, $n = 1, 2, \ldots$ be predictable processes. Define the functions ψ and ψ_n by $\psi(s) = \phi(\nu_\Gamma(s))$ and $\psi_n(s) = \phi_n(\nu_{\Gamma^n}(s))$, $s_1 \leq s \leq s_f$. Suppose*

(1) $X_1^{\Gamma_z}$ is an $L^{2,2}$-bounded process such that the induced function on $[s_1, s_f]$ given by $X_1^{\Gamma_z}(\nu_\Gamma(s)) = Y(s)$ is $L^{2,2}$-bounded with respect to α^Γ, and $X_1^{\Gamma^n}$ is $L^{2,2}$-bounded such that the process $Y_1^n(s) = X_1^{\Gamma^n}(\nu_{\Gamma^n}(s))$ is $L^{2,2}$-bounded with respect to α^{Γ^n},

(2) there exists a σ-additive function $\tau : \mathcal{B}([s_1, s_f]) \otimes \Sigma \to I\!\!R^+$ such that for all $A \in \mathcal{B}([s_1, s_f]) \otimes \Sigma$, $\alpha^{\Gamma^n}(A) \leq \tau(A)$, $\alpha^\Gamma(A) \leq \tau(A)$, and $(\alpha^{\Gamma^n} - \alpha^\Gamma)([s_1, s_f] \times \Omega) \to 0$ as $n \to \infty$,

(3) $\int_{[s_1, s_f] \times \Omega} \psi_n^2(s) \, d\tau < \infty$ and $\int_{[s_1, s_f] \times \Omega} \psi^2(s) \, d\tau < \infty$,

(4) $\lim_{n \to \infty} \int_{[s_1, s_f] \times \Omega} (\psi_n - \psi)^2 \, d\tau = 0$.

Then

$$\lim_{n \to \infty} \int_{\Gamma^n} \phi_n \partial_1 X = \int_\Gamma \phi \partial_1 X.$$

Proof. Since $\alpha^{\Gamma^n} \leq \tau$ and $\alpha^{\Gamma} \leq \tau$ it follows that the difference $(\alpha^{\Gamma^n} - \alpha^{\Gamma}) \leq 2\tau$. Consequently $d(\alpha^{\Gamma^n} - \alpha^{\Gamma}) \leq 2d\tau$, $d\alpha^{\Gamma^n} \leq d\tau$, and $d\alpha^{\Gamma} \leq d\tau$. Using this, it follows that:

$$\int_{\Gamma} \phi \partial_1 X - \int_{\Gamma^n} \phi_n \partial_1 X = \int_{s_1}^{s_f} \psi dY - \int_{s_1}^{s_f} \psi_n dY^n$$

$$= \int_{s_1}^{s_f} (\psi - \psi_n) \, dY + \int_{s_1}^{s_f} (\psi_n - \psi) \, d(Y - Y^n)$$

$$+ \int_{s_1}^{s_f} \psi d(Y - Y^n).$$

Considering the right-hand side term by term:

$$E\left\{\left(\int_{s_1}^{s_f} (\psi - \psi_n) \, dY\right)^2\right\} \leq \int_{[s_1,s_f]\times\Omega} (\psi - \psi_n)^2 \, d\alpha^{\Gamma}$$

$$\leq \int_{[s_1,s_f]\times\Omega} (\psi - \psi_n)^2 \, d\tau$$

$$\rightarrow 0,$$

$$E\left\{\left(\int_{s_1}^{s_f} (\psi_n - \psi) \, d(Y - Y^n)\right)^2\right\} \leq \int_{[s_1,s_f]\times\Omega} (\psi_n - \psi)^2 \, d\left(\alpha^{\Gamma} + \alpha^{\Gamma^n}\right)$$

$$\leq 2 \int_{[s_1,s_f]\times\Omega} (\psi - \psi_n)^2 \, d\tau$$

$$\rightarrow 0,$$

and

$$E\left\{\left(\int_{s_1}^{s_f} \psi d(Y - Y^n)\right)^2\right\} \leq \int_{[s_1,s_f]\times\Omega} \psi^2 d\left(\left|\alpha^{\Gamma} - \alpha^{\Gamma^n}\right|\right). \qquad (*)$$

Writing

$$\psi^2 = \psi^2 \chi_{\{|\psi| \leq m\}} + \psi^2 \chi_{\{|\psi| > m\}}$$

$$\leq m^2 + \psi^2 \chi_{\{|\psi| > m\}},$$

one has

$$(*) \leq m^2 \int_{[s_1,s_f]\times\Omega} d\left(\left|\alpha^\Gamma - \alpha^{\Gamma^n}\right|\right) + 2 \int_{[s_1,s_f]\times\Omega} \psi^2 \chi_{\{|\psi|>m\}} d\tau$$

$$= m^2 \left(\left|\alpha^\Gamma - \alpha^{\Gamma^n}\right|\right) ([s_1,s_f]\times\Omega) + 2 \int_{[s_1,s_f]\times\Omega} \psi^2 \chi_{\{|\psi|>m\}} d\tau.$$

Now first let $n \to \infty$, then $m \to \infty$, so that $(*)$ tends to 0 using condition (2) and the dominated convergence theorem. $\qquad\qquad\square$

A similar proof holds for $i = 2$ with Γ of type I or II'.

Remark 32. The above theorem includes Proposition (4.6) of Cairoli and Walsh [4] wherein for strong martingales $d\tau = d\langle Y_0 \rangle$ for Γ^0, with Γ^0 defined as a curve above Γ^n and Γ, $n = 1, 2, \ldots$, and

$$\int_{[s_1,s_f]\times\Omega} d\left(\left|\alpha^\Gamma - \alpha^{\Gamma^n}\right|\right) = \int_{[s_1,s_f]\times\Omega} d\langle Y - Y^n \rangle$$

$$= E\left\{M^2\langle A_n \rangle\right\}$$

$$\to 0 \text{ as } n \to \infty$$

where $A_n = \text{Area}(\Gamma^n \cup \Gamma)^\circ \longrightarrow 0$.

Remark 33. For any $L^{2,2}$-bounded process X with bounding measure α, by Proposition 25 the measure α can be used in place of τ in condition (2). The hypothesis of condition (2) then becomes

$$\int_{\Omega\times T} \left(\chi_{\overline{A^n}-\Gamma}\right)^2 d\alpha \longrightarrow 0 \quad \text{as} \quad n \longrightarrow \infty$$

where A^n is the open area between the curves Γ^n and Γ. If X does not charge the boundary of $\Gamma \cup \Gamma^n$ then the hypothesis follows from the dominated convergence theorem. For the above two cases, step $(*)$ can be computed as follows:

$$E\left\{\left(\int_{s_1}^{s_f}\phi d\left(Y-Y^n\right)\right)^2\right\}=E\left\{\left(\int^T\phi(X_\Gamma-X_{\Gamma^n})dX\right)^2\right\}$$

$$=E\left\{\left(\int_T\phi X_{\overline{A^n}-\Gamma}dX\right)^2\right\}\qquad(A)$$

$$\leq\int_{\Omega\times T}\phi^2 X_{\overline{A^n}-\Gamma}d\alpha$$

where step (A) requires some algebra and manipulation starting from simple functions ϕ. Then the hypotheses of the theorem are satisfied and the result holds.

2.4   The Mixed Integral

The above work has provided a well-defined line integral for a pure curve Γ. If the curve Γ is either a vertical or horizontal segment, i.e., V_{st_0} or $H_{s_0 t}$, then it is possible to define a mixed integral. First, in the case of $H_{s_0 t}$, set

$$I_{s_0 t}=\int_{H_{s_0 t}}\phi\partial_1 X$$

for $\mathcal{F}_z^1$ predictable ϕ. Then integrate $I_{s_0 t}$ over t relative to an increasing (Lebesgue-Stieltjes) measure $dF(t)$, such that $|F(z_0)|<\infty$, to get

$$\int_0^{t_0}\int_0^{s_0}\phi\partial_1 X dF(t)=\int_0^{t_0}I_{s_0 t}dF(t).$$

Clearly $I_{s_0 t}$ must be $(\mathcal{B}[0,t_0],dt)$ measurable for the integral to make sense. Now for Γ, an increasing curve (i.e., type I), the following property holds:

$$\int_\Gamma\phi dX^\Gamma=\int_\Gamma\phi\partial X$$

$$=\int_\Gamma\phi\partial_1 X+\int_\Gamma\phi\partial_2 X$$

$$=\int_\Gamma\phi dX_1^{\Gamma(z)}+\int_\Gamma\phi dX_2^{\Gamma(z)}.$$

Thus if Γ is the increasing curve $H_{s_0 t}$, (similarly for $V_{s t_0}$) then

$$\int_{H_{s_0 t}} \phi \, dX^{H_{s_0 t}} = \int_{H_{s_0 t}} \phi \partial X$$

$$= \int_{H_{s_0 t}} \phi \partial_1 X$$

where $\int_{H_{s_0 t}} \phi \partial_2 X = 0$ since there is no contribution by the vertical increment in $H_{s_0 t}$. In particular, if X is a 1-martingale one has the identity

$$X_1^{\Gamma(z)} = X \left(D_1^{\Gamma}(z) \right)$$

$$= E \left\{ D_z^{\Gamma}(z_f) | \mathcal{F}_z^1 \right\}$$

$$= X_1^{\Gamma}(z) \text{ an } \mathcal{F}_z^1 \text{ martingale.}$$

Similarly, 2-martingales satisfy $X_2^{\Gamma(z)} = X_2^{\Gamma}(z)$. To define the mixed integral it is assumed that the $L^{2,2}$-bounded process $\{X_z, \mathcal{F}_z\}_{z \in T}$ has the bounding measure $\alpha(\omega, s, t) = \mu(\omega, s, t) \otimes P(\omega)$ such that for any $\mathcal{F}_z^i$ predictable simple ϕ,

$$E \left\{ \left(\int_{\Gamma} \phi \partial_i X \right)^2 \right\} \leq \int_{\Gamma \times \Omega} \phi^2 \, d\alpha,$$

and if $t' \in [0, t_0]$ is fixed, then

$$\int_{H_{s_0 t'} \times \Omega} \phi(\omega, s, t) \, d\alpha(\omega, s, t) = \int_{[0, s_0] \times \Omega} \phi(\omega, s, t') \, d\alpha(\omega, s, t')$$

$$= \int_{\Omega \times [0, s_0]} \phi(\omega, s, t') \, d\mu(\omega, s, t') \otimes dP(\omega).$$

Integrating over $[0, t_0]$ by $F(t)$, for ϕ simple, it follows from Tonelli's Theorem,

$$\int_{[0, t_0]} \int_{\Omega \times [0, s_0]} \phi(\omega, s, t') \, d\mu(\omega, s, t') \otimes dP(\omega) \, dF(t')$$

$$= \int_{[0, t_0]} \int_{\Omega} \int_{[0, s_0]} \phi(\omega, s, t') \, d\mu(\omega, s, t') \, dP(\omega) \, dF(t')$$

$$= \int_{\Omega} \int_{[0, t_0]} \int_{[0, s_0]} \phi(\omega, s, t') \, d\mu(\omega, s, t') \, dF(t') dP(\omega).$$

Consequently, the Bochner bounding measure exists for the mixed integral generated by the measure $\mu(\omega, s, t') \otimes F(t') \otimes P(\omega)$.

Theorem 34. *Let $\{X_z, \mathcal{F}_z\}_{z \in T}$ be a separable $L^{2,2}$-bounded process such that for $\mathcal{F}_z^1$-predictable ϕ, X satisfies*

$$\int_\Omega \int_{[0,t_0]} \int_{[0,s_0]} |\phi(\omega, s, t')| \, d\mu(\omega, s, t') \otimes dF(t') \otimes dP(\omega) < \infty,$$

and let $F : [0, t_0] \to \mathbb{R}$ be a continuous increasing Lebesgue-Stieltjes measure with $0 < F(t_0) < \infty$. Then there exists a measurable process $\{I_z, z \prec z_0\}$ such that

(a)   *for a.e. $dF(t)$, with fixed $t' \leq t_0$*

$$P\left\{I_{st} = \int_{H_{st}} \phi \partial_1 X, \text{ for all } s \leq s_0\right\} = 1,$$

consequently, $\left\{I_{st}, \mathcal{F}_{st}^1\right\}_{s \in [0,s_0]}$ is a one-parameter right continuous martingale if X is a 1-martingale,

(b)   (i) *for X a 1-martingale,*

$$E\left\{\sup_{s < s_0} I_{st}^2\right\} < \infty \text{ a.e. } [dF(t)]$$

and I_{st} is integrable in t,

   (ii) *if I_{st} satisfies the Brennan-Doob inequality from Theorem 3 for $\phi(x) = x^2$ the same holds,*

(c) $E\left\{\left(\int_0^{t_0} I_{s_0 t} dF(t)\right)^2\right\}$

$$\leq F(t_0) \int_\Omega \int_0^{t_0} \int_0^{s_0} \phi^2(s, t') d\alpha(s, t') \otimes dF(t') \otimes dP$$

for all $\mathcal{F}_z^1$ predictable ϕ that are square integrable with respect to $d\alpha(\omega, s, t) \otimes dF(t) \otimes dP(\omega)$.

Proof. Let ϕ be an $\mathcal{F}_z^1$ predictable simple process. Observe that the process $I_{s_0t} = \int_{H_{s_0t}} \phi \partial_1 X$ satisfies for fixed $t' \le t_0$,

$$
\begin{aligned}
E\left\{(I_{s_0t'})^2\right\} &= E\left\{\left(\int_{H_{s_0t'}} \phi \partial_1 X\right)^2\right\} \\
&\le \int_{H_{s_0t'} \times \Omega} \phi^2 \, d\alpha \\
&= \int_\Omega \int_0^{s_0} \phi^2 \left(\omega, s, t'\right) d\alpha \left(\omega, s, t'\right) dP\left(\omega\right) \\
&< \infty
\end{aligned}
$$

and is integrable with respect to $F(t)$ by hypothesis. So it follows that

$$
\begin{aligned}
E&\left\{\left(\int_0^{t_0} I_{s_0t} dF(t)\right)^2\right\} \\
&= F(t_0)^2 E\left\{\left(\int_0^{t_0} I_{s_0t} \frac{dF(t)}{F(t_0)}\right)^2\right\} \\
&\le F(t_0)^2 E\left\{\int_0^{t_0} I_{s_0t}^2 \frac{dF(t)}{F(t_0)}\right\} \qquad \text{(a.e. } dF(t) \text{ by Jensen's)} \\
&= F(t_0) \int_0^{t_0} E\left\{I_{s_0t}\right\}^2 dF(t) \qquad\qquad \text{(by Tonelli's)} \\
&\le F(t_0) \int_0^{t_0} \int_\Omega \int_0^{s_0} \phi^2 \left(\omega, s, t'\right) d\alpha \left(\omega, s, t'\right) dP\left(\omega\right) dF(t') \\
&= F(t_0) E\left\{\int_0^{t_0} \int_0^{s_0} \phi^2 d\alpha dF(t)\right\} \qquad \text{(Tonelli's)}
\end{aligned}
$$

thus (c) holds given the truth of (a), since the extension for L^2-bounded ϕ (from the simple case) is routine.

Considering the measurability properties of I_{st}, let $\phi = \alpha \mathcal{X}_{(z,w]}$ for $\alpha \in \mathcal{F}_z^1$ and suppose the intersection of H_{st} with $(z, w]$ is nonempty and occurs at the points $\tilde{z}$ and $\tilde{w}$, so $\tilde{z} = (\max\{s, z_1\}, t)$ and $\tilde{w} = (\min\{s, w_1\}, t)$. Then

$$
\begin{aligned}
I_{st} &= \int_{H_{st}} \phi_s \partial_1 X \\
&= \int_{H_{st}} \alpha \mathcal{X}_{(z,w]} \partial_1 X
\end{aligned}
$$

$$= \int_{H_{st}} \alpha X_{(z,w]} dX_1^{H_{st}}$$

$$= \alpha \left(X(\tilde{w}) - X(\tilde{z}) \right)$$

where α is $\mathcal{F}_z^1 \subseteq \mathcal{F}_{\tilde{z}}^1 \subseteq \mathcal{F}_{st}^1$ adapted. For $\phi = \sum_{i=1}^{n} \alpha_i X_{(z_i, w_i]}$, with the same notation as before,

$$I_{st} = \sum_{i=1}^{n} \alpha_i \left(X(\tilde{w}_i) - X(\tilde{z}_i) \right)$$

is a $\mathcal{F}_{st}^1$-predictable process. Also, since the difference

$$X(\tilde{w}_i) - X(\tilde{z}_i) = X(\max\{s, w_1\}, t) - X(\min\{s, z_1\}, t),$$

I_{st} will be $(\mathcal{B}[0, t_0], dF(t))$ measurable whenever $X(\omega, s, t)$ is $\mathcal{B}[0, t_0]$ measurable for fixed (ω, s). This follows from the sections being measurable.

By the structure theorem for $\mathcal{F}_z^1$-predictable ϕ, there exists a sequence of $\mathcal{F}_z^1$-predictable simple functions $\{\phi^n\}_{n=1}^{\infty}$ such that $\phi^n \to \phi$. Thus letting $n \longrightarrow \infty$,

$$\int_{\Omega} \int_0^{t_0} \int_0^{s_0} (\phi^n - \phi)^2 \, d\alpha \otimes dF(t) \otimes dP \to 0$$

and

$$\int_0^{t_0} \int_{\Omega} \int_0^{s_0} (\phi^n - \phi)^2 \, d\alpha \otimes dP \otimes dF(t) \to 0.$$

By separability, the process X can be restricted to a separable dense subset $D \subset T$. For each t-section of $T \times \Omega = ([0, s_0] \times [0, t_0]) \times \Omega$ a subsequence of the ϕ^n can be found such that for a.a. t $[dF(t)]$,

$$\int_{\Omega} \int_0^{s_0} (\phi^n - \phi)^2 \, d_t\alpha \otimes dP < 2^{-n}$$

for large n. Hence for a.a. t $[dF(t)]$,

$$\int_{\Omega} \left(\int_{H_{st}} (\phi^n - \phi) \, \partial_1 X \right)^2 dP \leq \int_{\Omega} \int_0^{s_0} (\phi^n - \phi)^2 \, d_t\alpha dP$$

$$< 2^{-n}.$$

Thus,

$$\int_{H_{st}} \phi^n \partial_1 X \to \int_{H_{st}} \phi \partial_1 X$$

almost surely uniformly in s and in L^2. By separability, I_{st} can be defined by

$$I_{st} = \begin{cases} \lim_{n\to\infty} \int_{H_{st}} \phi^n \partial_1 X & \text{, if the limit exists} \\ 0 & \text{, otherwise.} \end{cases}$$

Thus (a) holds. When X is a 1-martingale it follows that

$$X_1^{\Gamma(z)} = X_1^{\Gamma}(z) = E\left\{ X\left(D_1^{\Gamma}(z_f)\right) | \mathcal{F}_z^1 \right\}$$

is a martingale. Thus

$$I_{st} = \int_{H_{st}} \phi \partial_1 X$$

is a martingale in s for fixed t and the classical maximal submartingale inequality applies, giving

$$E\left\{ \sup_{s \leq s_0} I_{st}^2 \right\} \leq 4E\left\{ I_{st}^2 \right\}$$

$$\leq 4E\left\{ \int_0^{s_0} \phi^2 d\langle X_1^{\Gamma} \rangle_s \right\}$$

$$< \infty.$$

Consequently, I_{st} is square integrable. For (ii), if I_{st} satisfies the hypothesis of the Brennan-Doob inequality for $\phi(x) = x^2$ then

$$E\left\{ \sup_{s \leq s_0} (I_{st})^2 \right\} = E\left\{ \sup_{s \leq s_0} \left(\int_{H_{st}} \phi \partial_1 X \right)^2 \right\}$$

$$= E\left\{ \sup_{s \leq s_0} \left(\int_{H_{st}} \phi dX^{H_{st}} \right)^2 \right\}$$

$$\leq E\left\{ \sup_{(s,t)\in T} \left(\int_T \phi \chi_{H_{st}} dX \right)^2 \right\}$$

$$\leq aE\left\{\left(\int_T \phi \chi_{H_{s_0 t_0}} dX\right)^2\right\} + b$$

$$\leq a\int_{T\times\Omega} \phi^2 \chi_{H_{s_0 t_0}} d\alpha + b$$

$$< \infty.$$

□

2.5   The Double Integral

An extension of the double integral, developed by Cairoli and Walsh, is given for $L^{2,2}$-bounded processes using the Bochner boundedness principle. Let $\{X_z, \mathcal{F}_z\}_{z\in T}$ be a right continuous $L^{2,2}$-bounded process. The complementary ordering to $\prec$ defined earlier by $(z_1, z_2) \, \lambda \, (w_1, w_2)$ if and only if $z_1 \leq w_1$ and $z_2 \geq w_2$ will be used in the following work. Let A and B be two rectangles in T. The notation $A \, \lambda \, B$ is used if $z \, \lambda \, w$ for all $z \in A$ and $w \in B$. Consider the process

$$Y_z = \alpha X(A\cap R_z)X(B\cap R_z), \quad z\in T$$

where α is a bounded $\mathcal{F}_{w_1 w_2}$-measurable process for an $A = ((z_1, z_2), (z_1', z_2')]$ and $B = ((w_1, w_2), (w_1', w_2')]$. It is shown in Cairoli and Walsh [4] that if X is a martingale, $\{Y_z, \mathcal{F}_z\}_{z\in T}$ is also. To begin, relations like the above between properties of X and those of Y are given.

Theorem 35. Let $A = (z_1, z_1'], \quad B = (z_2, z_2'],$ and $R_z = ((0,0), z]$ where $z_i = (z_i^1, z_i^2]$ such that $A \, \lambda \, B$. Then for

$$Y_z = \alpha X\left(A\cap R_z\right)X\left(B\cap R_z\right), \quad \alpha \in \mathcal{F}^1_{z_2^1 z_1^2}$$

the following assertions hold:

(1)   *if X is an i-martingale, then Y is also, $i = 1, 2$,*

(2)   *let X be a weak martingale with decomposition $X = X^{(1)} + X^{(2)}$ into the sum of adapted i-martingales for $i = 1, 2$. If $X^{(1)}$ and $X^{(2)}$ satisfy the relation $E\left\{X^{(1)}(C)X^{(2)}(D)|\mathcal{F}_w\right\} = 0$ for all $C \overset{\wedge}{\curlywedge} D$ where C and D are rectangles and*

$$w = (\inf\{s : (s,t) \in D\}, \inf\{t : (s,t) \in C\}),$$

then Y is a weak martingale.

Proof. Before the proof is begun it should be mentioned that strong martingales are an example of i-martingales, so the above concept applies and extends the previous work of Cairoli and Walsh [4]. Let $D = (z, z']$ where $z = (s,t) \prec\prec z' = (s', t')$, then

$$Y(D) = \alpha X\left(A \cap (R_{s't'} - R_{s't})\right) X\left(B \cap (R_{s't'} - R_{st'})\right)$$
$$= \alpha X(A')X(B'), \quad \text{(say)}.$$

Denote by z'' the lower left-hand corner of $A' = A \cap (R_{s't'} - R_{s't})$ and by z''' the lower left-hand corner of $B' = B \cap (R_{s't'} - R_{st'})$. Then $z'' \succ \inf(A) = z_1$, and also α is both $\mathcal{F}_{z''}^2$ and $\mathcal{F}_{z'''}^1$ measurable. For $i = 1$ it can be seen,

$$E\left\{Y(D)|\mathcal{F}_z^1\right\} = E\left\{\alpha X(A')X(B')|\mathcal{F}_z^1\right\}$$
$$= E\left\{\left\{\alpha X(A')X(B')|\mathcal{F}_{z'''}^1\right\}|\mathcal{F}_z^1\right\}$$
$$= E\left\{\alpha X(A')E\left\{X(B')|\mathcal{F}_{z'''}^1\right\}|\mathcal{F}_z^1\right\}$$
$$= 0.$$

The case for $i = 2$ is similar. For the second claim the conditional expectation with respect to $\mathcal{F}_z$ is computed,

$$E\left\{Y(D)|\mathcal{F}_z\right\} = E\left\{\alpha X(A')X(B')|\mathcal{F}_z\right\}$$

$$= E\left\{\alpha X^{(1)}(A')X^{(1)}(B')|\mathcal{F}_z\right\}$$
$$+ E\left\{\alpha X^{(1)}(A')X^{(2)}(B')|\mathcal{F}_z\right\}$$
$$+ E\left\{\alpha X^{(2)}(A')X^{(1)}(B')|\mathcal{F}_z\right\}$$
$$+ E\left\{\alpha X^{(2)}(A')X^{(2)}(B')|\mathcal{F}_z\right\},$$

the first, third, and last summands are zero by conditioning and using the properties of i-martingales, $(i = 1, 2)$. Using hypothesis (2) the second summand becomes

$$E\left\{\alpha X^{(1)}(A')X^{(2)}(B')|\mathcal{F}_z\right\} = E\left\{\alpha E\left\{X^{(1)}(A')X^{(2)}(B')|\mathcal{F}_w\right\}\mathcal{F}_z\right\}$$
$$= 0,$$

so Y is a weak martingale. $\square$

In constructing the Bochner bounding measure for the double integral the following lemma will be needed:

Lemma 36. Let $\{\mathcal{F}_z\}_{z\in T}$ be a right continuous filtration satisfying the (F4) condition on $T = [0, s_0] \times [0, t_0]$. If X is $\mathcal{F}_z^1$ measurable and Y is $\mathcal{F}_z^2$ measurable, then

$$E\left\{XY|\mathcal{F}_z\right\} = E\left\{X|\mathcal{F}_z\right\}E\left\{Y|\mathcal{F}_z\right\}.$$

Proof.

$$E\left\{XY|\mathcal{F}_z\right\} = E\left\{E\left\{XY|\mathcal{F}_z^1\right\}|\mathcal{F}_z^2\right\}, \quad \text{by (F4)}$$
$$= E\left\{XE\left\{Y|\mathcal{F}_z^1\right\}|\mathcal{F}_z^2\right\}, \quad (X \text{ is } \mathcal{F}_z^1 \text{ adapted})$$
$$= E\left\{XE\left\{E\left\{Y|\mathcal{F}_z^2\right\}|\mathcal{F}_z^1\right\}|\mathcal{F}_z^2\right\}, \quad (Y \text{ is } \mathcal{F}_z^2 \text{ adapted})$$
$$= E\left\{XE\left\{Y|\mathcal{F}_z\right\}|\mathcal{F}_z^2\right\}, \quad \text{by (F4)}$$
$$= E\left\{X|\mathcal{F}_z^2\right\}E\left\{Y|\mathcal{F}_z\right\}, \quad \text{since } \mathcal{F}_z \subseteq \mathcal{F}_z^2$$
$$= E\left\{E\left\{X|\mathcal{F}_z^1\right\}|\mathcal{F}_z^2\right\}E\left\{Y|\mathcal{F}_z\right\}, \quad (X \text{ is } \mathcal{F}_z^1 \text{ adapted})$$
$$= E\left\{X|\mathcal{F}_z\right\}E\left\{Y|\mathcal{F}_z\right\}, \quad \text{by (F4)}.$$

□

Remark 37. For $X^{(1)}(A)$ and $X^{(2)}(B)$ with $A \curlywedge B$ and $w = (\inf\{s : (s,t) \in B\}, \inf\{t : (s,t) \in A\})$, it follows that $X^{(1)}(A)$ is $\mathcal{F}_w^1$ measurable and $X^{(2)}(B)$ is $\mathcal{F}_w^2$ measurable, hence

$$E\left\{X^{(1)}(A)X^{(2)}(B)|\mathcal{F}_w\right\} = E\left\{X^{(1)}(A)|\mathcal{F}_w\right\}E\left\{X^{(2)}(B)|\mathcal{F}_w\right\}.$$

To construct the double integral a dyadic partition will be used on the parameter set $T = [0,1]^2$. Fix an integer n and divide up the region T into dyadic subrectangles $\Delta_{ij} = (z_{ij}, z_{i+1j+1}]$ where $z_{ij} = \left(\frac{i}{2^n}, \frac{j}{2^n}\right)$ for $0 \leq i, j \leq 2^n$, i, j, k, and l are positive integers satisfying

$$0 < i < k \leq 2^n - 1, \quad 0 < l < j \leq 2^n - 1.$$

Define the function

$$\psi_{ij,kl}(\eta, \xi) = \alpha \chi_{\Delta_{ij}}(\eta)\chi_{\Delta_{kl}}(\xi) \tag{2.24}$$

where α is bounded and $\mathcal{F}_{z_{kj}}$ measurable with $\Delta_{ij} \curlywedge \Delta_{kl}$ for all i, j, k, and l. Define the double integral with respect to X, an $L^{2,2}$-bounded process, by

$$\int_{R_z \times R_z} \psi_{ij,kl} dX dX = \alpha X\left(\Delta_{ij} \cap R_z\right) X\left(\Delta_{kl} \cap R_z\right).$$

A process $X \cdot X$ will be used which is defined by

$$X \cdot X : T \times T \to \mathbb{R}$$

such that

$$X \cdot X(\eta, \xi) = X(\eta)X(\xi).$$

The process $X \cdot X(\eta, \xi)$ will be called adapted if $X(\eta)X(\xi)$ is $\mathcal{F}_{\eta \wedge \xi}$ measurable with respect to $\{\mathcal{F}_z\}_{z \in T}$. Some conditions on $\{X_z, \mathcal{F}_z\}_{z \in T}$ are needed so the process $\{X \cdot X(\eta, \xi), \mathcal{F}_{\eta \wedge \xi}\}_{(\eta, \xi) \in T \times T}$ will satisfy the Bochner boundedness principle. The following lemma will be needed.

Lemma 38. *If* $\{X_z, \mathcal{F}_z\}_{z \in T}$ *is an* $L^{2,2}$*-bounded process with respect to the measure* $d\mu(\omega, z)dP(\omega)$ *on the class of simple predictable functions, then*

$$E\left\{X\left(\Delta_{ij}\right)^2 | \mathcal{F}_{kj}\right\} \leq \mu\left(\Delta_{ij}\right) \quad a.e. \ (\omega)$$

for $\Delta_{ij} \curlywedge \Delta_{kl}$.

Proof. Let $F \in \mathcal{F}_{ij} \subseteq \mathcal{F}_{kj}$, then by definition of conditioning

$$\int_F E\left\{X\left(\Delta_{ij}\right)^2 | \mathcal{F}_{kj}\right\} dP_{\mathcal{F}_{kj}} = \int_F X\left(\Delta_{ij}\right)^2 dP$$

$$= \int_F \left(\int_T \chi_{\Delta_{ij}} dX\right)^2 dP$$

$$= \int_\Omega \chi_F \left(\int_T \chi_{\Delta_{ij}} dX\right)^2 dP$$

$$= E\left\{\left(\int_T \chi_{\Delta_{ij} \times F} dX\right)^2\right\}$$

$$\leq \int_\Omega \int_T \chi_{\Delta_{ij} \times F} d\mu dP$$

($L^{2,2}$-bounded relative to $d\mu dP$ on simple functions)

$$= \int_F \int_T \chi_{\Delta_{ij}} d\mu dP$$

$$= \int_F \mu(\Delta_{ij}) dP, \quad a.e. \ \omega.$$

$\square$

Now let $\psi = \sum_{m=1}^n \psi_{ij,kl}^m$ be a sum of functions as in (2.24) such that the sets $\Delta_{ij} \times \Delta_{kl}$ are disjoint on $T \times T$. One can now define a class of processes $X \cdot X$ which will satisfy the Bochner boundedness principle.

Definition 39. $\{X \cdot X(\eta, \xi), \mathcal{F}_{\eta \wedge \xi}\}_{(\eta, \xi) \in T \times T}$ is said to satisfy the *cross-term domination* property if for every simple function $\psi = \sum_{m=1}^n \psi_{ij,kl}^m$,

$$\sum_{1 \leq m < p \leq n} E\left\{\alpha_m \alpha_p X\left(\Delta_{ij}^m\right) X\left(\Delta_{kl}^m\right) X\left(\Delta_{ij}^p\right) X\left(\Delta_{kl}^p\right)\right\}$$

$$\leq K \sum_{m=1}^n E\left\{\alpha_m^2 X\left(\Delta_{ij}^m\right)^2 X\left(\Delta_{kl}^m\right)^2\right\},$$

for some absolute constant $K \geq 0$.

It is immediate for the case when X is a strong martingale having four moments that X satisfies the cross-term domination condition. Thus the work by Cairoli and Walsh is extended in this formulation. Another class of processes which satisfies the cross-term domination property is the following. Call a process $X \cdot X : T \times T \to I\!\!R$ defined on $T \times T$ a *1-productmartingale* if for $s_1 \leq s_2$

$$E\left\{X \cdot X\left((s_2, t), (s_2, r)\right) | \mathcal{F}^1_{s_1}.\right\} = X \cdot X\left((s_1, t), (s_1, r)\right)$$

for each $0 \leq r, t \leq t_0$. Thus, if $X \cdot X$ is a 1-productmartingale and X is a 1-martingale, for $\Delta_{ij} \curlywedge \Delta_{kl}$ and $\Delta_{i'j'} \curlywedge \Delta_{k'l'}$ disjoint, it follows that

$$E\{\alpha\alpha' X \cdot X\left(\Delta_{ij} \times \Delta_{kl}\right) X \cdot X\left(\Delta_{i'j'} \times \Delta_{k'l'}\right)\}$$

$$= E\left\{\alpha\alpha' X\left(\Delta_{ij}\right) X\left(\Delta_{kl}\right) X\left(\Delta_{i'j'}\right) X\left(\Delta_{k'l'}\right)\right\}$$

$$= E\left\{\alpha\alpha' X\left(\Delta_{ij}\right) X\left(\Delta_{i'j'}\right) E\left\{X\left(\Delta_{kl}\right) X\left(\Delta_{k'l'}\right) | \mathcal{F}^1_{\min\{k,k'\}}\right\}\right\}$$

and if $k = k'$

$$E\{X\left(\Delta_{kl}\right) X\left(\Delta_{k'l'}\right) | \mathcal{F}^1_k\}$$

$$= E\left\{(X_{k+1l+1} - X_{k+1l})(X_{k+1l'+1} - X_{k+1l'}) | \mathcal{F}^1_k\right\}$$

$$+ E\left\{(X_{k+1l+1} - X_{k+1l})(X_{kl'+1} - X_{kl'}) | \mathcal{F}^1_k\right\}$$

$$+ E\left\{(X_{kl+1} - X_{kl})(X_{k+1l'+1} - X_{k+1l'}) | \mathcal{F}^1_k\right\}$$

$$+ E\left\{(X_{kl+1} - X_{kl})(X_{kl'+1} - X_{kl'}) | \mathcal{F}^1_k\right\}$$

$$= E\left\{(X_{k+1l+1} - X_{k+1l})(X_{k+1l'+1} - X_{k+1l'}) | \mathcal{F}^1_k\right\}$$

$$- (X_{kl+1} - X_{kl})(X_{kl'+1} - X_{kl'})$$

$$= E\left\{X_{k+1l+1} X_{k+1l'+1} | \mathcal{F}^1_k\right\} - E\left\{X_{k+1l+1} X_{k+1l'} | \mathcal{F}^1_k\right\}$$

$$- E\left\{X_{k+1l} X_{k+1l'+1} | \mathcal{F}^1_k\right\} + E\left\{X_{k+1l} X_{k+1l'} | \mathcal{F}^1_k\right\}$$

$$- X_{kl+1} X_{kl'+1} + X_{k+1l} X_{kl'} + X_{kl} X_{kl'+1} - X_{kl} X_{kl'}$$

$$= 0,$$

by the 1-productmartingale property. For $k \neq k'$ (supposing without loss of generality that $k < k'$) by conditioning with $\mathcal{F}_{k'.}^1$,

$$E\left\{ \alpha \alpha' X\left(\Delta_{ij}\right) X\left(\Delta_{i'j'}\right) X\left(\Delta_{kl}\right) E\left\{ X\left(\Delta_{k'l'}\right) | \mathcal{F}_{k'.}^1 \right\} \right\} = 0,$$

since X is a 1-martingale. The above discussion includes strong martingales, but it is not characteristic of strong martingales. Finally, the Bochner bounding measure for $X \cdot X$ is given by the following result:

Theorem 40. *Let* $\{X_z, \mathcal{F}_z\}_{z \in T}$ *be an* $L^{2,2}$*-bounded process with bounding measure* $d\mu(\omega, z) dP(\omega)$ *on the simple functions having four moments such that the process* $X \cdot X$ *is cross-term dominated. Then* $X \cdot X$ *is an* $L^{2,2}$*-bounded process with bounding measure* $d\mu_\eta \otimes d\mu_\xi \otimes dP$.

Proof. Let $\psi = \sum_{m=1}^{n} \psi_{ij,kl}^m$ where $\psi_{ij,kl}^m = \alpha_m \mathcal{X}_{\Delta_{ij}^m}(\eta) \mathcal{X}_{\Delta_{kl}^m}(\xi)$, α_m is $\mathcal{F}_{kj}$ measurable, $\Delta_{ij}^m \perp \Delta_{kl}^m$ being disjoint in $T \times T$. Then it follows that

$$E\left\{ \left(\int_{T \times T} \psi \, dX dX \right)^2 \right\} = E\left\{ \left(\sum_{m=1}^{n} \alpha_m X\left(\Delta_{ij}^m\right) X\left(\Delta_{kl}^m\right) \right)^2 \right\}$$

$$= E\left\{ \sum_{m=1}^{n} \alpha_m^2 X\left(\Delta_{ij}^m\right)^2 X\left(\Delta_{kl}^m\right)^2 \right\}$$

$$+ E\left\{ \sum_{1 \leq p < m \leq n} \alpha_m \alpha_p X\left(\Delta_{ij}^m\right) X\left(\Delta_{ij}^p\right) X\left(\Delta_{kl}^m\right) X\left(\Delta_{kl}^p\right) \right\}$$

$$\leq (K+1) \sum_{m=1}^{n} E\left\{ \alpha_m^2 X\left(\Delta_{ij}^m\right)^2 X\left(\Delta_{kl}^m\right)^2 \right\}$$

(by the cross-term domination)

$$= (K+1) \sum_{m=1}^{n} E\left\{ \alpha_m^2 E\left\{ X\left(\Delta_{ij}^m\right)^2 X\left(\Delta_{kl}^m\right)^2 | \mathcal{F}_{kj}^1 \right\} | \mathcal{F}_{kj}^2 \right\}$$

(by $F4$ and α_m being $\mathcal{F}_{kj}$ measurable)

$$= (K+1) \sum_{m=1}^{n} E\left\{\alpha_m^2 E\left\{X\left(\Delta_{ij}^m\right)^2 E\left\{X\left(\Delta_{kl}^m\right)^2 |\mathcal{F}_{kj}^1\right\} |\mathcal{F}_{kj}^2\right\}\right\}$$

(by Lemma 36)

$$= (K+1) \sum_{m=1}^{n} E\left\{\alpha_m^2 E\left\{X\left(\Delta_{ij}^m\right)^2 |\mathcal{F}_{kj}\right\} E\left\{X\left(\Delta_{kl}^m\right)^2 |\mathcal{F}_{kj}\right\}\right\}$$

(by Lemma 36)

$$\le (K+1) \sum_{m=1}^{n} E\left\{\alpha_m^2 \mu_\eta\left(\Delta_{ij}^m\right) \mu_\xi\left(\Delta_{kl}^m\right)\right\}$$

(by Lemma 38)

$$= (K+1)E\left\{\int_{T\times T} \sum_{m=1}^{n} \alpha_m^2 \chi_{\Delta_{ij}^m}(\eta)\chi_{\Delta_{kl}^m}(\xi)d\mu_\eta d\mu_\xi\right\}$$

$$= (K+1)E\left\{\int_{T\times T} \psi^2 d\mu_\eta d\mu_\xi\right\}.$$

$\square$

Thus for cross-term dominated $X \cdot X$, one has a double integral which is well defined and extends the work of Cairoli and Walsh. Now let $\mathcal{D}$ be the σ-algebra on $T \times T \times \Omega$ generated by the simple functions $\psi = \sum_{m=1}^{n} \psi_{ij,kl}^m$. This $\mathcal{D}$ is called the algebra of predictable sets. A process $\{\psi(\eta, \xi) : \eta, \xi \in T\}$ is said to be predictable if it is $\mathcal{D}$ measurable as a function of (η, ξ, ω). The process $\psi(\eta, \xi)$ is said to be adapted if it is $\mathcal{F}_{\eta \wedge \xi}$ measurable as before. Let $\mathcal{L}_{XX}^2(z_0)$ be the class of all processes satisfying:

(1) ψ is predictable,

(2) $\psi(\eta, \xi) = 0$ unless $\eta \overset{\wedge}{\lambda} \xi$,

(3) $\int_{T\times T\times\Omega} \psi^2(\eta, \xi)d\mu_\eta \otimes d\mu_\xi \otimes dP < \infty$.

A general stochastic Fubini theorem will now be established.

Theorem 41 (Stochastic Fubini Theorem). *Let* $\psi \in \mathcal{L}_{XX}^2(z_0)$ *and* X *a separable* $L^{2,2}$*-bounded process such that* $X \cdot X$ *is cross-term dominated.*

Then $\psi(\eta, \cdot)$ is $\mathcal{F}_\xi^1$ predictable and

$$\int_{T \times \Omega} \psi^2 (\eta, \xi) \, d\alpha < \infty$$

for $d\mu_\eta$ a.a. η where X has the bounding measure $\alpha(\xi, \omega) = \mu(\xi, \omega) \otimes dP(\omega)$.
Furthermore, a process $\{I(\eta), \eta \in T\}$ can be defined such that:

(a) $I(\eta)$ is $\mathcal{F}_\eta^2$ predictable,

(b)

$$I(\eta) = \int_T \psi(\eta, \xi) dX_\xi \quad d\mu_\eta \quad a.a \quad \eta$$

(c)

$$\int_T I(\eta) dX_\eta = \iint_{T \times T} \psi dX_\xi dX_\eta$$

in L^2. A similar equation for $I(\xi)$ holds.

Proof. Let $\psi_{ij,kl} = \alpha \chi_{\Delta_{ij}} \chi_{\Delta_{kl}}$ where $\Delta_{ij} \downarrow \Delta_{kl}$ and α is $\mathcal{F}_{kj}$-measurable.
Define for $\psi_{ij,kl}$ the process

$$I(\eta) = \alpha \chi_{\Delta_{ij}} X (\Delta_{kl}) .$$

Since $X(\Delta_{kl})$ and α are $\mathcal{F}_{kj}^2$-measurable it follows that $I(\eta)$ is $\mathcal{F}_\eta^2$-predictable,
because for all $\eta \in \Delta_{ij}$ we have $\mathcal{F}_{kj}^2 \subseteq \mathcal{F}_\eta^2$. So $I(\eta)$ is $\mathcal{F}_\eta^2$-predictable for
simple functions ψ and consequently, by taking limits, for general ψ. Com-
puting the integral of $I(\eta)$ it can be seen that, since $I(\eta)$ is $\mathcal{F}_\eta^2$-predictable,

$$\begin{aligned}
\int_T I(\eta) dX_\eta &= \int_T \alpha \chi_{\Delta_{ij}}(\eta) X (\Delta_{kl}) \, dX_\eta \\
&= \alpha X (\Delta_{ij}) X (\Delta_{kl}) \\
&= \iint_{T \times T} \psi_{ij,kl} (\eta, \xi) \, dX_\eta dX_\xi \\
&= \iint_{T \times T} \psi_{ij,kl} (\eta, \xi) \, dX_\xi dX_\eta .
\end{aligned}$$

(2.25)

This extends to finite sums of simple $\psi_{ij,kl}$. Using this, the following can be derived:

$$E\left\{\left(\int_T I(\eta)\,dX_\eta\right)^2\right\} = E\left\{\alpha^2 X\,(\Delta_{ij})^2\,X\,(\Delta_{kl})^2\right\}$$

$$\leq E\left\{\alpha^2 \mu_\eta\,(\Delta_{ij})\,\mu_\xi\,(\Delta_{kl})\right\} \text{ by Lemma 38}$$

$$= \int_\Omega \int_{T\times T} \psi_{ij,kl}^2\,d\mu_\eta\,d\mu_\xi\,dP$$

$$= \int_\Omega \int_{T\times T} \psi_{ij,kl}^2\,d\mu_\xi\,d\mu_\eta\,dP.$$

This extends to finite sums $\psi = \sum_{m=1}^n \psi_{ij,kl}^m$ since it was assumed $X \cdot X$ is cross-term dominated. Given $\psi \in \mathcal{L}_{XX}^2(z_0)$ we can find a sequence of $\mathcal{D}$ predictable simple functions such that $\psi_n \to \psi$ pointwise and

$$\int_\Omega \int_{T\times T} (\psi_n - \psi)^2\,d\mu_\xi \otimes d\mu_\eta \otimes dP \to 0$$

by dominated convergence. Using the separability conditions one can restrict the process to a separable subset $D \times D \subset T \times T$. Then given that $\beta = \mu_\eta \otimes \mu_\xi \otimes dP$ the η sections will be considered for $\eta \in D$. Thus for large enough n,

$$\int_\Omega \int_T (\psi_n(\eta,\xi) - \psi(\eta,\xi))^2\,d\beta_\xi < 2^{-n},$$

for $d\mu_\eta$ a.a. η. Now define for $\eta \in D$

$$I(\eta) = \begin{cases} \lim_{n\to\infty} \int_T \psi_n(\eta,\cdot)\,dX(\cdot) & \text{, if the limit exists,} \\ 0 & \text{, otherwise.} \end{cases}$$

Now by separability extend this definition to all of T.

Thus, given the fact that

$$\int_T I_n(\eta)\,dX_\eta = \iint_{T\times T} \psi_n(\eta,\xi)\,dX_\xi dX_\eta, \qquad (2.26)$$

where the integrand $I_n(\eta) = \int_T \psi_n(\eta,\cdot)dX(\cdot)$ is $\mathcal{F}_\eta^2$-predictable, it follows that the same holds for the limit

$$I(\eta) = \lim_{n\to\infty} I_n(\eta).$$

For part (c) one can do the following:

$$E\left\{\left(\int_T I_n(\eta)\,dX_\eta - \int_T I(\eta)\,dX_\eta\right)^2\right\}$$

$$= E\left\{\left(\int_T I_n(\eta) - I(\eta)\,dX_\eta\right)^2\right\}$$

$$= E\left\{\left(\int_T\left(\int_T \psi_n(\eta,\xi) - \psi(\eta,\xi)\,dX_\xi\right)dX_\eta\right)^2\right\}$$

$$\leq \int_{T\times T\times\Omega}(\psi_n(\eta,\xi) - \psi(\eta,\xi))^2\,d\mu_\xi\times d\mu_\eta\times dP$$

by (2.5)

$$\to 0 \quad \text{as } n\to\infty,$$

so

$$\int_T I_n(\eta)\,dX_\eta \xrightarrow{L^2} \int_T I(\eta)\,dX_\eta.$$

In addition,

$$\iint_{T\times T}\psi_n(\eta,\xi)\,dX_\xi dX_\eta \xrightarrow{L^2} \iint_{T\times T}\psi(\eta,\xi)\,dX_\xi dX_\eta.$$

Thus by (2.25),

$$\int_T I(\eta)\,dX_\eta = L^2 - \lim_{n\to\infty}\int_T I_n(\eta)\,dX_\eta$$

$$= L^2 - \lim_{n\to\infty}\iint_{T\times T}\psi_n(\eta,\xi)\,dX_\xi dX_\eta$$

$$= \iint_{T\times T}\psi(\eta,\xi)\,dX_\xi dX_\eta.$$

Therefore,

$$\int_T\left(\int_T \psi(\eta,\xi)\,dX_\eta\right)dX_\xi$$

$$= \iint_{T\times T}\psi(\eta,\xi)\,dX_\xi dX_\eta$$

$$= \int_T\left(\int_T \psi(\eta,\xi)\,dX_\xi\right)dX_\eta.$$

$\square$

Remark 42. If the process $\{X \cdot X(\eta, \xi), \mathcal{F}_{\eta \wedge \xi}\}_{(\eta, \xi) \in T \times T}$ has a Bochner bounding measure which on the simple functions is equal to $d(\mu_\eta \otimes \mu_\xi)dP$ with $(\mu_\eta \otimes \mu_\xi)$ being independent of ω, or it can be assumed that $d\mu_\eta \otimes d\mu_\xi \otimes dP \ll d\lambda \otimes d\nu \otimes dQ$ where $\lambda \times \nu \times Q : T \times T \times \Omega \to I\!\!R^+$ is a product measure, then $I(\eta)$ can be defined in the same manner as in the theorem. This includes the case for strong martingales with four moments that Cairoli and Walsh studied where they assumed $\langle M \rangle_z^i$ to be independent of ω.

2.6 The J_X Measure

The connection between the stochastic surface integral and the line integral is well expressed by the stochastic Green Theorem. The present development of this theorem will use the work of the previous sections and requires the definition of a measure J_X associated to the stochastic measure $\partial_1 X \partial_2 X$. This approach was devised by Cairoli and Walsh for the case of strong martingales and is used here with the new case of the extended Bochner boundedness principle in generalizing that theory. Another approach to the Green Theorem is through "path independent" stochastic partial derivatives, which was considered by Dozzi [6]. Related ideas with extensions to subclasses of $L^{2,2}$-bounded processes are also presented here.

To construct the desired measure, consider the indicator function

$$\psi(\eta, \xi) = \begin{cases} 1 & , \text{if } \eta \overset{\wedge}{\lambda} \xi \\ 0 & , \text{otherwise,} \end{cases}$$

and let

$$J_X(z) = \iint_{R_z \times R_z} \psi \, dX \, dX$$

for $z \prec z_0$. Divide the parameter set T into dyadic squares with corners $z_{ij} = (\frac{i}{2^n} s_0, \frac{j}{2^n} t_0)$ for $i, j = 0, 1, \dots, 2^n$, the squares being $\Delta_{ij} =$

$(z_{ij}, z_{i+1,j+1}]$. Set

$$\delta_{ij} = (z_{0j}, z_{ij+1}] \quad \text{and} \quad \epsilon_{ij} = (z_{i0}, z_{i+1,j}]$$

where $\delta_{ij} \overset{\wedge}{\lambda} \epsilon_{ij}$. Define

$$J_{ij}^n(z) = X\left(\delta_{ij} \cap R_z\right) X\left(\epsilon_{ij} \cap R_z\right),$$

where R_z is the rectangle with diagonal from $(0,0)$ to (z_1, z_2). Over the partition let

$$J_X^n(z) = \sum_{i,j=0}^{2^n-1} J_{ij}^n(z)$$

$$= \sum_{i,j=0}^{2^n-1} X\left(\delta_{ij}\right) X\left(\epsilon_{ij}\right).$$

Then for $\psi_{ij}^n(\eta, \xi) = X_{\delta_{ij}}(\eta) X_{\epsilon_{ij}}(\xi)$ the following equation holds,

$$J_X^n(z) = \iint_{R_z \times R_z} \sum_{i,j=0}^{2^n-1} X_{\delta_{ij}}(\eta) X_{\epsilon_{ij}}(\xi) dX_\eta dX_\xi.$$

On letting $n \longrightarrow \infty$ it is seen that

$$J_X^n(z) \longrightarrow J_X(z) \quad \text{a.e, for each } z.$$

It can be verified that

$$J_X^n(\Delta_{ij}) = X\left(\epsilon_{ij}\right) X\left(\delta_{ij}\right).$$

This is the connection between J_X and $\partial_1 X \partial_2 X$ where $X(\epsilon_{ij})$ is the increment of X over the interval $\overline{z_{ij} z_{ij+1}}$, similarly for $X(\delta_{ij})$. To complete the development of J_X the integral $\int_T X dX$ will be considered approximating X by X^n where $X_z^n = X_{z_{ij}}$ for $z \in \Delta_{ij}$ and $X^n = 0$ on the axes. Thus X^n can be written as

$$X_z^n = \sum_{i,j=0}^{2^n-1} X_{z_{ij}} X_{\Delta_{ij}}(z)$$

which is an $\mathcal{F}_z$-predictable function. The integral of X^n with respect to X satisfies the sum

$$\int_T X^n dX = \sum_{i,j=0}^{2^n-1} X_{z_{ij}} X(\Delta_{ij})$$

$$= \sum_{i,j=0}^{2^n-1} X_{z_{ij}} \left(X(\epsilon_{i,j+1}) - X(\epsilon_{i,j}) \right)$$

(by adding, subtracting $\sum_{i,j=0}^{2^n-1} X_{z_{i,j+1}} X(\epsilon_{i,j+1})$, and rearranging)

$$= \sum_{i,j=0}^{2^n-1} \left(X_{z_{ij+1}} X(\epsilon_{i,j+1}) - X_{z_{ij}} X(\epsilon_{i,j}) \right) + \sum_{i,j=0}^{2^n-1} \left(X_{z_{ij}} - X_{z_{ij+1}} \right) X(\epsilon_{i,j+1})$$

$$(2.27)$$

$$= \sum_{i=0}^{2^n-1} X_{i,2^n} X(\epsilon_{i,2^n}) - \sum_{i,j=0}^{2^n-1} X(\delta_{ij}) X(\epsilon_{i,j}) - \sum_{i,j=0}^{2^n-1} X(\delta_{ij}) X(\Delta_{ij}) \quad (2.28)$$

with the first summand of (2.27) being a telescoping sum in j, and with the substitution of $X(\epsilon_{i,j+1}) = X(\epsilon_{ij}) + X(\Delta_{ij})$ into the second summand. The first term of (2.28) can be written as a line integral (as defined in Section III) for $L^{2,2}$-bounded processes X.

$$\sum_{i=0}^{2^n-1} X_{i,2^n} X(\epsilon_{i,2^n}) = \sum_{i=0}^{2^n-1} X_{i,2^n} \left(X_{i+1,2^n} - X_{i,2^n} \right)$$

$$= \int_{H_{z_0}} \overline{X}^n_{st_0} \partial_1 X$$

$$(2.29)$$

where $\overline{X}^n$ is defined by

$$\overline{X}^n_{st_0} = \begin{cases} X_{\frac{i}{2^n} s_0, t_0} & , \text{ if } s \in (\frac{i}{2^n} s_0, \frac{i+1}{2^n} s_0], \\ 0 & , \text{ otherwise.} \end{cases}$$

If X is a 1-martingale $\{X_z, \mathcal{F}_z\}_{z \in T}$ then

$$X_{\frac{i+1}{2^n} s_0, t} - X_{\frac{i}{2^n} s_0, t} = E\left\{ X_{s_0 t_0} | \mathcal{F}^1_{\frac{i+1}{2^n} s_0, t} \right\} - E\left\{ X_{s_0 t_0} | \mathcal{F}^1_{\frac{i}{2^n} s_0, t} \right\}$$

$$= E\left\{ X(D^1_{s_0 t_0}) | \mathcal{F}^1_{\frac{i+1}{2^n} s_0, t} \right\} - E\left\{ X(D^1_{s_0 t_0}) | \mathcal{F}^1_{\frac{i}{2^n} s_0, t} \right\},$$

and

$$\sum_{i=0}^{2^n-1} X_{i,2^n}\left(X_{i+1,2^n} - X_{i,2^n}\right) = \int_{H_{z_0}} \overline{X}^n_{s,t_0}\, \partial X_1^\Gamma(z).$$

Using the fact that an $L^{2,2}$-bounded A-quasimartingale on any increasing path Γ is a one-parameter quasimartingale, it follows that on H_{z_0} the process X can be decomposed as follows:

$$X_{s,2^n} = M_{s,2^n} + A_{s,2^n}$$

where $\{M_{s,2^n}, \mathcal{F}_{s,2^n}\}_{s\in[0,s_0]}$ is an L^2-martingale and $\{A_{s,2^n}, \mathcal{F}_{s,2^n}\}_{s\in[0,s_0]}$ is a predictable process of bounded variation. Then

$$\sum_{i=0}^{2^n-1} X_{\frac{i}{2^n}s_0,t_0}\left(X_{\frac{i+1}{2^n}s_0,t_0} - X_{\frac{i}{2^n}s_0,t_0}\right) = \sum_{i=0}^{2^n-1} X_{\frac{i}{2^n}s_0,t_0}\left(\left(M_{\frac{i+1}{2^n}s_0,t_0} - M_{\frac{i}{2^n}s_0,t_0}\right)\right.$$
$$\left. + \left(A_{\frac{i+1}{2^n}s_0,t_0} - A_{\frac{i}{2^n}s_0,t_0}\right)\right)$$
$$= \int_{H_{z_0}} \overline{X}^n_{s,t_0}\, \partial_1 M + \int_{H_{z_0}} \overline{X}^n_{s,t_0}\, \partial_1 A.$$

In Section 4 it is shown that for an A-quasimartingale X restricted along a horizontal curve, i.e., $H_{s_0 t}$, satisfies

$$\int_{H_{z_0}} \phi \partial X = \int_{H_{z_0}} \phi \partial_1 X + \int_{H_{z_0}} \phi \partial_2 X$$
$$= \int_{H_{z_0}} \phi \partial_1 X$$
$$= \int_{H_{z_0}} \phi \partial_1 M + \int_{H_{z_0}} \phi \partial_1 A$$

for all $\mathcal{F}_z$-predictable ϕ. Thus for A-quasimartingales

$$\int_{H_{s t_0}} \overline{X}^n_{s,t_0}\, \partial_1 M + \int_{H_{s t_0}} \overline{X}^n_{s,t_0}\, \partial_1 A = \int_{H_{s t_0}} \overline{X}^n_{s,t_0}\, (\partial_1 M + \partial_1 A). \qquad (2.30)$$

Setting

$$\delta^n(z) = \begin{cases} X(\delta_{ij}) = X_{z_{i,j+1}} - X_{z_{ij}} & \text{, if } z \in \Delta_{ij}, \\[2mm] 0 & \text{, on the axes,} \end{cases}$$

and considering (2.28) and (2.29) for general $L^{2,2}$-bounded process X it follows that,

$$\int_T X^n dX = \int_{H_{z_0}} \overline{X}^n \partial_1 X - \int_T \phi dX^n dX^n - \int_T \delta^n dX$$

$$= \int_{H_{z_0}} \overline{X}^n \partial_1 X - J_X^n(z_0) - \int_T \delta^n dX.$$

Then solving for $J_X^n(z_0)$ the equation becomes:

$$J_X^n(z_0) = \int_{H_{z_0}} \overline{X}^n \partial_1 X - \int_T X^n dX - \int_T \delta^n dX. \tag{2.31}$$

Likewise for an A-quasimartingale by (2.30) and (2.28):

$$J_X^n(z_0) = \int_{H_{z_0}} \overline{X}^n (\partial_1 M + \partial_1 A) - \int_T X^n dX - \int_T \delta^n dX. \tag{2.32}$$

To complete the development of the J_X measure, the L^2-limits of the terms on the right-hand sides of (2.31) and (2.32) need to be considered. In order to establish convergence in $L^2(P)$, however, additional hypotheses and a maximal inequality are required. Here the generalized Doob maximal inequality for a class of separable H-quasimartingales, given in Theorem 3 from the Introduction to this chapter, will be needed. The generalization includes the case when X is a continuous strong martingale having four moments. So the work below extends the treatment of Cairoli and Walsh [4].

Theorem 43. *Let $\{X_z, \mathcal{F}_z\}_{z \in T}$ be a separable right continuous $L^{2,2}$-bounded process having four moments and satisfying the conditions of the Brennan-Doob Inequality for $\phi(x) = |x|^4$. Suppose the Bochner bounding measure is given by $d\alpha = d\mu(\omega, z) \otimes dP(\omega)$ on $\mathcal{F}_z^1$-predictable functions and that $E\{\mu(T)^2\} < \infty$. Given the nth dyadic partition of T determined by the set of points $\{z_{ij}\}_{1 \leq i,j \leq 2^n}$, if*

$$E \left\{ \sup_{i,j} \left(X_{z_{ij+1}} - X_{z_{ij}} \right)^4 \right\} \longrightarrow 0, \quad a.e. \ z, \quad as \ n \longrightarrow \infty \tag{2.33}$$

then

$$J_X^n \xrightarrow{L^2} J_X.$$

Proof. First, it should be noted that the approximation X^n of X inherits all the conditions of the hypothesis. Also, since these processes satisfy the framework of the Brennan-Doob inequality they are H-quasimartingales, consequently X^n is an A-quasimartingale. Thus the representation in (2.32) can be used for J_X^n. First, for $\{\overline{X}^n\}_{n=1}^{\infty}$,

$$E\left\{\left(\int_{H_{z_0}}\left(\overline{X}^n-\overline{X}^m\right)\partial_1 M\right)^2\right\}=\int_{\Omega\times H_{z_0}}\left(\overline{X}^n-\overline{X}^m\right)^2 d\alpha$$

$$=\int_{\Omega}\left(\int_{H_{z_0}}\left(\overline{X}^n-\overline{X}^m\right)^2 d\langle M\rangle\right)dP$$

(since Bochner's bounding measurefor a martingale is $d\langle M\rangle\otimes dP$)

$$\leq E\left\{\sup_s\left(\overline{X}^n-\overline{X}^m\right)^2\langle M\rangle_{st_0}(H_{z_0})\right\}$$

$$\leq E\left\{(\langle M\rangle_{s_0 t_0})^2\right\}^{\frac{1}{2}}$$

$$E\left\{\sup_z\left(\overline{X}^n-\overline{X}^m\right)^4\right\}^{\frac{1}{2}}$$

(by Hölder's)

since $\langle M\rangle_{st_0}$ is increasing in s. By the generalized maximal inequality,

$$E\left\{\sup_z\left(\overline{X}_z^n-\overline{X}_z^m\right)^2\right\}\leq a+bE\left\{\left(\overline{X}_{z_0}^n-\overline{X}_{z_0}^m\right)^2\right\}$$

$$<\infty.$$

Also, $E\{(\langle M\rangle_{s_0 t_0})^2\}^{\frac{1}{2}}<\infty$ since by the Burkholder Inequality for right continuous martingales

$$E\left\{(\langle M\rangle_{z_0})^2\right\}\leq KE\left\{M_{z_0}^4\right\}<\infty$$

for some $K>0$ where $\langle M\rangle_{st_0}$ is the natural increasing process attached to the martingale M_{st_0}. The term $\sup_z(\overline{X}_z^n-\overline{X}_z^m)^4$ is dominated by

$$\sup_z\left(\overline{X}_z^n-\overline{X}_z^m\right)^4\leq 8\sup_z\left(\overline{X}_z^n\right)^4+8\sup_z\left(\overline{X}_z^m\right)^4$$

$$\leq 16\sup_z(X_z)^4,$$

and again by the generalized maximal inequality

$$E\left\{\sup_z (X_z)^4\right\} \le a + bE\left\{X_{z_0}^4\right\} < \infty.$$

Thus by the dominated convergence theorem

$$\lim_{m,n\to\infty} E\left\{\sup_z \left(\overline{X}_z^n - \overline{X}_z^m\right)^4\right\} = E\left\{\lim_{m,n\to\infty}\sup_z \left(\overline{X}_z^n - \overline{X}_z^m\right)^4\right\}$$

$$= 0,$$

since $\overline{X}_z^n \longrightarrow X_z$ pointwise, hence

$$\int_{H_{z_0}} \overline{X}^n \partial_1 M \xrightarrow{L^2} \int_{H_{z_0}} X\partial_1 M.$$

For $\int_{H_{z_0}} \overline{X}^n \partial_1 A$ the same holds, using the fact that $E\{|A_{s_0 t_0}|^2\} < \infty$ for a process of bounded variation. So,

$$\int_{H_{z_0}} \overline{X}^n \partial_1 A \xrightarrow{L^2} \int_{H_{z_0}} X\partial_1 A.$$

Given that $X^n = \sum_{i,j=0}^{2^n-1} X_{z_{ij}} \chi_{\Delta_{ij}}(z)$ for all n, it follows that X^n is an $\mathcal{F}_z$-predictable simple function. Thus,

$$E\left\{\left(\int_T (X^n - X^m)\,dX\right)^2\right\} \le \int_{\Omega\times T} (X^n - X^m)^2\,d\alpha$$

$$= \int_\Omega \left(\int_T (X^n - X^m)^2\,d\mu\right)dP$$

$$\le E\left\{\mu(T)\sup_z (X_z^n - X_z^m)^2\right\}$$

$$\le E\left\{\mu(T)^2\right\}^{\frac{1}{2}} E\left\{\sup_z (X^n - X^m)^4\right\}^{\frac{1}{2}}$$

$$\longrightarrow 0$$

as before since $E\{\mu(T)^2\} < \infty$. Finally for the term $\int_T \delta^n(z)\,dX$ the extra hypothesis will be used. If X has continuous paths the condition (2.33) is immediate. Since

$$\delta^n(z) = \sum_{i,j=0}^{2^n-1} X(\delta_{ij})\chi_{\Delta_{ij}}(z)$$

is $\mathcal{F}_z^1$-predictable and $E\left\{\mu(T)^2\right\} < \infty$ it follows that,

$$E\left\{\left(\int_T \delta^n(z)dX\right)^2\right\} \le \int_{\Omega \times T} (\delta^n(z))^2 \, d\alpha$$

$$\le E\left\{\mu(T) \sup_z (\delta^n(z))^2\right\}$$

$$\le E\left\{\mu(T)^2\right\}^{\frac{1}{2}} E\left\{\sup_z (\delta^n(z))^2\right\}^{\frac{1}{2}}$$

$$\longrightarrow 0.$$

Thus for $\{J_X^n\}_{n=1}^\infty$,

$$E\left\{(J_X^n - J_X^m)^2\right\} = E\left\{\left(\left(\int_{H_{z_0}} \overline{X}^n \partial_1 M + \int_{H_{z_0}} \overline{X}^n \partial_1 A\right.\right.\right.$$

$$\left.- \int_T X^n dX - \int_T \delta^n dX\right)$$

$$- \left(\int_{H_{z_0}} \overline{X}^m \partial_1 M + \int_{H_{z_0}} \overline{X}^m \partial_1 A\right.$$

$$\left.\left.\left.- \int_T X^m dX - \int_T \delta^m dX\right)\right)^2\right\}$$

$$\le K_1 E\left\{\left(\int_{H_{z_0}} \left(\overline{X}^n - \overline{X}^m\right) \partial_1 M\right)^2\right\}$$

$$+ K_2 E\left\{\left(\int_{H_{z_0}} \left(\overline{X}^n - \overline{X}^m\right) \partial_1 A\right)^2\right\}$$

$$+ K_3 E\left\{\left(\int_T (X^n - X^m) dX\right)^2\right\}$$

$$+ K_4 E\left\{\left(\int_T \delta^n dX\right)^2\right\}$$

$$+ K_5 E\left\{\left(\int_T \delta^m dX\right)^2\right\}$$

$$\longrightarrow 0$$

where $\max\{K_1, K_2, K_3, K_4, K_5\} \le 8$. Therefore, $J_X^n(z_0) \xrightarrow{L^2} J_X(z_0)$. $\qquad \square$

Denote the limit as

$$J_X(z_0) = \int_{H_{z_0}} X\partial_1 M + \int_{H_{z_0}} X\partial_1 A - \int_T X dX. \qquad (2.34)$$

If $\{X_z, \mathcal{F}_z\}_{z \in T}$ is a square integrable continuous strong martingale having four moments, then $\{J_X(z), \mathcal{F}_z\}$ is a square integrable martingale (see Cairoli and Walsh [4]). Hence,

$$\langle J_X \rangle_z = \iint_{R_z \times R_z} \chi_{\{\eta \hat{\lambda} \xi\}} d\langle X \rangle_\eta d\langle X \rangle_\xi$$

so that for $\phi \in \mathcal{L}^2_{J_X}(z_0)$ one has

$$E\left\{ \left(\int_T \phi dJ_X \right)^2 \right\} = E\left\{ \int_T \phi^2 d\langle J_X \rangle \right\}$$

$$\stackrel{\text{def}}{=} E\left\{ \int_T \phi^2 d_s\langle X \rangle_{st} d_t\langle X \rangle_{st} \right\}$$

where $d_s\langle X \rangle_{st} d_t\langle X \rangle_{st} \leq d_s\langle X \rangle_{st_0} \otimes d_t\langle X \rangle_{s_0 t}$ (Cairoli and Walsh [4], page 149-150). In general, the hypothesis for the process J_X

$$E\left\{ \left(\int_T \phi dJ_X \right)^2 \right\} \leq \int_{\Omega \times T} \phi^2 d\alpha$$

$$\leq \int_{\Omega \times T} \phi^2 d\mu \otimes dP$$

for $\mathcal{F}_z$-predictable simple ϕ where $d\alpha$ is majorized by a product measure $d\mu \otimes d\lambda$ can be used. For the processes which satisfy the hypothesis of Theorem 43, the following result can be established.

Theorem 44. *Suppose the conditions of Theorem 43 hold for a process* $\{X_z, \mathcal{F}_z\}_{z \in T}$. *Then J_X obeys the Bochner boundedness principle, i.e., for the $\mathcal{F}_z$-predictable functions ϕ one has*

$$E\left\{ \left(\int_T \phi dJ_X \right)^2 \right\} \leq 2 \int_{\Omega \times T} \phi^2 d\beta$$

where $d\beta = \chi_{H_{z_0}} d\sigma + (X^2 d\alpha)$ with σ the Bochner bounding measure for a one-parameter semimartingale and the bounding measure α for X.

Proof. Since

$$J_X(z) = \int_{H_z} X\partial_1 X + \int_{H_z} X\partial A - \int_T X dX$$

the bounding measure can be constructed using the right-hand side. Since X is predictable the process defined for $z \in \Gamma_1$ by

$$\gamma(z) = \int_{H_z} X\partial_1 M + \int_{H_z} X\partial_1 A$$

is a one-parameter semimartingale with respect to $\{\mathcal{F}_z\}_{z \in H_z}$. Thus, the process $\gamma(z)$ has an $L^{2,2}$ bounding measure for all $\mathcal{F}_z$ predictable simple functions ϕ,

$$E\left\{ \left(\int_{H_{z_0}} \phi d\gamma \right)^2 \right\} \le \int_{\Omega \times H_{z_0}} \phi^2 d\sigma,$$

(see Rao [14]). Let $\kappa(z) = \int_{R_z} X dX$ and again consider the approximation of the integrand by

$$X^n(z) = \sum_{i,j=0}^{2^n-1} X_{z_{ij}} \chi_{\Delta_{ij}}(z)$$

where $X^n \longrightarrow X$ pointwise, by $\mathcal{F}_z$-simple functions. Thus by the $L^{2,2}$-boundedness of X

$$E\left\{ \left(\int_T X^n dX \right)^2 \right\} \le \int_{\Omega \times T} (X^n)^2 d\alpha$$

and, as in the previous theorem, it can be verified that

$$\int_T X^n dX \xrightarrow{L^2} \int_T X dX.$$

Let $\psi = \sum_{i=1}^n \alpha_i \chi_{(s_i,t_i]}$ be an $\mathcal{F}_z$-predictable simple function. Then

$$E\left\{ \left(\int_T \psi d\kappa \right)^2 \right\} = E\left\{ \left(\int_T \sum_{i=1}^n \alpha_i \chi_{(s_i,t_i]} d\kappa \right)^2 \right\}$$

$$= E\left\{ \left(\sum_{i=1}^n \alpha_i \int_{(s_i,t_i]} X dX \right)^2 \right\}$$

$$= E\left\{\left(\int_T \sum_{i=1}^n \alpha_i X_{(s_i,t_i]} X dX\right)^2\right\}$$

$$= E\left\{\left(\int_T (\psi X) dX\right)^2\right\} \quad \text{(where } \psi X \text{ is } \mathcal{F}_z\text{-predictable)}$$

$$\leq \int_{\Omega \times T} (\psi X)^2 \, d\alpha$$

$$= \int_{\Omega \times T} \psi^2 X^2 d\alpha.$$

Putting the two results together, it follows that the bounding measure for J_X satisfies

$$E\left\{\left(\int_T \psi dJ_X\right)^2\right\} \leq 2E\left\{\left(\int_{H_{z_0}} \psi d\gamma\right)^2\right\} + 2E\left\{\left(\int_T \psi d\kappa\right)^2\right\}$$

$$\leq 2\int_{\Omega \times H_{z_0}} \psi^2 d\sigma + 2\int_{\Omega \times T} \psi^2 X^2 d\alpha$$

$$= 2\int_{\Omega \times T} \psi^2 \left(\chi_{H_{z_0}} d\sigma + X^2 d\alpha\right)$$

$$= 2\int_{\Omega \times T} \psi^2 d\beta.$$

$\square$

2.7   A Stochastic Green Theorem

Before a Green theorem is presented, a simple proposition will be proven which is used in the theorem to follow.

Proposition 45. *If X is an $L^{2,2}$-bounded A-quasimartingale then $\partial_1 X = \partial_1 M + \partial_1 A$ on horizontal paths (∂_2 for vertical), where $X_{s\tilde{t}} = M_{s\tilde{t}} + A_{s\tilde{t}}$ is the Doob-Meyer decomposition of $\{X_{s\tilde{t}}, \mathcal{F}_{s\tilde{t}}\}_{s\in[0,s_0]}$, with $\tilde{t}$ fixed.*

Proof. By the Itô-formula for $f(x) = x$ with $\tilde{t}$ fixed,

$$f(X_{s\tilde{t}}) - f(X_{0\tilde{t}}) = \int_0^s \frac{\partial f}{\partial x}(X_{v\tilde{t}}) \, dM_{v\tilde{t}} + \int_0^s \frac{\partial f}{\partial x}(X_{v\tilde{t}}) \, dA_{v\tilde{t}}$$

$$+ \frac{1}{2} \int_0^s \frac{\partial^2 f}{\partial x^2} \left(X_{v\tilde{\imath}} \right) d_{v\tilde{\imath}} \langle M \rangle$$

$$= \int_0^s dM_{v\tilde{\imath}} + \int_0^s dA_{v\tilde{\imath}}$$

since $X = 0$ on the axes. Hence,

$$df \left(X_{s\tilde{\imath}} \right) = dX_{s\tilde{\imath}} = dM_{s\tilde{\imath}} + dA_{s\tilde{\imath}},$$

so $\partial_1 X_{s\tilde{\imath}} = \partial_1 M_{s\tilde{\imath}} + \partial_1 A_{s\tilde{\imath}}$. $\square$

Let $\Phi = \{\Phi_{st}\}_{(s,t) \in T}$ be a process which satisfies the equation

$$\Phi_{st} = \Phi_{s0} + \int_{V_{st}} \phi \partial_2 X + \int_{V_{st}} \psi d\nu \qquad (2.35)$$

for some ϕ and ψ $\mathcal{F}_{st}$-predictable processes. Here $\nu : [0, t_0] \longrightarrow \mathbb{R}$ is the Lebesgue measure and V_{st} is the vertical segment joining the s-axis with the point (s, t). This will be a derivative with respect to t. For a derivative with respect to s use H_{st} and Φ_{0t} instead. Also let the $L^{2,2}$-bounding measure be β for J_X, such that

$$\int_0^t |\phi_{sv}| d\beta(v) < \infty \text{ a.e. and } \int_0^t |\psi_{sv}| d\nu < \infty \text{ a.e..}$$

Definition 46. If (2.35) holds for fixed s and each $t \leq t_0$, then Φ is said to have a *stochastic partial derivative with respect to* (X, t) *along* V_{st_0}. If (2.35) holds along $s \leq s_0$, $t \leq t_0$, then Φ is said to have *stochastic partial with respect to* (X, t) *on* T.

Thus prepared, the stochastic version of a Green Theorem can be presented. With all the above preliminary work, it is possible to give a proof close to that of Cairoli and Walsh, and this is done.

Theorem 47. *Let* $\{X_z, \mathcal{F}_z\}_{z \in T}$ *be an* $L^{2,2}$-*bounded process satisfying the hypothesis of Theorem 43 with a bounding measure* $d\alpha = d\mu(\omega, z) \otimes dP(\omega)$. *Suppose* ϕ *and* ψ *are* $\mathcal{F}_z$-*predictable, and* Φ *is* $\mathcal{F}_z$-*predictable, such that* Φ *has* ϕ *and* ψ *as stochastic partials along* V_{st_0} *with respect to* (X, t), *a.a* t,

$s \leq s_0$. *Suppose also that*

(1)

$$\int_{\Omega \times T} \Phi_{s0}^2 d\alpha < \infty,$$

(2)   either,

(I)   for simple γ_n,

$$E\left\{\left(\int_{H_{st}} \gamma_n \partial_1 X\right)^2\right\} \leq E\left\{\int_{H_{st}} \gamma_n^2 d\mu_1\right\} \quad and$$

$$E\left\{\left(\int_{V_{st}} \gamma_n \partial_2 X\right)^2\right\} \leq E\left\{\int_{V_{st}} \gamma^2 d\mu_2\right\},$$

where $d\mu_1$ and $d\mu_2$ are independent of ω, or $d\mu_1$ and $d\mu_2$ are majorized by the product measure $d\mu_i(\omega, z)dP(\omega) \leq d\lambda(z) \otimes dP(\omega)$. Suppose further that,

$$E\left\{\int_0^{s_0} \int_0^{t_0} \phi^2 d\lambda(s_0, u)d\lambda(v, t_0)\right\} < \infty \quad and$$

$$E\left\{\int_0^{s_0} \int_0^{t_0} \psi_n(u, v)dv d\lambda(u, t_0)\right\} < \infty$$

where $\nu : [0, t_0] \longrightarrow \mathbb{R}$ is the Lebesgue measure in Definition 2.35, or

(II)   there exists a bounded function $k : T \times T \longrightarrow \mathbb{R}$ such that for a.a (ω, s)

$$\left(\int_{V_{st}} \phi \partial_2 X\right)^2 \leq \left(\int_{V_{st}} k(u, v)\phi(u, v)d\lambda(u, v)\right)^2$$

with $\lambda : \mathcal{B}(T) \longrightarrow \mathbb{R}$ as a finite Lebesgue-Stieltjes measure such that

$$\int_\Omega \int_0^{s_0} \int_0^{t_0} \psi_n^2(u, v)dv d\mu(u, t_0) \otimes dP < \infty \quad and$$

$$\int_\Omega \int_0^{s_0} \int_0^{t_0} \phi_n^2(u, v)d_v \lambda(u, v)d\mu(u, t_0) \otimes dP < \infty.$$

Then for any rectangle A

$$\int_{\partial A} \Phi \partial_1 X = \int_A \Phi dX + \int_A \phi dJ_X + \iint_A \psi \partial_1 X dt$$

where the integral is in the clockwise direction, and ∂A is the boundary of the rectangle A.

Proof. Let $A = (z_1, z_2]$ where $z_1 = (s_1, s_2) \prec\prec (s_2, t_2) = z_2$. It may be assumed that $\Phi = 0$ on the lower edge of A. Indeed, writing $\Phi_{st} = \Phi_{st_1} + (\Phi_{st} - \Phi_{st_1})$, $(s, t) \in A$, it follows, since Φ_{st_1} is fixed in the second coordinate, that

$$\int_A \Phi_{st_1} dX_{st} = \int_{s_1}^{s_2} \Phi_{st_1} d(X_{st_2} - X_{st_1})$$

$$= \int_{s_1}^{s_2} \Phi_{st_1} dM_{st_2} - \int_{s_1}^{s_2} \Phi_{st_1} dM_{st_1}$$

$$+ \int_{s_1}^{s_2} \Phi_{st_1} dA_{st_2} - \int_{s_1}^{s_2} \Phi_{st_1} dA_{st_1}$$

$$= \int_{s_1}^{s_2} \Phi_{st_1} (\partial_1 M_{st_2} + \partial_1 A_{st_2})$$

$$- \int_{s_1}^{s_2} \Phi_{st_1} (\partial_1 M_{st_1} + \partial_1 A_{st_1})$$

$$= \int_{s_1}^{s_2} \Phi_{st_1} \partial_1 X_{st_2} - \int_{s_1}^{s_2} \Phi_{st_1} \partial_1 X_{st_1}$$

$$= \int_{\partial A} \Phi_{st_1} \partial_1 X.$$

So what is to be proven will hold if and only if it holds for $\{(\Phi_{st} - \Phi_{st_1})\}$. Suppose ϕ and ψ are bounded simple functions. It is possible to write $A = \cup_{i=1}^n A_i$ where the A_i are subrectangles on which ϕ and ψ are constant. Thus since the integral satisfies

$$\int_{\partial A} \Phi \partial_1 X = \sum_{i=1}^n \int_{\partial A_i} \Phi \partial_1 X$$

where the line integrals cancel each other on the common boundaries of A_i, it can be assumed $\phi = \phi_1$ and $\psi = \psi_1$ are constant on A_i, $i = 1, 2, \ldots, n$.

By hypothesis

$$\Phi_{st} = \Phi_{s0} + \int_{V_{st}} \phi \partial_2 X + \int_{V_{st}} \psi d\nu$$

$$= 0 + \int_{V_{st}} \phi_1 \partial_2 X + \int_{V_{st}} \psi_1 d\nu$$

$$= \phi_1 \int_{V_{st}} \partial_2 X + \psi_1 \int_{V_{st}} d\nu$$

$$= \phi_1 X_{st} + \psi_1 t.$$

Thus for Φ_{st}, by the previous comment,

$$\Phi_{st} = \phi_1 (X_{st} - X_{st_1}) + \psi(t - t_1), \quad (s,t) \in A.$$

Since $A = R_{s_2 t_2} - R_{s_2 t_1} - R_{s_1 t_2} + R_{s_1 t_1}$ it follows that for the process J_X,

$$J_X(A) = J_X(R_{s_2 t_2}) - J_X(R_{s_2 t_1}) - J_X(R_{s_1 t_2}) + J_X(R_{s_1 t_1})$$

where

$$J_X(R_{s_2 t_2}) = J_X(s_2, t_2) - J_X(0, t_2) - J_X(s_2, 0) + J_X(0, 0)$$

$$= J_X(s_2, t_2)$$

since $X = 0$ on the axes. So

$$\int_A \phi dJ_X = \int_A \phi_1 dJ_X = \phi_1 J_X(A)$$

$$= \phi_1 \left(J_X(s_2, t_2) - J_X(s_2, t_1) - J_X(s_1, t_2) + J_X(s_1, t_1) \right)$$

$$= \phi_1 \left(\int_{H_{s_2 t_2}} X \partial_1 X - \int_{R_{s_2 t_2}} X dX - \int_{H_{s_2 t_1}} X \partial_1 X + \int_{R_{s_2 t_1}} X dX \right)$$

$$- \phi_1 \left(\int_{H_{s_1 t_2}} X \partial_1 X - \int_{R_{s_1 t_2}} X dX - \int_{H_{s_1 t_1}} X \partial_1 X + \int_{R_{s_1 t_1}} X dX \right)$$

(using (2.34))

$$= \phi_1 \int_{\partial A} X \partial_1 X - \phi_1 \int_A X dX$$

$$= \int_{\partial A} \phi_1 (X_{st} - X_{st_1}) \partial_1 X - \int_A \phi_1 (X_{st} - X_{st_1}) dX.$$

Now using the Itô formula for a semimartingale $\{N_t\}_{t\in[0,t_0]}$ with $N_t = M_t + A_t$ and $f(x,t) = (t-t_1)x$ it follows that

$$f(N_t, t) - f(N_0, 0) = \int_0^t \frac{\partial f}{\partial x}(N_s, s)dM_s + \int_0^t \frac{\partial f}{\partial t}(N_s, s)ds$$
$$+ \int_0^t \frac{\partial f}{\partial t}(N_s, s)dA_s,$$

so that,

$$(t-t_1)N_t = \int_0^t (s-t_1)dM_s + \int_0^t N_s ds + \int_0^t (s-t_1)dA_s.$$

Thus,

$$d\left((t-t_1)N_t\right) = (t-t_1)dM_t + N_t dt + (t-t_1)dA_t$$

$$= (t-t_1)dN_t + N_t dt.$$

Setting $N_t = X_{s_2 t} - X_{s_1 t}$ and using the above formula in the second equality,

$$\int_A \psi(t-t_1)dX_{st} = \psi_1 \int_{t_1}^{t_2} (t-t_1)d(X_{s_2 t} - X_{s_1 t})$$
$$= \psi_1 \int_{t_1}^{t_2} d\left((t-t_1)(X_{s_2 t} - X_{s_1 t})\right) - \psi_1 \int_{t_1}^{t_2} (X_{st} - X_{s_1 t})dt$$
$$= \psi_1(t_2 - t_1)(X_{s_2 t_2} - X_{s_1 t_2}) - \int_{t_1}^{t_2}\int_{s_1}^{s_2} \psi_1 d_s X_{st} dt$$
$$= \int_{\partial A} \psi_1(t-t_1)\partial_1 X - \int_{t_1}^{t_2}\int_{s_1}^{s_2} \psi_1 d_s X_{st} dt.$$

Therefore the following equations are true:

$$\Phi_{st} = \phi_1(X_{st} - X_{st_1}) + \psi_1(t-t_1), \qquad (2.36)$$

$$\int_A \phi_1 dJ_X = \int_{\partial A} \phi_1(X_{st} - X_{st_1})\partial_1 X - \int_A \phi_1(X_{st} - X_{st_1})dX, \qquad (2.37)$$

and

$$\int_A \psi_1(t-t_1)dX_{st} = \int_{\partial A} \psi_1(t-t_1)\partial_1 X - \int_{t_1}^{t_2}\int_{s_1}^{s_2} \psi_1 d_s X_{st} dt. \qquad (2.38)$$

Integrating (2.36) by $\partial_1 X$ over the boundary ∂A

$$\int_{\partial A} \Phi_{st}\partial_1 X = \int_{\partial A} \phi_1\left(X_{st} - X_{st_1}\right)\partial_1 X + \int_{\partial A} \phi_1(t - t_1)\partial_1 X. \qquad (2.39)$$

Then using (2.37) for the first term on the right-hand summand and (2.38) for the left-hand summand of the right-hand side of (2.39)

$$(2.39) = \int_A \phi_1 dJ_X + \int_A \phi_1\left(X_{st} - X_{st_1}\right) dX$$
$$+ \int_A \psi_1(t - t_1) dX_{st} + \int_{t_1}^{t_2}\int_{s_1}^{s_2} \psi_1 d_s X_{st} dt,$$

and integrating (2.36) by dX over A

$$\int_A \Phi dX = \int_A \phi_1\left(X_{st} - X_{st_1}\right) dX + \int_A \psi_1(t - t_1) dX.$$

Thus making substitutions,

$$\int_{\partial A} \Phi_{st}\partial_1 X = \int_A \phi_{st} dJ_X + \int_A \Phi_{st} dX - \int_A \psi_1(t - t_1) dX$$
$$+ \int_A \psi_1(t - t_1) dX + \int_{t_1}^{t_2}\int_{s_1}^{s_2} \psi_1 d_s X_{st} dt$$
$$= \int_A \phi_1 dJ_X + \int_A \Phi_{st} dX + \int_A \psi_1 d_s X_{st} dt.$$

Consequently, for simple ϕ_n and ψ_n with

$$\Phi_{st}^n = \int_{V_{st}} \phi_n \partial_2 X + \int_{V_{st}} \psi_n d\nu$$

it follows that

$$\int_{\partial A} \Phi_{st}^n \partial_1 X = \int_A \phi_{st}^n dJ_X + \int_A \Phi_{st}^n dX + \int_A \psi^n d_s X_{st} dt. \qquad (2.40)$$

It is at this point that the hypotheses I and II will be used in order to establish the L^2-convergence of both sides of (2.40). It has already been shown for $\phi_{st}^n \in \mathcal{L}^2_{J_X}(z_0)$ that

$$E\left\{\left(\int_T \phi^n dJ_X\right)^2\right\} \leq \int_{\Omega \times T} \phi_{st}^2 d\beta, \qquad \text{(by Theorem 44)}$$

and for the mixed integral

$$E\left\{\left(\int_0^{s_0}\int_0^{t_0}\psi^n d_s X_{st}dt\right)^2\right\} \le t_0\int_{\Omega\times T}(\psi^n)^2 d\alpha\otimes dt\otimes dP$$

where $F(t) = t$ in Theorem 34. To complete the argument the $L^{2,2}$-bounding measures for the following integrals need to be considered:

(i) $\displaystyle\int_T\left(\int_{V_{st}}\phi_n\partial_2 X\right)dX,$

(ii) $\displaystyle\int_T\left(\int_{V_{st}}\psi_n d\nu\right)dX,$ (where ν is Lebesgue),

(iii) $\displaystyle\int_{H_{st}}\left(\int_{V_{st}}\phi_n\partial_2 X\right)\partial_1 X,$

(iv) $\displaystyle\int_{H_{st}}\left(\int_{V_{st}}\psi_n d\nu\right)\partial_1 X.$

Since the hypothesis gives Φ_n as predictable it follows that both the terms $\int_{V_{st}}\psi_n d\nu$ and $\int_{V_{st}}\phi_n\partial_2 X$ are also predictable. It may be noted that only the Lebesgue measure ν is used here, whereas the mixed integral was defined in Section 4 with respect to the more general Lebesgue-Stieltjes measures. [The extension to this class seems likely. However, it will require an investigation of the Itô formula for functions such as $f(x, g(x))$ where $g(x)$ is a right continuous function of bounded variation. In this case, the second derivatives in the Itô formula for semimartingales will be non-zero in general, as well as a multitude of higher-order derivatives. This will require some further development. Thus ν is restricted to be a Lebesgue measure here.] By Theorem 29 of Section 3 for simple $\mathcal{F}_z^1$-predictable γ_n it follows that

$$E\left\{\left(\int_{H_{st}}\gamma_n\partial_i X\right)^2\right\} \le \int_{\Omega\times T}\gamma_n^2 d\alpha$$
$$= \int_{\Omega\times T}\gamma_n^2 d\mu\otimes dP$$

for $i = 1, 2$, with $d\mu\otimes dP$ the $L^{2,2}$ bounding measure of X. Using hypothesis (I), $d\mu(\omega, z)\otimes dP(\omega)$ is majorized by a Cartesian product measure $d\lambda(z)\otimes$

$dP(\omega)$ on $\mathcal{B}(T) \otimes \Sigma$, it now follows that

$$E\left\{\left(\int_{H_{st}} \gamma_n \partial_1 X\right)^2\right\} \leq \int_{\Omega \times H_{st}} \gamma_n^2 d\mu \otimes dP$$

$$\leq \int_\Omega \int_{H_{st}} \gamma_n^2 d\lambda dP$$

$$= \int_0^{s_0} E\{\gamma_n^2\} d\lambda. \qquad (2.41)$$

Thus,

$$E\left\{\left(\int_{V_{st}} \phi_n \partial_2 X\right)^2\right\} \leq \int_\Omega \int_{V_{st}} \phi_n^2 d\lambda dP$$

$$= \int_0^{t_0} E\{\phi_n^2\} d\lambda \qquad (2.42)$$

and

$$E\left\{\left(\int_{V_{st}} \psi_n d\nu\right)^2\right\} = E\left\{\left(\int_0^t \psi_n d\nu\right)^2\right\}$$

$$\leq t_0 \int_0^{t_0} E\left\{\psi_n^2\right\} d\nu \qquad (2.43)$$

using Jensen's inequality and Tonelli's theorem for product measures. Using the majorization $d\lambda \times dP$ for $\mathcal{F}_z$-predictable Y and the results of Section 3, one has

$$E\left\{\left(\int_T Y dX\right)\right\} \leq \int_T \int_\Omega Y^2 dP d\lambda \qquad (2.44)$$

$$E\left\{\left(\int_{H_{z_0}} Y \partial_1 X\right)^2\right\} \leq \int_0^{s_0} \int_\Omega Y^2 dP d\lambda. \qquad (2.45)$$

Putting (2.42), (2.43), (2.44), and (2.45) together, the desired result is obtained:

(i)

$$E\left\{\left(\int_T \left(\int_{V_{st}} \phi_n \partial_2 X\right) dX\right)^2\right\} \le \int_T \int_\Omega \left(\int_{V_{st}} \phi_n \partial_2 X\right)^2 dP d\lambda$$

$$\text{(by (2.44))}$$

$$= \int_T E\left\{\left(\int_{V_{st}} \phi_n \partial_2 X\right)^2\right\} d\lambda$$

$$\le \int_T \left(\int_0^{t_0} \int_\Omega \phi_n^2 dP d\lambda(s_0, v)\right) d\lambda(z)$$

$$\text{(by (2.42))}$$

$$= \int_0^{t_0} \int_0^{t_0} E\left\{\phi_n^2\right\} d\lambda(s_0, v) d\lambda(u, t_0),$$

(ii)

$$E\left\{\left(\int_T \left(\int_{V_{st}} \psi_n d\nu\right) dX\right)^2\right\} \le \int_T E\left\{\left(\int_{V_{st}} \psi_n d\nu\right)^2\right\} d\lambda \quad \text{(by (2.44))}$$

$$\le \int_T \left(t_0 \int_0^{t_0} E\left\{\psi_n^2\right\} d\nu\right) d\lambda(z)$$

$$\text{(by (2.43))}$$

$$= t_0 \int_0^{s_0} \int_0^{t_0} E\left\{\psi_n^2(u, v)\right\} d\nu d\lambda(u, t_0),$$

(iii)

$$E\left\{\left(\int_{H_{st}} \left(\int_{V_{st}} \phi_n \partial_2 X\right) \partial_1 X\right)^2\right\} \le \int_0^{s_0} E\left\{\left(\int_{V_{st}} \phi_n \partial_2 X\right)^2\right\} d\lambda(z)$$

$$\text{(by (2.45))}$$

$$\le \int_0^{s_0} \int_0^{t_0} E\left\{\phi_n^2\right\} d\lambda(s_0, v) d\lambda(u, t_0),$$

$$\text{(by (2.42))}$$

(iv)

$$E\left\{\left(\int_{H_{st}} \left(\int_{V_{st}} \psi_n d\nu\right) \partial_1 X\right)^2\right\} \le \int_0^{s_0} E\left\{\left(\int_{V_{st}} \psi_n d\nu\right)^2\right\} d\lambda(u, t_0)$$

$$\text{(by (2.41))}$$

$$\leq t_0 \int_0^{s_0} \int_0^{t_0} E\left\{\psi_n^2\right\} d\nu d\lambda(u, t_0).$$

Then by hypothesis, since the right-hand sides of the Equations (i) through (iv) above are finite, there exist sequences $\psi_n \longrightarrow \psi$ and $\phi_n \longrightarrow \phi$ such that the left-hand and right-hand sides of (2.40) converge in L^2 to

$$\int_{\partial A} \Phi \partial_1 X = \int_A \phi dJ_X + \int_A \Phi dX + \int_A \psi d_s X_{st} dt.$$

For hypothesis (II) the following can be done:

(i)

$$E\left\{\left(\int_T (\int_{V_{st}} \phi_n \partial_2 X) dX\right)^2\right\} \leq \int_{\Omega \times T} \left(\int_{V_{st}} \phi_n \partial_2 X\right)^2 d\mu dP$$

$$\leq \int_{\Omega \times T} \left(\int_{V_{st}} k(u, v) \phi_n(u, v) d\lambda_v(u, v)\right)^2 d\mu dP$$

$$\leq a^2 t_0 \int_\Omega \int_0^{s_0} \int_0^{t_0} \phi_n^2(u, v) d\lambda_v(u, v) d\mu(u, t_0) dP,$$

(ii)

$$E\left\{\left(\int_T (\int_{V_{st}} \psi_n(u, v) d\nu) dX\right)^2\right\} \leq \int_{\Omega \times T} \left(\int_{V_{st}} \psi_n(u, v) d\nu\right)^2 d\mu dP$$

$$\leq t_0 \int_\Omega \int_0^{s_0} \int_0^{t_0} \psi_n^2(u, v) d\nu d\mu(u, t_0) dP,$$

(iii)

$$E\left\{\left(\int_{H_{st}} (\int_{V_{st}} \phi_n(u, v) \partial_2 X) \partial_1 X\right)^2\right\}$$

$$\leq \int_\Omega \int_0^{s_0} \left(\int_{V_{st}} \phi_n(u, v) \partial_2 X\right)^2 d\mu(u, t_0) dP$$

$$\leq \int_\Omega \int_0^{s_0} \left(\int_{V_{st}} k(u, v) \phi_n(u, v) d\lambda_v(u, v)\right)^2 d\mu(u, t_0) dP$$

$$\leq a^2 t_0 \int_\Omega \int_0^{s_0} \int_0^{t_0} \phi_n^2(u, v) d\lambda_v(u, v) d\mu(u, t_0) dP,$$

(iv)

$$E\left\{\left(\int_{H_{st}}\left(\int_{V_{st}}\psi_n d\nu\right)\partial_1 X\right)^2\right\} \le \int_\Omega \int_0^{s_0}\left(\int_{V_{st}}\psi_n d\nu\right)^2 d\mu(u,t_0)dP$$

$$\le t_0 \int_\Omega \int_0^{s_0}\int_0^{t_0}\psi_n^2(u,v)d\nu d\mu(u,t_0)dP.$$

Then, as in the case for hypothesis (I), there exist sequences $\phi_n \longrightarrow \phi$ and $\psi_n \longrightarrow \psi$ such that both sides of (2.40) converge in L^2 to the asserted result.   □

A similar argument holds for $\partial_2 X$ to get

$$-\int_{\partial A}\Phi\partial_2 X = \int_A \Phi dX + \int_A \widehat\phi dJ_X + \iint_A \widehat\psi \partial_2 X dt$$

with the associated hypotheses (I) and (II). The above theorem is for rectangular regions A. As usual, an extension to general regions with piecewise pure boundaries is done by approximating it with rectangles and taking the L^2-limits. The argument is outlined. It is very similar to the strong martingale case of Cairoli and Walsh [4].

Theorem 48. *Let $D \subset T$ be a region with piecewise pure boundary ∂D. Suppose that the $L^{2,2}$-bounded process X does not charge the boundary ∂D (i.e., $\int_B dX = 0$ a.e. for all $B \subset \partial D$ where B is Borel) and that Φ, ϕ, and ψ satisfy the conditions of the stochastic Green Theorem above. Then*

$$\int_{\partial D}\Phi\partial_1 X = \int_D \Phi dX + \int_D \phi J_X + \iint_D \psi\partial_1 X dt \qquad (2.46)$$

where the integral is taken in the clockwise direction.

Proof. Express ∂D as a finite number of curves Γ_i, $i = 1,2,\ldots,p$, each of which is of type I, II, I', II', and approximate each Γ_i by stepped paths Γ_i^n of the same type as Γ_i with the same final and initial points. Select the curves in such a way that they intersect at endpoints only. Let D^n be the region bounded by $\cup_{i=1}^p \Gamma_i^n$, and express D^n as a union of disjoint rectangles

A_j^n. Apply the last theorem to each A_j^n, where

$$\int_{\partial D^n} \Phi \partial_1 X = \sum_j \int_{\partial A_j^n} \partial_1 X$$

since the terms cancel each other on the mutual boundaries of the rectangles A_j^n. Then adding over Γ^i, the Equation (2.46) holds for D^n. For each $i = 1, \ldots, p$, let B_i^n be the region enclosed by $\Gamma_i^n \cup \Gamma_i$ such that $\Gamma_i^n \longrightarrow \Gamma_i$ as in Section 3. Then by Remark 33,

$$\lim_{n \to \infty} \int_{\Gamma_i^n} \Phi \partial_1 X = \int_{\Gamma_i} \Phi \partial_1 X \text{ in } L^2(P).$$

Consequently,

$$\lim_{n \to \infty} \int_{\partial D^n} \Phi \partial_1 X = \int_{\partial D} \Phi \partial_1 X \text{ in } L^2(P)$$

with ∂D^n being the sum of the Γ_i^n. The right-hand side converges in L^2 as well, hence the result follows. $\qquad\qquad\qquad\qquad\qquad\qquad\qquad\qquad \square$

Bibliography

[1] D. Bakry. Une remarque sur les semimartingales. In *Séminaire de Probabilités XV*, number **850** in Lecture Notes in Mathematics, pages 671–672, 1981.

[2] S. Bochner. Stationarity, boundedness, almost periodicity of random valued functions. *Proc. Third Berkeley Symp. Math. Statist. & Prob.*, 2:7–27, 1956.

[3] M. D. Brennan. "Planar Semimartingales". *J. of Multivar. Anal.*, 9:465–486, 1979.

[4] R. Cairoli and J. Walsh. Stochastic integrals in the plane. *Acta Math.*, **134**:111–183, 1975.

[5] C. Doléans-Dade. Existence du processus croissant natural associé à un potentiel de la class(D). *Z. Wahrs.*, 9:309–314, 1968.

[6] M. Dozzi. *Stochastic Processes with Multidimensional Parameter*. Research Notes in Mathematics. Longman Scientific & Technical, England, 1989.

[7] M. Dozzi and B. G. Ivanoff. Doob-Meyer decomposition for set-indexed submartingales. *J. of Theor. Probab.*, 7(3):499–525, 1994.

[8] M. Emry. *Stochastic Calculus in Manifolds*. Springer-Verlag, New York, 1989.

[9] M. L. Green. Inequalities for random fields. *Stochastic Anal. Appl.*, (to appear).

[10] B. G. Ivanoff, E. Merzbach, M. Dozzi, and I. Schiopu-Kratina. Predictability and stopping on lattices of sets. *Probability Th. Relat. Fields*, 97:433–446, 1993.

[11] M. Métivier and J. Pellaumail. *Stochastic Integrals*. Academic Press, New York, 1980.

[12] M. M. Rao. *Stochastic Processes and Integration*. Sijthoff & Noordhoff, Alphen aan den Rijn, The Netherlands, 1979.

[13] M. M. Rao. *Measure Theory and Integration*. Wiley-Interscience, New York, 1987.

[14] M. M. Rao. Stochastic integration: a unified approach. *C.R. Acad. Sci. Paris, Sér I*, 314:629–633, 1992.

[15] M. M. Rao. *Stochastic Processes: General Theory*. Kluwer Academic Publishers, Dordrecht, Netherlands, 1995.

[16] E. Wong and M. Zakai. Weak martingales and stochastic integrals in the plane. *The Annals of Probability*, 4(4):570–586, 1976.

Chapter 3

Probability Metrics and Limit Theorems in AIDS Epidemiology

S. T. Rachev, V. Haynatzka, and G. Haynatzki

Metrics have been used in probability theory since the early 1930's as a powerful tool for estimating closeness between distributions, and the works of A.N. Kolmogorov, P. Lévy, Fortét and Mourier, L.V. Kantorovich, Yu.V. Prokhorov, A.V. Skorokhod, and V. Strassen are the prime examples. Although fundamental relationships between various metrics in probability theory (the so-called *probability metrics*) were already established by the late 1960's, the theory of probability metrics as such has been developed by V.M. Zolotarev and his students who introduced and studied main concepts and relationships. Theirs, for example, are the concepts of simple and compound metrics, minimality and protominimality, and ideality, that date from the 1970's and 1980's (**cf.** Zolotarev [96], [98], [101], and Rachev [56], [57]). Today, after probability metrics and the method based on them have been successfully applied to many theoretical and applied problems such as obtaining the rate of convergence in limit theorems, stability of queuing models, financial and insurance mathematics, and environmental processes, among others (**cf.** Rachev [62], [34], [50]), there is still much left to be done in this field. In this chapter we shall review some basic concepts and well-known relationships as well as recently obtained limiting results with applications to the spread of AIDS – the first that have been established by means of probability metrics. The two models being discussed are quite different in nature and, at the same time, complement each other in two ways:

- One evolves in discrete time and is described through a stochastic recursion, while the other does so in continuous time and is characterized by a system of stochastic differential equations;

- One describes the dynamics of the AIDS epidemic within a distinct group of intravenous drug users, while the other does the same within a large number of communities, each having the above group as one of its major four transmission groups.

We shall follow the exposition in Haynatzki [29] and Gani et al. [20], [21], formulating the results from these papers and only sketching the proofs.

3.1   Introduction

The notion of probability metrics comes from the basic notions of metric space and metric as studied in real and functional analysis [16], [36], [15].

Definition 3.1.1 *A set (S, ρ) is said to be a **metric space** with a **metric** ρ if ρ is a mapping from $S \times S$ to $[0, \infty)$ having the following properties. For each $x, y, z \in S$,*

(1) $\rho(x, y) = 0 \Leftrightarrow x = y$,

(2) $\rho(x, y) = \rho(y, x)$,

(3) $\rho(x, y) \leq \rho(x, z) + \rho(z, y)$.

If (1) *is substituted by*

(1') $\rho(x, y) = 0 \Leftarrow x = y$, *then ρ and (S, ρ) are called a **semimetric** and a **semimetric space**, respectively.*

If, instead of (3), *holds the more general*

(3') $\rho(x, y) \leq \kappa[\rho(x, z) + \rho(z, y)]$ *for some $\kappa \geq 1$,*

*then ρ is a **probability distance**.*

Following this avenue, Zolotarev [96], [101] (cf. also Rachev [62]) defined *probability metric* as follows.

Let U be a complete separable metric space, $U^k = U \times \ldots \times U$ — the k-fold Cartesian product of U, and $\mathcal{P}_k = \mathcal{P}_k(U)$ — the space of all probability measures defined on the σ algebra $\mathcal{B}_k = \mathcal{B}_k(U)$ of Borel subsets of U^k. For any set $\{\alpha, \beta, \ldots, \gamma\} \subseteq \{1, 2, \ldots, k\}$ and for any $P \in \mathcal{P}_k$, let us define the marginal of P on the coordinates $\alpha, \beta, \ldots, \gamma$ by $T_{\alpha, \beta, \ldots, \gamma} P$. Let B be the operator in U^2 defined by $B(x, y) := (y, x)$, $(x, y) \in U^2$.

Definition 3.1.2 *A mapping μ defined on $\mathcal{P}_2$ and taking values in the extended interval $[0, \infty]$ is said to be a **probability metric (semimetric)** in $\mathcal{P}_2$, if it possesses the three properties listed below.*

(1) *If $P \in \mathcal{P}_2$ and $P(\bigcup_{x \in U} \{(x, x)\}) = 1$ then $\mu(P) = 0$.*

(1′) *If $P \in \mathcal{P}_2$, then $P(\bigcup_{x \in U} \{(x, x)\}) = 1 \Leftrightarrow \mu(P) = 0$.*

(2) *If $P \in \mathcal{P}_2$ then $\mu(P \circ B^{-1}) = \mu(P)$.*

(3) *If $P_{1,3}, P_{1,2}, P_{2,3} \in \mathcal{P}_2$ and there exists a law $Q \in \mathcal{P}_3$ such that the following "consistency" condition holds:*

if $T_{13} Q = P_{13}$, $T_{12} Q = P_{12}$ and $T_{23} Q = P_{23}$, then $\mu(P_{13}) \leq \mu(P_{12}) + \mu(P_{23})$.

By the established tradition, it is common to talk about "metrics" even for mappings satisfying only **(1)**, **(2)**, and **(3)**, i.e., in the case of "semimetrics".

Let $\mathcal{X}(U)$ be the set of all r.v.s on a given probability space $(\Omega, \mathcal{A}, \text{Pr})$ taking values in $(U, \mathcal{B})$. By $\mathcal{LX}_2$ we denote the space of all joint distributions $\text{Pr}_{X,Y}$ generated by the pairs $X, Y \in \mathcal{X}$. Consider a mapping μ on the subset $\mathcal{LX}_2 \subseteq \mathcal{P}_2$.

When the probability space $(\Omega, \mathcal{A}, \text{Pr})$ has no atoms then $\mathcal{LX}_2 \equiv \mathcal{P}_2$, and Definition 3.1.2 is equivalent to the following more natural definition.

Definition 3.1.3 *A mapping μ defined on $\mathcal{LX}_2$ and taking values in $[0, \infty]$ is said to be a **probability semimetric (metric)** if, for all r.v.s $X, Y, Z \in \mathcal{X}(U)$:*

(1) $\text{Pr}(X = Y) = 1 \Rightarrow \mu(\text{Pr}_{XY}) = 0$;

(1′) $\text{Pr}(X = Y) = 1 \Leftrightarrow \mu(\text{Pr}_{XY}) = 0$ *(identity property)*;

(2) $\mu(\text{Pr}_{XY}) = \mu(\text{Pr}_{YX})$ *(symmetry)*;

(3) $\mu(\text{Pr}_{XY}) \leq \mu(\text{Pr}_{XZ}) + \mu(\text{Pr}_{ZY})$ *(triangle inequality)*.

The metric μ from Definition 3.1.3 can be considered defined on $\mathcal{X} \times \mathcal{X}$ (according to Definition 3.1.2) through the equation $\mu(X, Y) := \mu(\text{Pr}_{XY})$.

Definition 3.1.4 *If $\mu(X,Y)$ is a probability metric whose value is determined by the pair of marginal distributions P_X and P_Y, i.e., $\mu(P) = \mu(Q)$ for all $P, Q \in \mathcal{P}_2$ such that $T_i P = T_i Q, i = 1, 2$, then the metric μ is called* **simple;** *otherwise it is called* **compound.**

When μ is a simple metric, it is actually defined on the set $\mathcal{P}_1 \times \mathcal{P}_1$; so the following equivalent notation makes sense:

$$\mu(P_{XY}) = \mu(X,Y) = \mu(P_X, P_Y) = \mu(F_X, F_Y).$$

For any simple metric the following holds:

(1') $P_1 = P_2 \Rightarrow \mu(P_1, P_2) = 0$;

(2) $\mu(P_1, P_2) = \mu(P_2, P_1)$;

(3) $\mu(P_1, P_3) \leq \mu(P_1, P_2) + \mu(P_1, P_3)$.

The analogy for (1') from Definition 3.1.3 for the case of simple metrics is

(1) $P_1 = P_2 \Leftrightarrow \mu(P_1, P_2) = 0$.

As the definition of probability metric shows, the typical case is that of compound probability metrics. Unfortunately, a compound metric, unlike a simple one, is not a metric on $\mathcal{P}_1$ since it is not a function of pairs of probability measures (or, respectively, random variables). On the other hand, if its value is small then the measure is concentrated near the diagonal and, therefore, its marginals (and the respective random variables) are close to each other. The next definition gives us a procedure for deriving simple metrics from compound ones.

Definition 3.1.5 *Let $\mu(X,Y)$ be a probability metric in $\mathcal{X}(U)$ and consider the mapping*

$$\hat{\mu}(X,Y) = \hat{\mu}(P_X, P_Y) := \inf\{\mu(X,Y) : P_{XY} \in \mathcal{P}_2; P_X, P_Y \text{ - fixed}\}.$$

This mapping can be shown to be a probability metric in $\mathcal{X}$ and is called the **minimal metric** *for μ.*

Note that if μ is a simple metric then $\hat{\mu} = \mu$, and obviously each minimal metric $\hat{\mu}$ is simple. In some cases a simple metric is considerably more tangible than a corresponding compound one. It is then natural to try to obtain simple metrics that are similar or analogous (in a certain sense) to compound ones.

Let us note that the properties of the derived simple metrics through the minimization procedure largely match the properties of the original compound metrics. This is important since it is more convenient and easier to obtain restrictions of some kind (usually in the form of inequalities) for the stochastic model of interest in terms of the original compound metrics. Then the minimization procedure will enable us to switch to restrictions in terms of simple metrics. It turns out that when we do that, in many relevant cases not only the qualitative properties of the model (existence of some kind of convergence) but the accompanying quantitative (upper or lower) bounds are "inherited" as well.

The problem of giving a description of the minimal metrics is central in studying the relation between the two main classes of probability metrics – *compound* and *simple*. Moreover, it is important to know how to find the explicit form of the minimal metric in focus. This question has now been resolved for many widely encountered compound metrics, and the corresponding minimal ones are available. For example, minimal for the $\mathcal{L}_1$ metric is the Kantorovich metric and, more generally, the l_p metric is minimal for the $\mathcal{L}_p$ metric (**cf.** Section 2), for the Ky Fan compound metric minimal is the Lévy-Prokhorov metric (Strassen's theorem in [93]), for the indicator metric minimal is the distance in variation (Dobrushin's theorem in [14]). It is unrealistic to believe, though, that some universal algorithm for producing minimal metrics exists in the general case.

We discussed above the relationship between simple, compound and minimal metrics. To continue further, we have to describe the general situation where the *method of probability metrics* arises in a natural way and plays an important role.

Questions concerning the bounds within which stochastic models can be applied (as in probabilistic limit theorems) can only be answered by investigation of the qualitative and quantitative stability of the corresponding approximation. Such stability is very often convenient to express in terms of some metric, and any concrete stochastic approximation problem requires an "appropriate" or "natural" (Zolotarev calls it "ideal") metric having properties which are helpful in solving the problem. If the solution of the approximation problem has to be expressed in terms of another metric, the transition is carried out by using general relationships between metrics. This two-stage approach is the basis of the method of probability metrics, and we shall consider several examples in the next four sections, but let us now give a more detailed description of the method as we are going to apply it to two problems from AIDS epidemiology in Sections 4 and 5 (**cf.** Haynatzki [29] and Gani et al. [20], [21]).

Studying the development of a stochastic process (e.g., an epidemic), we sometimes arrive at a stochastic equation – recursive in the discrete case and SDE in the continuous one – whose solution characterizes the process of interest. Unfortunately, such stochastic equations are difficult to deal with directly and one resorts to studying the limiting properties of their solution. There are several possible approaches to accomplish this task:

- The moment-generating functions approach;

- The martingale method;

- The method of branching processes (approximation by Brownian excursions);

- The stochastic approximation approach.

The probability metrics method, as we noted above, falls in the last category and can be summarized, in the case of a stochastic equation, as follows:

1. Find the correct normalization of the equation;

2. Determine the recursion for the normalized equation;

3. Determine the limiting form of the normalized equation;

4. Choose an appropriate ("ideal") metric μ such that the transformation A, determined by the limiting equation, has good contraction properties with respect to μ;

5. Establish the convergence of the equation to the fixed point of A;

6. Investigate the convergence in distribution, in probability, in moments, etc., that follows from convergence in μ.

What we called "ideal" metric in point **4** will vary from model to model and, although ideal metrics have so far been identified for a large class of models, the classification problem for ideal metrics is a difficult task. Zolotarev [96], [97], [100], Maejima and Rachev [47], Rachev and Yukich [83], [84], de Haan and Rachev [25], and Omey and Rachev [54] considered "ideal" metrics for the basic schemes in probability theory: *summation* and *maxima* of i.i.d. r.v.s. Ideal metrics for noncommutative operations between random elements (as, e.g., random motions) are studied in Rachev and Yukich [85]. A problem of Zolotarev (**cf.** Zolotarev [100], p. 300) concerning the existence of "doubly ideal" metrics, which enjoy the properties of $(r, +)$- and $(r, \bigvee)$-ideal metrics, was resolved by Rachev and Rüschendorf [68].

However, another problem of Zolotarev [100], the rate of convergence problem with respect to the "general summation scheme", is still far from its solution. It can be probably approached by introducing ideal metrics with respect to the operation "generalized sum" and by using the structure of so-called "metrics of convolution type" (**cf.** Rachev and Yukich [83], [84]). Rachev and Yukich [85] introduced two types of ideal metrics in the space of random motions, which provided a refined rate of convergence in the integral and local CLT for random motions in the d-dimensional Euclidean space. It may be possible to investigate the structure of ideal metrics with respect to other noncommutative operations.

Rachev and Rüschendorf [67] applied ideal metrics to de Finetti type theorems (**cf.** Diaconis and Freedman [12]). In their seminal paper, Diaconis and Freedman proved that the first k-components of a point uniformly distributed on the 2-sphere in R^n are close, with respect to the variation distance, to k independent standard normal random variables; the distance being of *exact* order k/n. Similar result was obtained for the 1-sphere. Rachev and Rüschendorf [67] considered this problem for every $p > 0$ and a meaningful extension for $p = \infty$ as well. The class F_p of distributions, arising in this way, is an exponential class of distributions connecting the exponential distribution ($p = 1$) with the normal ($p = 2$) and uniform ($p = \infty$) distributions. The stability of characterizations of F_p-distributions using different metrics has been also studied in that paper; this leads to immediate applications to de Finetti type theorems.

The scope of recent research on *ideal metrics* (**cf.** [48], [70], [73], [29], [20], [21]) includes the following:

(1) The structure of ideal metrics: description of classes of ideal metrics which arise from mass-transshipment problems (**cf.** Sections 2 and 3).

(2) Ideal metrics and limit theorems for i.i.d. random vectors.

(3) Ideal metrics and dependence: application to CLT for martingale differences.

(4) Ideal metrics and contractions: application to limit theorems for recursive equations.

We discussed in detail some of the possible directions for future research on ideal metrics, which is a rather specialized topic in the theory of probability metrics. More general research directions in that theory are the following (**cf.** Rachev [62]).

Direction 1. Description of the basic structures of probability metrics.

Direction 2. Analysis of the topologies in the space of probability measures, generated by different types of probability metrics.

The applied areas and problems that have been successfully handled by the method of probability metrics are

(1) Limit theorems and rate of convergence (**cf.** Zolotarev [101], Maejima and Rachev [48], [47], Rachev and Rüschendorf [70], Rachev and Yukich [83]);

(2) Mass-transportation problems (**cf.** Rachev and Rüschendorf [74], [73], Rachev and Schief [76], Rachev [63]);

(3) Queuing theory (**cf.** Kalashnikov and Rachev [34], [33], Rachev [62], [61]);

(4) Environmental processes (**cf.** Rachev and Samorodnitsky [75], Rachev and Todorovich [79]);

(5) Rounding proportions (**cf.** Balinski et al. [6], Balinski and Rachev [7]);

(6) Computer tomography (**cf.** Klebanov and Rachev [40], [41]);

(7) Mathematical finance and insurance (**cf.** Mittnik and Rachev [49], [50], [51], Rachev and SenGupta [77], Rachev [65], Beirlant and Rachev [5]);

(8) Carcinogenic risk estimation (**cf.** Dimitrov et al. [13], Rachev and Yakovlev [81], [82], Kadyrova et al. [32], Hanin et al. [26], Klebanov et al. [38], [39], Rachev et al. [80]).

It is obvious that we cannot cover in detail the whole area of probability metrics here, therefore we shall concentrate our efforts in the next four sections on certain "representative" subareas.

Section 2 is dedicated to mass-transportation and mass-transshipment problems, namely, the Monge-Kantorovich and Kantorovich-Rubinstein problems, and their various versions. Probability metrics naturally connected with the primal and dual solutions of these problems are also considered. It may be noted that the mass-transportation and mass-transshipment problems give rise to a main metric structure in probability theory, the so-called Monge-Kantorovich class of metrics.

Section 3 is focused on ideal metrics, with plenty of examples: Zolotarev's metric, the L_p-version of Zolotarev's metric, convolution type metrics, and those metrics that can be represented in the form of a Kantorovich-Rubinstein functional; max-ideal and double ideal metrics are also discussed.

In Sections 4 and 5 we review some recent results obtained by the method of probability metrics: the limiting distribution for two problems from the spread of AIDS. We describe in Section 4 (**cf.** Haynatzki [29] and Gani et al. [20]) a model for the spread of AIDS among a group of intravenous drug users, a stochastic recursion from which we obtain limit theorems for the number of infectives produced by the group that evolves in discrete time. Then, in Section 5 (**cf.** Gani et al. [21]), we consider the spread of the AIDS epidemic in continuous time within N communities (cities), each having at least one of the major four HIV transmission groups: (i) homosexual/bisexual men; (ii) blood transfusion recipients; (iii) intravenous drug users; and (iv) heterosexuals. Our model consists of a system of $4N$ stochastic differential equations (SDEs). We show that, as $N \to \infty$, the number of infectives in each community converges to the unique solution of a Liouville type SDE.

3.2 The Monge-Kantorovich Problem and Probability Metrics

Let $A \subset U$ be the initial position set and $B \subset U$ the final position set, after a transportation of some mass. Let the probability functions $P_1, P_2 \in \mathcal{P}$ describe the densities of the sets A and B, i.e., for $a \subset A$,

$$P_1(a) = \frac{\text{mass of } a}{\text{mass of } A},$$

and for $b \subset B$,

$$P_2(b) = \frac{\text{mass of } b}{\text{mass of } B}.$$

Suppose that the cost of transferring a unit mass from point x to point y is $c(x, y)$, $c > 0$. Any probability function $P(x, y) \in \mathcal{P}_2$ with marginals $P_1(x)$ and $P_2(y)$ will be called an **admissible transportation (transference) plan**. $P(dx, dy)$ can be interpreted as proportional to the amount of mass from a dx neighborhood of x transferred into a dy neighborhood of y. Then, under an admissible plan P, the cost of shipping is

$$\int_{U \times U} c(x, y) dP(x, y). \tag{3.2.1}$$

Now we can state the **Monge-Kantorovich Problem** (MKP).

Let (U, d) be a metric space, P_1 and P_2 be two probability measures defined on its Borel σ-algebra, and let $\mathcal{P}^{(P_1, P_2)}$ be the set of all probability measures defined on the product σ-algebra with marginal distributions P_1 and P_2, respectively. The problem is to compute the functional

$$\mathcal{A}_c(P_1, P_2) := \inf \left\{ \int_{U \times U} c(x, y) dP(x, y) : P \in \mathcal{P}^{(P_1, P_2)} \right\}, \tag{3.2.2}$$

where the function $c : U \times U \to R_+$ is measurable.

The solution P^* of (3.2.2) (when existing) is called an **optimal transference plan** (OTP).

The first variant of the MKP was introduced in 1781 by Monge and considers transfers of soil from one location to another, although it (called the *Monge problem*) requires minimization of (3.2.1) not over the whole $\mathcal{P}^{(P_1, P_2)}$, but only over its subset consisting of one-to-one transformations (cf. [11]).

Monge has made the assumption that soil consists of small grains, and the problem is to give the final location of every grain in such a way that the costs of transportation are as low as possible, while it is not possible to divide the grains, mass sharing the same initial location must also share the same final one.

Often the function $c(x, y)$ has the following form. Denote by $\mathcal{H}$ the set of all nondecreasing continuous functions $H : R_+ \to R_+$ with $H(0) = 0$ and such that Orlicz's condition

$$\sup_{t>0} \frac{H(2y)}{H(t)} < \infty \tag{3.2.3}$$

holds. For such $H \in \mathcal{H}$, set $c(x, y) := H[d(x, y)]$.

Let us note that $\mathcal{L}_H(P) := \int_{U^2} H(d(x, y)) P(dx, dy)$ is a compound distance called H-**average compound distance**. If $H(t) = t^p, p > 0$, then $\mathcal{L}_p(P) = \mathcal{L}_H^{p'}(P)$ with $p' = \min(1, 1/p)$ becomes the p-**average metric**. So the MKP for this case requires actually finding the minimal metric of $\mathcal{L}_p(P)$.

The discrete case of MKP for $U = R$ turns out to be a well-known transportation problem:

$$\begin{aligned} &\text{minimize } \textstyle\sum_{i=1}^{m} \sum_{j=1}^{n} c_{ij} x_{ij} \text{ subject to} \\ &\textstyle\sum_{j=1}^{n} x_{ij} = a_i, \sum_{i=1}^{m} x_{ij} = b_j, \qquad x_{ij} \geq 0, \end{aligned} \tag{3.2.4}$$

where $\sum_{i=1}^{m} a_i = \sum_{j=1}^{n} b_j, \;\; a_i, b_j \geq 0$.

When $m = n$ and $a_i = b_j = 1$, (3.2.4) becomes the assignment problem, requiring minimization of the cost in assigning n persons to n jobs with cost c_{ij} for the ith person doing the jth job.

The MKP can be generalized as follows:

Let (U, d) be a separable metric space (s.m.s.), let $\tilde{P} = (P_1, P_2, \ldots, P_N)$, where $P_i, i = 1, \ldots, N$ are probability measures on the Borel σ-algebra of U, and let $\mathcal{B}(\tilde{P})$ be the set of all Borel probability measures on U^N having marginals $P_1, \ldots, P_N$, respectively. The multidimensional MKP looks for

$$A_c(\tilde{P}) = \inf \left\{ \int_{U^N} c\, dP : P \in \mathcal{B}(\tilde{P}) \right\}, \tag{3.2.5}$$

where c is a given continuous function on U^N.

3.2.1   Dual Representation of the MKP

The results discussed in this section are from [60], and [62]. Dual representations are an important step towards the solution of the MKP since:

- They help in the construction of algorithms for computing the optimal values of $A_c(P_1, P_2)$ for given discrete measures P_1 and P_2 — one of the most important cases in practice;

- They are efficient in deriving explicit representations and finding estimates for these optimal values.

In [60] is shown the existence of the optimal plan and obtained the dual representation for the general multidimensional case.

Consider a separable metric space (s.m.s.) (U, d) with $\mathcal{P}$ the set of all Borel probability measures on it. Let $\| \cdot \|$ be a seminorm in $R^m, m = \binom{N}{2}$, such that if $0 < b'_i \leq b''_i$, $i = 1, \ldots, m$, then $\|\mathbf{b}'\| \leq \|\mathbf{b}''\|$.

For any $x = (x_1, \ldots, x_N) \in U^N$, let

$$\mathcal{D}(x) = \|d(x_1, x_2), d(x_1, x_3), \ldots, d(x_1, x_N), d(x_2, x_N), \ldots, d(x_{N-1}, x_N)\|,$$

$$\mathcal{H}^* = \{H \in \mathcal{H} : H \text{ convex}\}.$$

Define

$$A_D(\tilde{P}) := \inf \left\{ \int_{U^N} D dP : P \in \mathcal{B}(\tilde{P}) \right\}, \tag{3.2.6}$$

where $D(x) := H[\mathcal{D}(x)]$ for $H \in \mathcal{H}^*$.

Let $\mathcal{P}^H$ be the space of all measures in $\mathcal{P}$ for which $\int_U H(d(x, a)) P(dx) < \infty$, $a \in U$. For any $U_0 \subseteq U$ define the class $\text{Lip}(U_0) = \bigcup_{\alpha > 0} \text{Lip}_{1,\alpha}(U_0)$, where $\text{Lip}_{1,\alpha}(U_0) = \{f : U \to R : |f(x) - f(y)| \leq \alpha d(x, y)$ for all $x, y \in U_0, \sup_{x \in U_0} |f(x)| < \infty\}$.

Define the class

$$\mathcal{G}(U_0) = \left\{ \mathbf{f} = (f_1, \ldots, f_N) : \sum_{i=1}^{N} f_i(x_i) \leq D(x_1, \ldots, x_N) \right\}$$

for $x_i \in U_0, f_i \in \text{Lip}(U_0)$, $i = 1, \ldots, N\}$. For any class $\mathcal{U}$ of vectors $\mathbf{f} = (f_1, \ldots, f_N)$ of measurable functions, let $\mathbf{K}(\tilde{P}, \mathcal{U}) := \sup\{\sum_{i=1}^{N} \int_U f_i dP_i : \mathbf{f} \in \mathcal{U}\}$, assuming that $P_i \in \mathcal{P}^H$ and f_i is P-integrable.

Theorem 3.2.1 *For any s.m.s. (U, d) and for any set $\tilde{P} = (P_1, \ldots, P_n)$, $P_i \in \mathcal{P}^H$, $i = 1, \ldots, N$, we have*

$$A_D(\tilde{P}) = \mathbf{K}(\tilde{P}, \mathcal{G}(U)). \tag{3.2.7}$$

If the set $\tilde{P}$ consists of tight measures, then the infimum in (3.2.6) is attained.

There are different kinds of results dealing with the dual representation of MKP and some are more general than the above theorem; **cf.** [44], [87], [86]. But Theorem 2.1 and its Corollary give a description of an important class of probability metrics — those minimal with respect to $\mathcal{L}_p$.

For the case $N = 2$, Theorem 2.1 gives

Corollary 3.2.1 Let $c = H \circ d$ for $H \in \mathcal{H}^*$. Let

$$Lip^c(U) = \left\{ (f,g) \in \bigcup_{\alpha > 0} [Lip_{1,\alpha}(U)]^2 : f(x) + g(y) \le c(x,y), x, y \in U \right\},$$

$$\mathcal{B}_c(P_1, P_2) = \sup \left\{ \int_U f dP_1 + \int_U g dP_2 : (f,g) \in Lip^c(U) \right\},$$

$$\mathcal{B}_d(P_1, P_2) = \sup \left\{ | \int_U f d(P_1 - P_2)| : f \in Lip_{1,1}(U) \right\}.$$

(i) If $\int_U c(x,a)(P_1 + P_2)(dx) < \infty$ for some $a \in U$, then

$$\mathcal{A}_c(P_1, P_2) = \mathcal{B}_c(P_1, P_2). \tag{3.2.8}$$

(ii) If $\int_U d(x,a)(P_1 + P_2)(dx) < \infty$ for some $a \in U$, then

$$\mathcal{A}_d(P_1, P_2) = \mathcal{B}_d(P_1, P_2). \tag{3.2.9}$$

If P_1 and P_2 are tight measures, then the infimum in (3.2.2) is attained.

Let $P_1, P_2 \in \mathcal{P}$ and define:

- for $p = 0$,

$$l_0 := \sup \left\{ | \int_U f d(P_1 - P_2)| : |f(x) - f(y)| \le 1 \ \forall x, y \in U, f \text{ bounded} \right\}$$

$$= \sup \left\{ |P_1(A) - P_2(A)| : A \text{ Borel} \right\}$$

- for $p \in (0, 1]$,

$$l_p := \sup \left\{ | \int_U f d(P_1 - P_2)| : |f(x) - f(y)| \le d^p(x,y), f \text{ bounded} \right\};$$

- for $p \in (1, \infty)$,

$$l_p^p := \sup \left\{ \int_U f dP_1 + \int_U g dP_2 : f, g \in \text{Lip}(U), \atop f(x) + g(y) \le d^p(x,y) \ \forall x, y \in U) \right\};$$

- for $p = \infty$,

$$l_\infty := \inf \{ \varepsilon > 0 : P_1(A) \le P_2(x : d(x,A) < \varepsilon), \forall A \text{ Borel set} \}.$$

An important dual representation for l_p follows from Corollary 2.1 (the representation for l_∞ comes from the Strassen-Dudley theorem):

Corollary 3.2.2 *Let* (U, d) *be a uniformly measurable s.m.s. Then, for* $0 \leq p \leq \infty$, l_p *are simple metrics on the space* $\mathcal{P}^p$ *of Borel probability measures,*

$$
\mathcal{P}^p = \begin{cases} \mathcal{P} & \text{for } p = 0; \\ P \in \mathcal{P} : \int d^p(x, a) dP < \infty & \text{for } p \in (0, \infty); \\ P \in \mathcal{P} \text{ with finite support} & \text{for } p = \infty, \end{cases}
$$

and

$$
l_p(P_1, P_2) = \widehat{\mathcal{L}_p}(P_1, P_2) \tag{3.2.10}
$$

for all $P_1, P_2 \in \mathcal{P}^p$.

3.2.2 Explicit Representation, Upper Bounds for, and Topological Properties of $\mathcal{A}_c(P_1, P_2)$

Explicit formulas for the value of $\mathcal{A}_c(P_1, P_2)$ as well as for the optimal transference plan are available only in special cases, and these are well known results; **cf.** [9], [30].

Theorem 3.2.2 *Let* $U = R$, *and let* $\mathcal{F}(F_1, F_2)$ *be the set of all distribution functions on* $\mathcal{X}(R^2)$ *with marginals* F *and* G *correspondingly. Then* $\overline{F}(x, y) \in \mathcal{F}(F_1, F_2)$ *if and only if*

$$
\max\{F(x) + G(y) - 1, 0\} \leq \overline{F}(x, y) \leq \min\{F(x), G(y)\}. \tag{3.2.11}
$$

Definition 3.2.1 *The continuous function* $c(x, y)$, $x, y \in R$, *is called an* **antitone function** *if and only if, for any* $x, y \in R$ *and* $d_x, d_y > 0$, *the following inequality holds:*

$$
c(x + d_x, y + d_y) - c(x + d_x, y) - c(x, y + d_y) + c(x, y) \leq 0. \tag{3.2.12}
$$

Examples of antitone functions are $H(x - y)$ with H as nonnegative convex, $\max(x, y)$, or any concave function on R^2.

Theorem 3.2.3 *Let* $U = R$ *and* $d(x, y) = |x - y|$, *and* $c(x, y)$ *is an antitone symmetric function. Let* $F(x)$ *and* $G(x)$ *be the distribution functions of some probability measures* P_1 *and* P_2 *on* $\mathcal{X}(R)$ *with* $\int c(x, a) d(F(x) + G(x)) < \infty$, $a \in R$, *and* $F^{-1}(x)$ *and* $G^{-1}(x)$ *their inverses, defined by* $F^{-1}(x) = \sup\{t : F(t) \leq x\}$. *Then*

(i)

$$
\mathcal{A}_c(P_1, P_2) = \int_0^1 c(F^{-1}(x), G^{-1}(x)) dx. \tag{3.2.13}
$$

(ii) *Furthermore,* (X^*, Y^*), *where* $X^* = F^{-1}(U)$ *and* $Y^* = G^{-1}(U)$ *with* U *being uniformly distributed on* $(0, 1)$, *is an optimal transference plan with distribution function* $\min\{F(x), G(y)\}$.

The optimal solution is not necessarily unique. Any probability measure with distribution function $\overline{F}$ that satisfies (3.2.11), and such that $\overline{F}(x,x) = \min\{F(x), G(x)\}$ for all $x \in R$, is also an OTP. The form of the distribution function of the OTP shows that the so-called "greedy" algorithms can lead to the solution in the discrete case.

Corollary 3.2.3 *Let $U = R$, $d(x,y) = |x - y|$ and $c = H \circ d$, $H \in \mathcal{H}^*$. Then if $\int H(|x|)dF < \infty$, $\int H(|x|)dG < \infty$,*

$$A_c(P_1, P_2) = \int_0^1 H(|F^{-1}(x) - G^{-1}(x)|)dx. \qquad (3.2.14)$$

So, for $p \geq 1$,

$$l_p = \left[\int_0^1 |F^{-1}(t) - G^{-1}(t)|^p dt \right]^{\frac{1}{p}};$$

$$l_\infty = \sup_{0 \leq t \leq 1} |F^{-1}(t) - G^{-1}(t)|.$$

Lately, a complete description of l_2 was obtained in [91] for the case of Hilbert space.

Theorem 3.2.4 *Let $U = R^k$ and $d(x,y) = \|x - y\|_2^2 = \sum_{i=1}^k (x_i - y_i)^2$. Let $P_1, P_2 \in \mathcal{P}$ with $\int |x|^2 dP_1(x) < \infty$, $\int |x|^2 dP_2(x) < \infty$.*

(i) *There exists an OTP of (3.2.2).*

(ii) *Let $X_1 \overset{d}{\sim} P_1$ and $X_2 \overset{d}{\sim} P_2$. Then (X_1, X_2) is a solution of (3.2.2) if and only if $X_2 \in \partial f(X_1)$ a.s. for some lower semicontinuous convex function f, where*

$$\partial f(x) = \{y : f(x') - f(x) \geq \langle y, x' - x \rangle, \text{ for all } x' \text{ in the domain of } f\}.$$

This is a convenient result since semicontinuous convex functions are comparatively well investigated, so different kinds of OTP are obtained this way (**cf.** [90], [89]), but even for such (U, d) finding of OTP is difficult.

An immediate application of Theorem 3.2.4 is to the case of Gaussian probability measures (**cf.** [22], [42]).

Proposition 3.2.1 *Let P_1 and P_2 be two n-dimensional Gaussian probability measures with means μ_1 and μ_2 and with covariance matrices Σ_1 and Σ_2, respectively (regular or not), then*

$$l_2(P_1, P_2) = \|\mu_1 - \mu_2\|_2^2 + \text{trace} \left[\Sigma_1 + \Sigma_2 - 2 \left(\Sigma_1^{1/2} \Sigma_2 \Sigma_1^{1/2} \right)^{1/2} \right]. \qquad (3.2.15)$$

Moreover, if X is a r.v. with distribution P_1, Σ_1 is non-singular and

$$A = \left(\Sigma_1^{1/2}\right)^{-1} \left(\Sigma_1^{1/2}\Sigma_2\Sigma_1^{1/2}\right)^{1/2} \left(\Sigma_1^{1/2}\right)^{-1},$$

then (X, AX) is an OTP for P_1 and P_2.

Complete characterization for other l_p does not exist. Some bounds on l_p are helpful in estimating its values, though; cf. [60].

Proposition 3.2.2 *Let $U = R^k$ and $d(x,y) = \|\cdot\|_1$.*

(i) *If P_1, P_2 have densities $p_1(x)$ and $p_2(x)$ respectively, then*

$$l_1(P_1, P_2) \leq \int_{R^k} \left[\sum_{i=1}^{k} |x_i|\right] \left|\int_0^1 (p_1 - p_2)(x_1/t, \ldots, x_k/t) \, t^{-m-1} dt\right| dx. \tag{3.2.16}$$

(ii) *If the system of differential equations*

$$\frac{\partial g(x)}{\partial x_i} = \operatorname{sgn} x_i \int_0^1 (p_1 - p_2)(x_1/t, \ldots, x_k/t) \, dt, \quad i = 1, \ldots, k,$$

has a solution with respect to some function $g(x)$, then (3.2.16) becomes an equality.

Important characterization of l_p-type metrics are their topological properties. The next theorem (**cf.** [60]) shows that for a special case of the cost function, the convergence in $\mathcal{A}_c$ is equivalent to weak convergence plus convergence in moments.

Theorem 3.2.5 *Let $c(x, y) = H[d(x, y)]$, where $H \in \mathcal{H}$, and let*

$$\int_U c(x, a) dP_n < \infty, \quad n = 0, 1, \ldots,$$

for some $a \in U$. Then $\{\mathcal{A}_c(P_0, P_n)\}$ converges to zero if and only if $\{P_n\}$ converges weakly to P_0 and

$$\lim_n \int_U c(x, b) d(P_n - P_0) = 0$$

for some (and therefore for all) $b \in U$.

3.2.3   The Kantorovich-Rubinstein Mass Transshipment Problem

Let (U, d) be a s.m.s., and let P_1 and P_2 be two probability measures in $\mathcal{P}$. Denote by $\mathcal{Q}^{(P_1, P_2)}$ the set of all finite nonnegative Borel measures on U^2 (not necessarily probability measures) such that, for any Borel set $A \in U$, the following balancing condition holds,

$$Q(A \times U) - Q(U \times A) = P_1(A) - P_2(A). \tag{3.2.17}$$

The functional

$$\mathcal{A}'_c(P_1, P_2) := \inf \left\{ \int_{U^2} c(x, y) dQ : Q \in \mathcal{Q}^{(P_1, P_2)} \right\} \tag{3.2.18}$$

is called the **Kantorovich-Rubinstein functional**, since it was first introduced by Kantorovich and Rubinstein in 1958. Here $c : U \times U \to R$ is a continuous nonnegative function.

The dual representation is as important for KRP as it is for MKP, and much has been written on that subject starting from Kantorovich and Rubinstein's paper in 1958, proving the theorem for the case when (U, d) is a compact space and $c = d$. The result from [78] states the following.

Theorem 3.2.6 *Under the following conditions imposed on function c:*
(C1) $c(x, y) = 0$ *iff* $x = y$;
(C2) $c(x, y) = c(y, x)$ *for all* $x, y \in U$;
(C3) There exists nonnegative measurable function $\lambda : U \to R$ *that maps bounded sets to bounded sets;*
(C4) $c(x, y) \le \lambda(x) + \lambda(y)$ *for* $x, y \in U$;
(C5) $\sup\{c(x, y) : d(x, a) < R, d(y, a) < r, d(x, y) \le \delta\}$ *tends to zero as* $\delta \to 0$ *for any* $a \in U$ *and* $R > 0$,
we have

$$\mathcal{A}'_c(P_1, P_2) = \sup \left\{ \int f d(P_1 - P_2) : f : U \to R, f(x) - f(y) \le c(x, y) \right\} \tag{3.2.19}$$

holds for any $P_1, P_2 \in \mathcal{P}$ *such that* $\int_U \lambda dP_1 < \infty$, $\int_U \lambda dP_2 < \infty$.
Moreover, there exists a function f^*, *such that* $f^*(x) - f^*(y) \le c(x, y)$ *and for which the supremum in (3.2.19) is attained.*

Conditions (C1)-(C5) are not too restrictive. For instance, functions $c(x, y) = H[d(x, y)]$, for $H \in \mathcal{H}$ or

$$c(x, y) = d(x, y) \max\{1, h[d(x, a)], h[d(y, a)]\}$$

with $h : R_+ \to R_+$ nondecreasing continuous, satisfy them.

By means of the above theorem, an explicit representation is obtained in the case $U = R$, $d(x,y) = |x - y|$ and

$$c(x,y) = |x - y| \max(h(|x - a|), h(|y - a|)), \qquad (3.2.20)$$

for h – nondecreasing continuous function, and $a \in R$.

Theorem 3.2.7 *If $X, Y \in \mathcal{X}(R)$ are such that $E[\lambda(X)] + E[\lambda(Y)] < \infty$ for $\lambda(x) = 2|x|h(|x - a|)$, then*

$$\mathcal{A}'_c(X,Y) = \int_{-\infty}^{\infty} h(|x - a|)|F_X(x) - F_Y(x)|dx. \qquad (3.2.21)$$

The optimal transshipment plan has not been found yet, except for the case $c = d$. Taking $h(x) = \max(1, x^{p-1})$, $p > 1$ in (3.2.21), we get

$$\mathcal{A}'_c(X,Y) = \int_{-\infty}^{\infty} \max(1, |x - a|^{p-1})|F_X(x) - F_Y(x)|dx$$

for the cost function $c(x,y) = |x - y| \max(1, |x - a|^{p-1}, |y - a|^{p-1})$. On the other hand, from (3.2.13),

$$\mathcal{A}_c(X,Y) = \int_0^1 c(|F_X^{-1}(x), F_Y^{-1}(x)|)dx.$$

So, in general, $\mathcal{A}'_c(X,Y) \neq \mathcal{A}_c(X,Y)$, but in the case when $c(x,y) = d(x,y)$ we always have $\mathcal{A}'_d(X,Y) = \mathcal{A}_d(X,Y)$.

Rachev and Rüschendorf in [73] generalized Theorem 3.2.7 as follows.

Theorem 3.2.8 *Let $c(x,y) = |x - y|\xi(x,y)$, $x, y \in R$ with $\xi(t,t) \leq \xi(x,y)$ for $x < t < y$, and $\xi(x,y)$ is symmetric, continuous, and locally bounded on the diagonal. Then, under the conditions of the duality theorem,*

$$\mathcal{A}'_c(X,Y) := \int \xi(t,t)|F_X(x) - F_Y(x)|dx. \qquad (3.2.22)$$

There is an interesting relationship between the Monge-Kantorovich and Kantorovich-Rubinstein functionals, which gives a practical meaning to the latter. Define, for a symmetric nonnegative cost function $c(x,y)$,

$$\tilde{c}(x,y) = \inf \left\{ \sum_{i=0}^{n-1} c(x_i, x_{i+1}) : n \in N, x_0 = x, x_n = y \right\}.$$

Comparing the dual representations of both functionals and using the fact that $\tilde{c}$ is symmetric and satisfies the triangle inequality — i.e., $\tilde{c}(x,y)$ is the largest semimetric dominated by $c(x,y)$, it can be shown that

$$\mathcal{A}'_c(P_1, P_2) = \mathcal{A}_{\tilde{c}}(P_1, P_2). \qquad (3.2.23)$$

So the Kantorovich-Rubinstein functional gives a minimum cost when arbitrary number of transfers are allowed. Moreover, from (3.2.23) can be concluded that while A_c do not necessarily satisfy the triangle inequality, A_c' always does.

Although different, both functionals have similar topological properties under certain conditions on the cost function $c(x, y)$. Namely, let

$$c(x, y) = d(x, y)k[d(x, a), d(y, a)] \qquad (3.2.24)$$

for $k(t, s)$ - symmetric continuous nondecreasing on both arguments $t \geq 0, s \geq 0$, and for which

C1) $\alpha = \sup\limits_{s \neq t} \dfrac{|tk(t, t) - sk(s, s)|}{|t - s|k(t, s)}$,

C2) $\beta = k(0, 0) > 0$,

C3) $\gamma = \sup\limits_{t \geq 0, s \geq 0} \dfrac{k(2t, 2s)}{k(t, s)} < \infty$.

Let $\lambda(x) = d(x, a)k[d(x, a), d(x, a)]$, and let $\mathcal{P}_\lambda$ be the set of all probability measures with finite λ-moments on a s.m.s. (U, d).

Theorem 3.2.9 *If $P_n, P \in \mathcal{P}_\lambda$ then the following are equivalent:*

(i) $A_c(P_n, P) \to 0$,

(ii) $A_c'(P_n, P) \to 0$,

(iii) $P_n \overset{d}{\to} P$ and $\int \lambda d(P_n - P) \to 0$ as $n \to \infty$.

3.2.4   Related Problems

Optimal Quality Usage

Suppose that the quality of an item of a product is described by a collection $x = (x_1, \ldots, x_m)$ of its characteristics, where m is a required number of quality characteristics and x_i is the value of the ith characteristic. The quality distribution is described by a Borel probability measure μ on R^m, and $\mu(A)$, $A \in \mathcal{B}^m$ is the proportion of all items with quality characteristics in A. The usage of the production is described by another Borel probability measure ν, where $\nu(B)$, $A \in \mathcal{B}^m$ is the proportion from the total consumption of all consumed items with quality characteristics in B.

Since usually for at least some $A \in \mathcal{B}^m$, $\mu(A) \neq \nu(A)$, let us denote by $\phi(x, y)$, $x, y \in R^m$ the loss function that takes a positive value whenever an item with quality x is used in place of an item with required quality y.

The question is how to distribute the total production so as to have minimal losses. The Borel measure $\theta(A, B)$, defined on R^{2m}, is an admissible distribution plan if

$$\theta(A, R^m) = \mu(A); \ \theta(R^m, B) = \nu(B) \text{ for all } A, B \in \mathcal{B}^m. \qquad (3.2.25)$$

Let $\Theta(\mu, \nu)$ denote the collection of all admissible plans for given μ and ν.

The usual case is that the distributions μ and ν are not known. What is known are the one-dimensional marginal quality and consumption distributions. If the ith marginal measure of production quality is denoted by μ_i and the jth marginal measure of consumption is denoted by ν_j, then the distribution plan is called **weakly admissible** if

$$\theta(R^{i-1} \times A_i \times R^{m-i}, R^m) = \mu_i(A_i), \ A_i \in \mathcal{B}^1, \ i = 1, \ldots, m, \quad (3.2.26)$$

and

$$\theta(R^m, R^{j-1} \times B_j \times R^{m-j}) = \nu_j(B_j), \ B_j \in \mathcal{B}^1, \ j = 1, \ldots, m. \quad (3.2.27)$$

Denote by $\overline{\Theta}(\mu_1, \ldots, \mu_m; \nu_1, \ldots, \nu_m)$ the collection of all weakly admissible plans, and let

$$\tau(\theta) := \int_{R^{2m}} \phi(x, y) \theta(dx, dy).$$

A distribution plan θ^* is called **optimal** if

$$\tau(\theta^*) = \inf_{\theta \in \Theta} \tau(\theta),$$

and the plan θ^0 is called **weakly optimal** if

$$\tau(\theta^0) = \inf_{\theta \in \overline{\Theta}} \tau(\theta).$$

These two problems are identical to the MKP and multidimensional MKP correspondingly. Based on the general theory of Monge-Kantorovich functionals, Rachev [66] obtains explicit representations and upper bounds for $\tau(\theta^*)$ and $\tau(\theta^0)$ in some particular cases. Kellerer [37] and Rüschendorf [87], [88] deal with different aspects of the same problem.

Stability of a Stochastic Process

Let[1] $C \subset R^m$ be a nonempty, closed set, $f : R^m \times R^s \to R \bigcup \{-\infty, +\infty\}$ be a Borel measurable, lower semicontinuous with respect to $z \in R^s$ for all fixed $x \in C$, and continuous on C. Let μ be a Borel probability measure on R^s. Consider the following stochastic optimization problem:

$$\varphi(\mu) = \inf \left\{ \int_{R^s} f(x, z) \mu(dz) : x \in C \right\}. \quad (3.2.28)$$

---

[1]The results in this subsection represent the main part of the lecture "Quantitative Stability of Stochastic Programs via Probability Metrics", by S.T. Rachev and W. Römisch given at the 3rd Int. Conf. on "Approximation and Optimization" in the Caribbean, Puebla (Mexico), Oct. 8-13, 1995.

Denote by $\psi(\mu) = \arg\{\varphi(\mu)\}$.

The question about the stability of this problem is closely related to the so-called Fortét-Mourier metric. Let (U, d) be a s.m.s. Choose $\Theta \in U$, an element that is playing the role of an origin. Let $H : R_+ \to R_+$ be a nondecreasing function with $H(0) = 0$. For any $h : U \to R$ and for any $r > 0$, define the Lipschitz norm of h by

$$\mathrm{Lip}_r(h) := \sup\left\{ \frac{|h(z) - h(\tilde{z})|}{d(z, \tilde{z})} : z \neq \tilde{z}, d(z, \Theta) \leq r, d(\tilde{z}, \Theta) \leq r \right\},$$

and the seminorm with respect to the fixed H by

$$\|h\|_H := \sup\left\{ \frac{\mathrm{Lip}_r(h)}{\max\{1, H(r)\}} : r > 0 \right\}.$$

The Fortét-Mourier metric is defined on the space

$$\mathcal{P}_H(U) := \left\{ P \in \mathcal{P}(U) : \int_U c_H(z, \Theta) P(dz) < \infty \right\}$$

by the formula

$$FM_H(P, Q) := \sup\left\{ \left| \int_U h(z)(P - Q)(dz) \right| : \|h\|_H \leq 1 \right\},$$

where $c_H(z, \Theta) = d(z, \Theta) \max\{1, H[d(z, \Theta)]\}$.

The stability of the optimization problem (3.2.28) with respect to the original distribution P can be evaluated in terms of the Fortét-Mourier metric.

Theorem 3.2.10 *Let $H : R_+ \to R_+$ be a nondecreasing function with $H(0) = 0, P \in \mathcal{P}_H(R^s)$ and $\psi(P)$ be nonempty and bounded. Assume that*

(i) *the function $f(\cdot, \xi)$ is convex for each $\xi \in R^s$, and*

(ii) *there exists an open, bounded subset V of R^m and a constant $L_0 > 0$ such that $\psi(P) \subset V$ and*

$$|f(x, \xi) - f(x, \tilde{\xi})| \leq L_0 \max\{1, H(\max\{\|\xi\|, \|\tilde{\xi}\|\})\} \ \|\xi - \tilde{\xi}\|$$

whenever $x \in V$ and $\xi, \tilde{\xi} \in R^s$.

Then there exist $\delta > 0$ such that

$$|\varphi(P) - \varphi(Q)| \leq FM_H(P, Q) \text{ whenever } Q \in \mathcal{P}_H(R^s), FM_H(P, Q) < \delta.$$

From the duality representation of the Kantorovich-Rubinstein functional it is obvious that

$$FM_H(P, Q) = \mathcal{A}'_{c_H}(P, Q).$$

So, from Theorem 3.2.9 it follows that if

$$\Delta_H := \sup_{t \neq s} \frac{|\, t \max\{1, H(t)\} - s \max\{1, H(s)\}\,|}{|\, t - s\,|\, \max\{1, H(t), H(s)\}} < \infty,$$

then

$$FM_H(P_n, P) \to 0 \text{ for } P_n, P \in \mathcal{P}_H(Z)$$

if and only if

$$P_n \overset{d}{\to} P\,, P_n, P \in \mathcal{P}_H(U) \text{ and } \int_U c_H(z, \Theta)(P_n - P)(dz) \to 0. \qquad (3.2.29)$$

The value of $\varphi(P_n)$ for a sequence P_n that satisfies (3.2.29) is an approximation for $\varphi(P)$.

When P_n are the "empirical measures"

$$P_n := \frac{1}{n}(\delta_{\xi_1} + \ldots + \delta_{\xi_n}),$$

where $\xi_1, \xi_2, \ldots$ are i.i.d. with distribution P, then

$$\varphi(P_n) = \min\left\{ \frac{1}{n} \sum_{i=1}^{n} f(x, \xi_i) : x \in C \right\}$$

and the bounds for distance between $\varphi(P)$ and $\varphi(P_n)$ provided by Theorem 3.2.10 can be used to estimate the sensitivity of a portfolio of asset returns having minimal risk with preassigned mean returns.

Problems With Relaxed Marginal Constraints

Let F_1, F_2 be given distribution functions. Consider the sets

$$\mathcal{H}(F_1, F_2) := \{F : F \text{ is a d.f. on } R^2 \text{ with marginals } \widetilde{F_1}, \widetilde{F_2},$$

$$\text{where } \widetilde{F_1} \leq F_1, \widetilde{F_2} \geq F_2\} \qquad (3.2.30)$$

and

$$\mathcal{F}^+(F_1, F_2) := \{F : F \text{ is a d.f. on } R^2 \text{ with marginals } \widetilde{F_1}, \widetilde{F_2},$$

$$\text{where } \widetilde{F_1}(x) + \widetilde{F_2}(x) = F_1(x) + F_2(x)\}. \qquad (3.2.31)$$

In [73] are investigated optimization transportation problems with bounds on the marginal distribution functions of the form

$$\text{minimize} \int c(x, y)dF(x, y), \text{ subject to } F \in \mathcal{H}(F_1, F_2), \qquad (3.2.32)$$

and

$$\text{minimize} \int c(x, y)dF(x, y), \text{ subject to } F \in \mathcal{F}^+(F_1, F_2). \qquad (3.2.33)$$

The corresponding discrete problems are as follows.
For Problem (3.2.32):

$$
\begin{aligned}
\text{minimize} \quad & \sum_{i=1}^{n} \sum_{j=1}^{n} c_{ij} x_{ij} \\
\text{subject to:} \quad & x_{ij} \geq 0 \\
& \sum_{s=1}^{j} \sum_{r=1}^{n} x_{rs} \geq \sum_{s=1}^{j} b_s =: G_j \\
& \sum_{r=1}^{j} \sum_{s=1}^{n} x_{rs} \leq \sum_{r=1}^{i} a_r =: F_i,
\end{aligned}
\qquad (3.2.34)
$$

where the sum of the "demands" $\sum_{s=1}^{n} b_s$ equals the sum of the "supplies" $\sum_{r=1}^{n} a_r$. The restrictions describe production and consumption processes based on priorities such that what remains in stage i of the production (or consumption) process can be transferred to some of the subsequent stages $i + 1, \ldots, n$.

For Problem (3.2.33):

$$
\begin{aligned}
\text{minimize} \quad & \sum_{i=1}^{n} \sum_{j=1}^{n} c_{ij} x_{ij} \\
\text{subject to:} \quad & x_{ij} \geq 0 \\
& \sum_{i=1}^{n} x_{ik} + \sum_{j=1}^{n} x_{kj} = h_k.
\end{aligned}
\qquad (3.2.35)
$$

These restrictions can be interpreted as a flow in a network with n nodes with fixed sum h_k of inflow and outflow for each node k.

Both problems (3.2.32) and (3.2.33) are solved explicitly in [73].

Theorem 3.2.11 *Suppose that $c(x, y)$ is a symmetric, antitone function such that $c(x, x) = 0$ for all x, and define*

$$H^*(x, y) := \min\{F_1(x), \max\{F_1(y), F_2(y)\}\}.$$

Then $H^* \in \mathcal{H}(F_1, F_2)$ and H^* solves the relaxed transportation problem (3.2.32). Furthermore,

$$\int cdH^* = \int_0^1 c\left(F_1^{-1}(u), \min\left(F_1^{-1}(u), F_2^{-1}(u)\right)\right) du.$$

Based on Theorem 3.2.11, a greedy algorithm is proposed for this problem when $c_{ii} = 0$, $c_{ij} = c_{ji}$ and $c_{ij} + c_{i+1,j+1} - c_{i,j+1} - c_{i+1,j} \leq 0$. Denote

$$H_i := \max\{F_i, G_i\}, 1 \leq i \leq n, \quad \delta_1 := H_1, \delta_{i+1} := H_{i+1} - H_i, i \leq n - 1,$$

then (3.2.34) is equivalent to the standard transportation problem

$$\begin{array}{c} \text{minimize } \sum\sum c_{ij}x_{ij} \text{ subject to} \\ \sum_{j=1}^{n} x_{ij} = a_i, \sum_{i=1}^{n} x_{ij} = \delta_j, \quad x_{ij} \geq 0, \end{array} \qquad (3.2.36)$$

and the North-West corner rule applied to these new equality restrictions solves (3.2.34).

Next theorem deals with problem (3.2.33).

Theorem 3.2.12 *Let* $G(x) = \frac{1}{2}[F_1(x) + F_2(x)]$, *and* U *is a uniformly* $[0, 1]$ *distributed random variable.*

(i) *If* $c \geq 0$ *is symmetric and antitone, or*

(ii) *if* c *is an antitone and* $x_1 \leq y \leq x_2$ *implies* $c(x_1, x_2) \geq c(y, y)$,

then

$$\inf\left\{\int c(x, y)dF(x, y) : F \in \mathcal{F}^+(F_1, F_2)\right\} = \int_0^1 c(G^{-1}(u), G^{-1}(u))du.$$

The optimal solution is given by $(G^{-1}(U), G^{-1}(U))$.

Closeness of Distributions With Increasing Number of Marginal Constraints

This problem originates from the field of computer tomography in the following context. Suppose an X-ray with an input intensity I_i passes through an object (e.g., a human body) and consequently its output intensity changes to I_o. Under certain conditions imposed on the object and the intensity of the input ray, it is known that

$$\log\left(\frac{I_i}{I_o}\right) = \int_{-\infty}^{\infty} p(x + \theta t)dt,$$

where the integral of the density $p : R^m \to R$ at point x is taken along the straight line $x + \theta t$ which the X-ray follows through the body. The goal – to reconstruct the body density $p(x)$ based on the input and output intensities, was solved as early as the 1920s by Radon. But this unique inversion of the Radon transformation holds only when all marginals are known. In reality, though, the data are incomplete. How to reconstruct the density function based on n marginals? The problem is close to the mass transshipment problems, since here also are imposed marginal restrictions.

Let P, Q be probability measures on R^m with identical marginal distributions in a finite number n of directions $\theta_1, \ldots, \theta_n$. Unfortunately, it was shown that w.r.t. the uniform metric these distributions may differ considerably. This is the so-called "computer tomography paradox", according to which the density reconstruction is impossible on the basis of finite data. But this paradox is solved in [40], [41] and, under some moment-type conditions, a measure of closeness between P and Q is obtained with respect to a metric that is weaker than the uniform metric.

The two-dimensional case $m = 2$ is easiest for consideration. Define

$$\lambda(P, Q) := \min_{T > 0} \max \left\{ \max_{\|t\| \leq T} \left| \int e^{i\langle t, x \rangle} (P - Q)(dx) \right|, \frac{1}{T} \right\}.$$

It is not difficult to see that λ metrizes the weak convergence.

Theorem 3.2.13 *(cf. [40])*
Let P, Q be probability measures on R^2 which have the same marginals in directions $\theta_1, \ldots, \theta_n$, no two of which are collinear. Suppose that P has support in the unit disc, then

$$\lambda(P, Q) \leq \left(\frac{2}{s!} \right)^{\frac{1}{s+1}},$$

with $s := 2 \left[\frac{n-1}{2} \right]$.

Moreover, if the distributions P and Q are with bounded support then they are λ-close even if their marginals (or moments) are not identical, but only λ-close, a more realistic setting.

Theorem 3.2.14 *(cf. [41])*
Let $\theta_1, \ldots, \theta_n$ be $n \geq 2$ directions in R^2 no two of which are collinear. Suppose that the supports of the measures P and Q lie in the unit disc, where they have ε-coinciding marginals with respect to the directions θ_j, i.e.,

$$\lambda(P^{(\theta_j)}, Q^{(\theta_j)}) := \min_{T > 0} \max \left\{ \max_{\|\tau\| \leq T} \left| \int e^{i\tau \langle \theta_j, x \rangle} (P - Q)(dx) \right|, \frac{1}{T} \right\} \leq \varepsilon,$$

$j = 1, \ldots, n$. Then there exists a constant C depending on the directions θ_j, $j = 1, \ldots, n$, such that for sufficiently small $\varepsilon > 0$ holds

$$\lambda(P, Q) \leq C \left(\frac{1}{\log(1/\varepsilon)} + \frac{1}{s} \right),$$

with $s := 2 \left[\frac{n-1}{2} \right]$.

Both theorems can be generalized for the multidimensional case $R^m, m \geq 2$ (**cf.** [41]), but it requires constraints over the directions $\theta_1, \ldots, \theta_n$.

3.3 Ideal Metrics

3.3.1 Definitions and Examples

We next define the class of **ideal metrics** since they possess desirable contraction properties.

Definition 3.3.1 *A compound metric μ in $\mathcal{X} \times \mathcal{X}$ is said to be an* **ideal metric** *of order $r > 0$ if*

(i) μ *is a regular functional, i.e.,*

$$\mu(X + Z, Y + Z) \leq \mu(X, Y) \text{ for any } X, Y, Z \in \mathcal{X};$$

(ii) μ *is a homogeneous functional of order r, i.e.,*

$$\mu(cX, cY) \leq |c|^r \mu(X, Y) \text{ for any } X, Y \in \mathcal{X} \text{ and } c \text{ is a non-zero constant.}$$

A simple metric μ in $\mathcal{X} \times \mathcal{X}$ is said to be an **ideal metric** *of order $r > 0$ if it is a homogeneous function of order r and (i) holds for any $X, Y, Z \in \mathcal{X}$ such that Z is independent of X and Y.*

From regularity and homogeneity directly follows the next subadditive property of ideal metrics.

Proposition 3.3.1 *If $X_1, \ldots, X_n, Y_1 \ldots, Y_n$ are independent r.v.s and μ is ideal of order r, then*

$$\mu\left(\sum_{i=1}^{n} c_i X_i, \sum_{i=1}^{n} c_i Y_i \right) \leq \sum_{i=1}^{n} |c_i|^r \mu(X_i, Y_i). \tag{3.3.1}$$

The last property makes ideal metrics a convenient tool for estimating closeness between sums of the form $\sum_{i=1}^{n} c_i X_i$ and $\sum_{i=1}^{n} c_i Y_i$.

Proposition 3.3.2 *An r-ideal metric with $r > 1$ assumes values either 1 or ∞.*

While there do not exist compound ideal metrics of order greater than 1, examples of simple r-ideal metrics with $r > 1$ do exist. Several examples are given below.

Zolotarev's Metric ζ_r

$$\zeta_r(X, Y) := \sup\{|E[f(X) - f(Y)]| : |f^{(m)}(x) - f^{(m)}(y)|$$
$$\leq \|x - y\|^\beta \quad a.e. \quad on \quad U\}, \tag{3.3.2}$$

where the supremum is taken over all continuous and bounded functions $f : U \to R$, $m = 0, 1, 2, \ldots$, and $\beta \in (0, 1]$ is such that $m + \beta = r$; $f^{(m)}$ denotes the mth Fréchet derivative of f (**cf.** [96], [97]).

Proposition 3.3.3 ζ_r *is an ideal metric of order r.*

For $X, Y \in \mathcal{X}(R)$, ζ_r is characterized by the following two propositions.

Proposition 3.3.4 *If r is an integer, then*

$$\zeta_r(X, Y) = \int_{-\infty}^{+\infty} \left| \int_{-\infty}^{x} \frac{(x-t)^{r-1}}{(r-1)!} d(F_X(t) - F_Y(t)) \right| dx. \qquad (3.3.3)$$

Proposition 3.3.5 *(An upper bound for ζ_r) Let $E(X^j - Y^j) = 0$ for $j = 1, \ldots, m$, then*

$$\zeta_r(X, Y) \le \frac{\kappa_r(X, Y)}{\Gamma(1+r)}. \qquad (3.3.4)$$

L_p-Version of Zolotarev's Metric $\zeta_{s,\delta}$

$$\zeta_{s,\delta}(X, Y) := \sup \left\{ |E[f(X) - f(Y)]| : \|f^{(s)}\|_\varepsilon := \left(\int |f^{(s)}|^\varepsilon \right)^{\frac{1}{\varepsilon}} \le 1 \right\}, \qquad (3.3.5)$$

where the supreme is taken over all continuous and bounded functions f having sth derivative $f^{(s)}$; here $s \in N$, $\delta \ge 1, 1/\varepsilon + 1/\delta = 1$, and the case $\delta = 1, \varepsilon = \infty$ corresponds to $\|f\|_\infty = \text{ess sup} \|f\|$ (cf. [47], [31]).

Proposition 3.3.6 *If $s - 1 + 1/\delta > 0$, then $\zeta_{s,\delta}$ is ideal of order $r = s - 1 + 1/\delta$.*

Proposition 3.3.7 *If $s - 1 + 1/\delta > 0$ and $\zeta_{s,\delta}(X, Y)$ is finite, then*

(i) $E(X^j - Y^j) = 0$ *for $j = 1, \ldots, s - 1$;*

(ii)

$$\zeta_{s,\delta}(X, Y) = \left[\int_{-\infty}^{+\infty} \left| \int_{-\infty}^{x} \frac{(x-t)^{s-1}}{(s-1)!} d(F_X(t) - F_Y(t)) \right|^\delta dx \right]^{\frac{1}{\delta}}. \qquad (3.3.6)$$

Remark : Unlike (3.3.3), (3.3.6) holds also for *noninteger*: $r = s - 1 + 1/\delta$.

Proposition 3.3.8 *(Upper bounds for $\zeta_{s,\delta}$)*
    If $E(X^j - Y^j) = 0$ for $j = 1, \ldots, s - 1$, then:

(i) *For any $r > 0$, $\zeta_{s,\delta}(X, Y) \le \zeta_r(X, Y)$;*

(ii) *For $0 < r \le 1$,*

$$\zeta_{s,\delta}(X, Y) \le \{\kappa_1(X, Y)\}^{\frac{1}{\delta}},$$

    and for $r > 1$,

$$\zeta_{s,\delta}(X, Y) \le \frac{\Gamma(1 + 1/\delta)}{\Gamma(1 + r)} \kappa_r(X, Y);$$

(iii) *For $1 \le \delta' \le \delta$,*

$$\zeta_{s,\delta}(X,Y) \le \{\max(E|X|^{s-1}, E|Y|^{s-1})\}^{1-\delta'/\delta} \{\zeta_{s,\delta'}(X,Y)\}^{\delta'/\delta}.$$

Proposition 3.3.9 *(Lower bounds for $\zeta_{s,\delta}$) For any $r > 0$,*

$$\mathbf{L}^{r+1}(X,Y) \le C(r)\zeta_{s,\delta}(X,Y), \tag{3.3.7}$$

where $C(r) = \dfrac{(2s)!\sqrt{2s+1}}{s!\sqrt{3-2/\delta}}$.

Moreover,

$$\zeta_{1,\delta}(X,Y) \ge [\mathbf{L}(X,Y)]^{1+\frac{1}{\delta}}, \tag{3.3.8}$$

$$\zeta_{2,\delta}(X,Y) \ge 1/4[\mathbf{L}(X,Y)]^{2+\frac{1}{\delta}}. \tag{3.3.9}$$

Convolution type ideal metrics

In the sequel, the following metrics will be used:

- Uniform metric between densities:

$$l(X,Y) := \text{ess sup}_x |p_X(x) - p_Y(x)|; \ X,Y \in \mathcal{X}(R^k). \tag{3.3.10}$$

- Uniform metric between characteristic functions:

$$\chi(X,Y) := \sup_{t \in R} |\psi_X(x) - \psi_Y(x)|; \ X,Y \in \mathcal{X}(R). \tag{3.3.11}$$

- Total variation metric:

$$\text{Var}(X,Y):$$
$$= \sup\{|Ef(X) - Ef(Y)| : f : U \to R \text{ is measurable and } \|f\|_\infty \le 1\}$$
$$= 2\sup_{A \in \mathcal{B}(U)} |P(X \in A) - P(Y \in A)|; \ X,Y \in \mathcal{X}(U). \tag{3.3.12}$$

In $\mathcal{X}(R^k)$, $\text{Var}(X,Y) = \int |d(F_X - F_Y)|$.

The l metric is ideal of order -1 and metrics χ and Var are ideal of order 0. By using these metrics, though, the metrics of any order r can be obtained.

- Let $\theta \in \mathcal{X}(U)$, $r > 0$. Then, as in [83], define for $X,Y \in \mathcal{X}(U)$:

$$\nu_{\theta,r}(X,Y) := \sup_{h \in R} |h|^r \text{Var}(X + h\theta, Y + h\theta). \tag{3.3.13}$$

- Let $\theta \in \mathcal{X}(R^k)$, $r > 0$. Then define for $X, Y \in \mathcal{X}(R^k)$:

$$\mu_{\theta,r}(X,Y) := \sup_{h \in R} |h|^r l(X + h\theta, Y + h\theta). \qquad (3.3.14)$$

- Let $r \in N_+$. Then for $X, Y \in \mathcal{X}(R)$:

$$\chi_r(X,Y) := \sup_{t \in R} |t|^{-r} |\phi_X(t) - \phi_Y(t)|. \qquad (3.3.15)$$

Proposition 3.3.10 (i) *For all $\theta \in \mathcal{X}$ and $r > 0$, $\nu_{\theta,r}$ is an ideal metric of order r.*

(ii) *For all $\theta \in \mathcal{X}$ and $r > 0$, $\mu_{\theta,r}$ is an ideal metric of order $r - 1$.*

(iii) *For all $r \in N_+$, χ_r is an ideal metric of order r.*

Below, the index θ for the metrics μ and ν will be omitted when it does not cause confusion.

The good features of these ideal metrics come from their smoothing properties with respect to the corresponding uniform metrics when θ is a α-stable random variable with $\alpha \in (0, 2]$.

Proposition 3.3.11 (i) *For any $X, Y \in \mathcal{X}(U)$ and $\sigma > 0$,*

$$Var(X + \sigma\theta, Y + \sigma\theta) \le \sigma^{-r} \nu_r(X,Y).$$

(ii) *For any $X, Y \in \mathcal{X}(R)$, $\sigma > 0$ and $r > \alpha$,*

$$\chi(X + \sigma\theta, Y + \sigma\theta) \le \left(\frac{r}{\alpha e}\right)^{\frac{r}{\alpha}} \sigma^{-r} \chi_r(X,Y).$$

(iii) *For any $X, Y \in \mathcal{X}(R^k)$ and $\sigma > 0$,*

$$l(X + \sigma\theta, Y + \sigma\theta) \le \sigma^{-r} \mu_r(X,Y).$$

When θ is α-stable with $\alpha \in (0, 2]$, the convolution metrics μ_r and ν_r are topologically equivalent to the rth pseudomoment κ_r, as the following proposition shows.

Proposition 3.3.12 *Let k be an odd integer and $X_n, Y \in \mathcal{X}(R)$ with $EX_n^j = EY^j$, $j = 1, \ldots, k - 2$. Then the following are equivalent:*

(i) $X_n \xrightarrow{d} U$ *and* $EX_n^{k-1} = EY^{k-1}$,

(ii) $\kappa_{k-1}(X_n, U) \to 0$,

(iii) $\mu_{k-1}(X_n, U) \to 0$,

(iv) $\nu_{k-1}(X_n, U) \to 0$.

3.3.2   Ideal Metrics and Minimality

In [66], [28], and [27], a whole class of r-ideal metrics is obtained as dual
solutions to some variations of the mass-transshipment problem.

Let $M = M(R^k)$ be the space of finite signed Borel measures on R^k such
that:

a) $m(R^k) = 0$ and $|m| = m^+ + m^- < \infty$;

b) $\int_{R^k} (x_1 \ldots x_k)^j m(dx) = 0, \qquad j = 1, \ldots, n$;

c) $\int_{R^k} |x_1 \ldots x_k|^n |m|(dx) < \infty$.

For any integer $n \geq 0$ and $m \in M$, define a signed measure m_n as
follows: its distribution function is

$$F_{m_n}(x_1, \ldots, x_k) := \int_{-\infty}^{x_1} \cdots \int_{-\infty}^{x_k} \prod_{i=1}^{k} \frac{(x_i - t_i)^n}{n!} m(dt_1, \ldots, dt_k)$$

for $x_i \leq 0$, and its survival function is

$$\overline{F}_{m_n}(x_1, \ldots, x_k) := (-1)^k \int_{x_1}^{\infty} \cdots \int_{x_k}^{\infty} \prod_{i=1}^{k} \frac{(x_i - t_i)^n}{n!} m(dt_1, \ldots, dt_k)$$

for $x_i > 0$.

For $x \in R^k$, define $J \subset \{1, \ldots, k\}$ to consist of $j : x_j > 0$ and $A_j = (x_j, \infty)$ for $j \in J$, $A_j = (-\infty, x_j]$ for $j \in \overline{J}$ (the complement of J).

Let

$$F_n^{(J)}(x_1, \ldots, x_k) := m_n(A_1, \ldots, A_k)$$

$$= \int_{A_1} \cdots \int_{A_k} \prod_{i=1}^{k} \frac{(x_i - t_i)^n}{n!} m(dt_1, \ldots, dt_k).$$

Then for any $n \geq 1$, we have $m \in M$ and m_n is absolutely continuous with
density

$$p_{m_n}(x) = F_{n-1}^{(J)}(x).$$

Moreover, m_n is a measure of total mass zero and finite variation.

Let $B_n(m)$ be the set of all nonnegative Borel measures on $R^k \times R^k$
satisfying $b(A \times R^k) - b(R^k \times A) = m_n(A)$ for all Borel sets $A \subset R^k$.
Consider the following variant of the Kantorovich-Rubinstein problem:

$$\|m\|_n := \inf \left\{ \int_{R^{2k}} c \, db : b \in B_n(m) \right\}, \tag{3.3.16}$$

where the cost function $c \geq 0$ satisfies the following conditions:

(a) $c(x, y) = 0$ iff $x = y$;

(b) $c(x, y) = c(y, x)$;

(c) $c(x,y) \leq \lambda(x) + \lambda(y)$, where λ is a measurable function that maps bounded sets to bounded sets;

(d) $\sup\{c(x,y) : \|x\| < a, \|x-y\| < \delta\}$ tends to zero as $\delta \to 0$, for any $a > 0$.

It is easily seen that the measures $b \in B_n(m)$ have absolutely continuous marginal difference $\Delta b(\cdot) = b(\cdot \times R^k) - b(R^k, \cdot)$. Its density $p_{\Delta b}$ is such that $\dfrac{\partial^{(n-1)k}}{\partial x_1^{(n-1)} \ldots \partial x_k^{(n-1)}} p_{\Delta b}$ is equal to the distribution function of m. That is, (3.3.16) can be viewed as an infinite-dimensional network flow problem, and $\|m\|_0$ is exactly KRP.

Let L be the space of functions $g : R^k \to R$ having Lipschitz norm

$$\|g\|_{L,c} := \sup_{x \neq y} |G(x) - g(y)|/c(x,y).$$

Let L_n be the space of integrals

$$g_n(x) = \int_0^x \prod_{j=1}^k \frac{(x_j - t_j)^{n-1}}{(n-1)!} g(t) dt x \in R^k, \quad n \geq 1,$$

and, for $n = 0$, let

$$g_0 = g,$$

where g belongs to L.

Theorem 3.3.1 *(Duality theorem for $\|m\|_n$). For $m \in M$ with $\int |\lambda| d|m| < \infty$,*

$$\|m\|_n = \sup_{f \in L_n} \left| \int f dm \right|. \tag{3.3.17}$$

Corollary 3.3.1 *For $k = 1$ and $c(x,y) = |x-y| \max(h(|x-a|), h(|y-a|))$, where $h(t) \geq 0$ is increasing on $t \geq 0$ with $h > 0$ for $t > 0$,*

$$\|m\|_n = \int_R \left| \int_{-\infty}^x \frac{(x-t)^n}{n!} dF_m(t) \right| h(|x-a|) dx. \tag{3.3.18}$$

So for $h \equiv 1$, $\|m\|_n$ is the Zolotarev metric of order $n+1$. In [31] is shown that ζ_k is a solution of a mass transportation problem with fixed marginals only for $k = 1$. For larger values of k, ζ_k does not have a dual representation similar to that in the Strassen-Dudley Theorem (**cf.** [15], [93]). But in (3.3.18) ζ_k is a solution of the Kantorovich-Rubinstein transshipment problem.

For $k \geq 1$, even in the simplest case $n = 0$, finding an explicit formula for the Kantorovich-Rubinstein norm

$$\|m\|_0 = \sup_{f \in L} \left| \int f dm \right|$$

is an open problem (**cf.** [14], [45], [60]). But the dual representation (3.3.17) provides an upper bound for $\|m\|_n$.

Corollary 3.3.2

$$\|m\|_n \leq \int_{R^k} c(x, \overline{0})|x_1 \ldots x_k|^n |m|(dx).$$

Corollary 3.3.3 *If $c(x, y) = \|x - y\|$, then $\|m\|_n$ defines an ideal metric of order $r = kn + 1$ by $Z_n(X, Y) = \|P^X - P^Y\|_n$.*

An inconvenience of Corollary 3.3.3 is that the order of ideality of the defined metric depends on the space dimension. In [27], a new ideal metric is defined that does not have this flaw.

Consider M – the set of finite signed Borel measures on R^n, and denote by M_r^0 the set of all $\mu \in M$ for which $\int_{R^n} \|x\|^r d|\mu|(x)$ is finite.

For $\mu \in M_{k-1}^0$ define Γ_μ as the set of all signed Borel measures on R^{2n} such that

$$\int_{R^n} f d\mu = \int_{R^{2n}} \Delta_h^k f(x) d\psi(x, h)$$

for all f on R^n – the functions with bounded support possessing all derivatives.

($\Delta_h^k f(x) = \sum_{i=0}^k (-1)^{k-i} \binom{k}{i} f(x + ih)$ is the kth finite difference of f).

Let $\varphi : R_+ \to R_+$, $\varphi(0+) = \varphi(0) = 0$, $\varphi(t) > 0$ for $t > 0$, be a nondecreasing function and $k \in N$ be fixed. Then in [27] the following modification of the Kantorovich-Rubinstein norm is proposed:

$$\|\mu\|_{k,\varphi} = \inf_{\psi \in \Gamma_\mu} \int_{R^{2n}} \varphi(\|h\|) d|\psi|(x, h); \quad \mu \in M_{k-1}^0.$$

The goal is to find a dual representation of $\|\mu\|_{k,\varphi}$. Define Λ_φ^k to be the set of all locally bounded functions f on R^n such that for some $C \geq 0$,

$$\Delta_h^k f(x) \leq C\varphi(\|h\|), \quad x, h \in R^n,$$

and define $\|f\|_{\Lambda_\varphi^k} = \inf C$.

Theorem 3.3.2 *Let $r = n + k - 1$. For any $\mu \in M_r^0$,*

$$\inf_{\psi \in \Gamma_\mu} \int_{R^{2n}} \varphi(\|h\|) d|\psi|(x, h) = \sup \left\{ \int_{R^n} f d\mu : \|f\|_{\Lambda_\varphi^k} \leq 1 \right\}.$$

Theorem 3.3.3 *For any $\mu \in M_r^0$, $r = n + k - 1$, there is some $f \in \Lambda_\varphi^k$ with $\|f\|_{\Lambda_\varphi^k} \leq 1$ such that*

$$\inf_{\psi \in \Gamma_\mu} \int_{R^{2n}} \varphi(\|h\|) d|\psi|(x, h) = \int_{R^n} f d\mu.$$

Now for $\varphi(t) = t^r$ and $k - 1 \leq r \leq k$ the following analog of the Kantorovich metric is an r-ideal metric:

$$K_r(P,Q) = \|P - Q\|_{k,\varphi} = \sup\left\{\left|\int_{R^n} f d(P-Q)\right| : \|f\|_{\Lambda^k_\varphi} \leq 1\right\}.$$

Theorem 3.3.4 (i) $K_r(P,Q)$ *is an ideal metric of order* r *and for* $k - 1 < r \leq k$, $c_1\zeta_r \leq K_r \leq c_2\zeta_r$ *for some positive constants* c_1 *and* c_2;

(ii) *if* $P - Q \in M^0_{n+k-1}$, *then* $K_r(P,Q)$ *admits the dual representation*

$$K_r(P,Q) = \inf_{\psi \in \Gamma_{P-Q}}\left\{\int_{R^{2n}} \|h\|^r d|\psi|(x,h)\right\},$$

and moreover,

$$K_r(P,Q) \leq A\int_{R^n} \|x\|^r d|p - q|(x) < \infty.$$

The order of ideality r of K_r does *not* depend on the dimension n.

3.3.3   Method of Metric Distances

In this subsection we shall explore the method of metric distances which we discussed already briefly in Section 1. When estimating some kind of "closeness" between two distributions, it is important to follow these two steps (**cf.** [96], [97], [100], [101]).

- Choose the most suitable (or, "ideal") for this concrete problem metric – the metric that makes the given problem most tractable.

- Using various relationships among probability metrics, transfer the relationship found during the first step into a relationship that satisfies our practical needs.

A lot of results, otherwise very difficult to achieve, were obtained by this method. Here we shall consider several concrete examples.

Rate of Convergence in the CLT

Using the convolution type metrics (3.3.14), (3.3.13), (3.3.15), Rachev and Yukich obtained in [83] the rate of convergence, with respect to the metrics $\chi(X,Y)$ and $l(X,Y)$, of the sum $S_n = 1/n^{1/\alpha}(X_1 + \cdots + X_n)$ of i.i.d. $X_1, X_2, \ldots$, to an α-stable law Y_α, $\alpha \in (0,2)$. The variables X_i are subject to tail restrictions. The main result states that

$$\chi(S_n, Y_\alpha) \leq 2^{-\frac{r}{\alpha}} n^{1-\frac{r}{\alpha}},   \qquad (3.3.19)$$

$$l(S_n, Y_\alpha) \le 2^{-\frac{r}{\alpha}} n^{1-\frac{r}{\alpha}}, \tag{3.3.20}$$

where $r > \alpha$ is such that X_i are "close" to Y_α in the χ_r metric for (3.3.19) and in μ_r and ν_r for (3.3.20).

A similar problem concerning convergence rate to an α-stable motion is solved by Rachev and Yukich in [85].

Application to Queuing Models

Let a $G/G/1/\infty$ queuing model be described by the sequences $\bar{e} = (e_1, e_2, \ldots)$ and $\bar{s} = (s_1, s_2, \ldots)$ of interarrival and service times, respectively, and let $v_i = s_i - e_i$. Let $\bar{w} = (w_1, w_2, \ldots)$ be the sequence of waiting times. Anastassiou and Rachev [1] estimate the Prokhorov distance between $\bar{w}$ and $\bar{w}^*$, where $\bar{w}^*$ is the sequence of waiting times of a $D/D/1/\infty$ queuing model with constant interarrival and service times, respectively, $\bar{e}^*$ and $\bar{s}^*$ that are close to $\bar{e}$ and $\bar{s}$ in the sense that for some $\epsilon_{1j} > 0, \epsilon_{2j} > 0$ and continuous functions f_j, g_j and, for $j = 1, 2, \ldots$,

$$e_j \in S_{e_j^*}(\epsilon_{1j}, \epsilon_{2j}, f_j, g_j), \tag{3.3.21}$$

$$s_j \in S_{s_j^*}(\epsilon_{1j}, \epsilon_{2j}, f_j, g_j), \tag{3.3.22}$$

or

$$v_j \in S_{v_j^*}(\epsilon_{1j}, \epsilon_{2j}, f_j, g_j). \tag{3.3.23}$$

Here

$$S_\alpha(\epsilon_1, \epsilon_2, f, g) = \{X \in R^\infty : f(X), g(X) \text{ are finite and}$$
$$|Ef(X) - Ef(\alpha)| \le \epsilon_1, |Eg(X) - Eg(\alpha)| \le \epsilon_2\}.$$

In order to obtain the estimation, the authors use the fact that the Prokhorov metric is minimal with respect to the Ky Fan metric $K(X, Y)$, and that both metrics are regular. For the most general case of dependent $\bar{e}$ and $\bar{s}$, the estimate is

$$\frac{1}{2}\pi(\bar{w}, \bar{w}^*) \le \sum_{j=1}^{\infty} \pi(v_j, v_j^*) = \sum_{j=1}^{\infty} K(v_j, v_j^*), \tag{3.3.24}$$

and

$$\frac{1}{2}\pi(\bar{w}, \bar{w}^*) \le \sum_{j=1}^{\infty} \pi(e_j, e_j^*) + \sum_{j=1}^{\infty} \pi(s_j, s_j^*) = \sum_{j=1}^{\infty} K(e_j, e_j^*) + \sum_{j=1}^{\infty} K(s_j, s_j^*). \tag{3.3.25}$$

The almost sharp upper bounds for the right-hand side of (3.3.24) and (3.3.25), i.e., $\sup_{X \in S_\alpha(\epsilon_1, \epsilon_2, f, g)} K(X, \alpha)$, are given for some special cases of f and g.

3.3.4   The Rate of Convergence to a Self-Similar Process

The following is another illustration of the method of metric distances.

Definition 3.3.2 *The process* $\{X(t), t \geq 0\}$ *is called* **self-similar with stationary increments and parameter** $H > 0$, *if*

(i) *The process* $X(c\cdot)$ *has the same distribution as* $c^H X(\cdot)$ *for any constant* $c > 0$;

(ii) $X(\cdot + b) - X(b)$ *has the same distribution as* $X(\cdot) - X(0)$ *for any* $b > 0$.

Definition 3.3.3 *A stationary sequence* $Y_1, Y_2, \ldots$ *of r.v.s is said to* **belong to the domain of attraction** *of a self-similar process* $X(\cdot)$ *if, for some slowly varying function* L,

$$n^{-H} L(n)^{-1} \sum_{j=1}^{[nt]} Y_j \overset{d}{=} X(t).$$

In [47], a sequence $\{X_j, j \in Z\}$ of i.i.d. random variables belonging to the domain of attraction of a stable law with characteristic function $\exp\{-|z|^\alpha\}$, $\alpha < 2$, is considered.

Let $\beta \in (1/\alpha - 1, 1/\alpha)$, $H \in (0, 1)$, and define

$$Y_k := \sum_{j \in Z} c_j X_{k-j}, k = 1, 2, \ldots,$$

where $c_j = |j|^{-\beta-1} \mathrm{sgn} j$, and for $t \in [0, 1]$,

$$\Delta_n(t) := |\beta| n^{-H} \left(\sum_{k=1}^{[nt]} Y_k + (nt - [nt]) Y_{[nt]+1} \right).$$

Let $\{Z_+(t), t \geq 0\}$ and $\{Z_-(t), t \geq 0\}$ be two independent stable processes, having the same characteristic function $\exp\{-t|z|^\alpha\}$.

Define the fractional stable process $\{\Delta(t), 0 \leq t \leq 1\}$ by

$$\Delta(t) := \int_{-\infty}^{\infty} (|t - s|^{-\beta} - |s|^{-\beta}) Z(s),$$

where $\Delta(0) := 0$ a.s. and $Z(s) = Z_+(s) I_{[s \geq 0]} - Z_-(-s + 0) I_{[s < 0]}$.

In [46] is shown that $\Delta(t)$ is self-similar with parameter H and proved the following

Theorem 3.3.5 $\Delta_n(t) \overset{d}{\to} \Delta(t)$ *as* $n \to \infty$.

The goal of [47] is to find the rate of convergence with respect to the uniform metric ρ. The next step is typical for that kind of calculations. Let $\{X_j^*, j \in Z\}$ be an i.i.d. sequence of symmetric α-stable random variables, and define $\Delta_n^*(t)$ in the same way as $\Delta_n(t)$, with X_j^* instead of X_j.

From the triangle inequality,

$$\rho(\Delta_n(t), \Delta(t)) \leq \rho(\Delta_n(t), \Delta_n^*(t)) + \rho(\Delta_n^*(t), \Delta(t)).$$

Since $\Delta_n(t)$, $\Delta_n^*(t)$ can be expressed as infinite weighted sums of $\{X_j\}$, $\{X_j^*\}$, respectively, the r-ideality of $\zeta_{s,\delta}$ helps to estimate $\zeta_{s,\delta}(\Delta_n(t), \Delta_n^*(t))$. Now (3.3.7) and (3.3.24) show that in order to estimate $\rho(\Delta_n(t), \Delta_n^*(t))$, it is sufficient to do so with the r-ideal metric $\zeta_{s,\delta}$, $r = s - 1 + 1/\delta$.

Theorem 3.3.6 *Consider the following cases:*
a) $0 < \alpha < 1$, $\frac{2\alpha}{(\alpha+1)} < r \leq 1$, $0 < H < 1$;
b) $0 < \alpha < 1$, $r > 1$, $1 - 1/r < H < 1$;
c) $1 \leq \alpha < 2$, $r > \alpha$, $1 - 1/r < H < 1$ $(H \neq 1/\alpha)$.
Then for each $t \in [0,1]$,

$$\zeta_{s,\delta}(\Delta_n(t), \Delta_n^*(t)) \leq C n^Q D(n) \zeta_{s,\delta}(X_0, X_0^*),$$

where C is a positive constant depending on H, r and t,

$$Q = -Hr + (r-1)^+ + \{2 - (1 - H + 1/\alpha)r\}^+$$

and

$$D(n) = \begin{cases} 1, & \text{if} \quad H \neq 1/\alpha + 1 - 2/r, \\ \log n, & \text{if} \quad H = 1/\alpha + 1 - 2/r. \end{cases}$$

Restrictions a) to c) assure that $C < \infty$ and $Q < 0$.

Theorem 3.3.7

$$\rho(\Delta_n^*(t), \Delta(t)) \leq (K\pi)^{-1} \psi(n),$$

where

$$K = \int_{-\infty}^{\infty} \left| |t - s|^{-\beta} - |s|^{-\beta} \right|^\alpha ds < \infty,$$

$$\psi(n) = \begin{cases} C_1 n^{-1}, & \text{for} \quad \beta < 0, \\ C_2(n,t) n^{-H\alpha}, & \text{for} \quad \beta > 0. \end{cases}$$

Here C_1 depends only on α and β, $C_2(n,t) < C_3$ if $nt > C_4$, and C_3, C_4 depend only on α and β.

From the above theorems, the next theorem easily follows.

Theorem 3.3.8 *If $\zeta_{s,\delta}(X_0, X_0^*) < \infty$ and if the conditions a) to c) of Theorem (3.3.6) hold, then*

$$\rho(\Delta_n(t), \Delta(t)) \leq (1 + C_5)\{C(r)\zeta_{s,\delta}(\Delta_n(t), \Delta_n^*(t))\}^{\frac{1}{r+1}} + (K\pi)^{-1}\psi(n),$$

for each $t \in [0,1]$, where C_5 is a positive constant that depends only on α and β.

3.3.5   Max-Ideal Metrics and the Rate of Convergence to a Max-Stable Law

The metrics that are ideal with respect to the summation scheme are convenient for comparing closeness between sums of random variables. In order to make conclusions about random variables of the form $\bigvee c_j X_j$, $X \bigvee Y := \max(X, Y)$, the notion of max-ideal metrics exists.

Definition 3.3.4 *A compound metric μ in $\mathcal{X} \times \mathcal{X}$ is said to be a **max-ideal metric of order** $r > 0$ if μ is a homogeneous functional of order r and*

$$\mu\left(X \bigvee Z, Y \bigvee Z\right) \leq \mu(X, Y) \text{ for any } X, Y, Z \in \mathcal{X}. \tag{3.3.26}$$

*A simple metric μ in $\mathcal{X} \times \mathcal{X}$ is **max-ideal of order** $r > 0$ if it is a homogeneous functional of order r and (3.3.26) holds only for such $X, Y, Z \in \mathcal{X}$ that Z is independent of X and Y.*

Proposition 3.3.13 *The weighted Kolmogorov metric*

$$\rho_p(X, Y) := \sup\{M^p(x)|F_X(x) - F_Y(x)| : x \in R^\infty\}, \tag{3.3.27}$$

$X, Y \in \mathcal{X}(R^\infty)$, where $M(x) = \inf_{i \geq 1}|x^{(i)}|$, $x \in R^\infty$, is a simple max-ideal metric of order $p > 0$.

Proposition 3.3.14 *For $U = R$ and $r > 0$, $p > 0$,*

$$\Delta_{r,p}(X, Y) = \left(\int_{-\infty}^{+\infty} \phi_{X,Y}^p(x)|x|^{rp-1}dx\right)^q, \tag{3.3.28}$$

where $q = \min(1, 1/p)$ and $\phi_{X,Y}(x) := \Pr(X \leq x < Y) + \Pr(Y \leq x < X)$ is a compound max-ideal metric of order $r \times \min(1,p)$ (see [69]).

It is important to note that, unlike sum-ideal metrics, there are nontrivial compound max-ideal metrics of order greater than one.

Let $\{X_j, j \geq 1\}$ be a sequence of i.i.d. random variables, taking values in R_+^∞, i.e., a sequence of i.i.d. discrete-time random processes with nonnegative coordinates (**cf.** [25], [54]). Let $\mathcal{C}$ be the the class of all sequences

$$\mathbf{C} = \left\{c_j(n); \ j, n = 1, 2, \cdots; \ c_1(n) > 0; \ c_j(n) \geq 0, \ j = 2, 3, \ldots;\right.$$

$$\sum_{j=1}^{\infty} c_j(n) = 1; \alpha_p(n) := \sum_{j=1}^{\infty} c_j^p(n) \to \infty \text{ as } n \to \infty \text{ for any } p > 1 \Big\}.$$

Take

$$\overline{X}_n := \bigvee_{j=1}^{\infty} c_j(n) X_j$$

and

$$\overline{Y}_n := \bigvee_{j=1}^{\infty} c_j(n) Y_j$$

with Y_j - i.i.d. with components having extreme-value distribution function $G(x) = e^{-1/x}$, $x \geq 0$.

We have $\overline{Y}_n \overset{d}{=} Y_1$ for any $n \geq 1$ and any $\mathbf{C} \in \mathcal{C}$, i.e. Y_j is a simple max-stable process (**cf.** [24]). The requirement for X_j to have nonnegative components is not restrictive since

$$\rho(X_n, Y) = \rho\Big(\bigvee_{j=1}^{\infty} c_j(n) \tilde{X}_j, Y \Big),$$

where $\tilde{X}_j^{(k)} := (X_j^{(k)})_+$.

For any $a, x \in R^{\infty}$, denote $a \circ x := (a^{(1)} x^{(1)}, a^{(2)} x^{(2)}, \ldots)$.

Proposition 3.3.15 *If $X, Y \in \mathcal{X}(R_+^{\infty})$ and Y has bounded marginal densities $p_{Y^{(i)}}, i = 1, 2, \ldots$, with $A_i := \sup_{x \in R} p_{Y^{(i)}} < \infty$, $p > 0$, $A := \sum_{i=1}^{\infty} A_i$ and $\Lambda(p) := (1 + p) p^{-\frac{p}{(1+p)}}$, then*

$$\rho(X, Y) \leq \Lambda(p) A^{\frac{p}{(1+p)}} \rho_p^{\frac{1}{(1+p)}}(X, Y).$$

By means of p max-ideality of the weighted Kolmogorov metric and the proposition above, the following theorem is proven in [25].

Theorem 3.3.9 *Let $\gamma > 0$ and $a \in R_+$ be such that $A(a, \gamma) := \sum_{k=1}^{\infty} (a^{(k)})^{-\frac{1}{\gamma}} < \infty$. Then for any $p > 1$ there exists a constant $c = c(\alpha, p, \gamma)$ such that*

$$\rho(\overline{X}_n, Y) \leq c \alpha_p(n)^{\frac{1}{(1+p\gamma)}} \rho_p(a \circ X, a \circ Y)^{\frac{1}{(1+p\gamma)}}. \tag{3.3.29}$$

So, the rate of convergence in (3.3.29) is $\alpha_p(n)^{\frac{1}{(1+p\gamma)}}$. Under additional assumptions over $\mathbf{C} \in \mathcal{C}$, it is proven that actually the convergence in (3.3.29) is of order $\alpha_p(n)$. For

$$c_j(n) = \begin{cases} 1/n & j = 1, \ldots, n; \\ 0 & j = n+1, n+2, \ldots, \end{cases}$$

i.e., in the case analogous to the CLT, these assumptions are met. Thus, we get the following

Corollary 3.3.4 *Let* $a \in R_+^\infty$ *is such that* $\sum_{i=1}^\infty a^{(i)} < \infty$. *Denote* $\overline{\lambda}_p :=$ $\max\{\rho(a \circ X, a \circ Y)\}$. *Then there exist constants* $\underline{C}$ *and* $\underline{D}$ *such that* $\alpha_p(n)^{\frac{1}{(1+p\gamma)}}$

$$\overline{\lambda}_p \le \underline{C} \Longrightarrow \rho\left(\frac{1}{n} \bigvee_{k=1}^n X_k, Y\right) \le \underline{D}\overline{\lambda}_p n^{1-p}. \tag{3.3.30}$$

Let us note that the convergence in (3.3.30) is uniform, which is not always the case for convergence of sums of random variables on R^∞. Moreover, the rate of convergence is almost the same as in the univariate case (**cf.** [54]).

Combining sum- and max-ideal metrics, we obtain *double ideal metrics* that have been investigated in [69].

Definition 3.3.5 *Let* U *be a Banach space with maxima operation* $\bigvee$. *A probability metric* μ *on* $\mathcal{X}(U)$ *is called*

(i) (r, I)**-ideal,** *if* μ *is compound* $(r, +)$*-ideal and compound* $(r, \bigvee)$*-ideal;*

(ii) (r, II)**-ideal,** *if it is compound* $(r, \bigvee)$*-ideal and simple* $(r, +)$*-ideal;*

(iii) (r, III)**-ideal,** *if it is simple* $(r, \bigvee)$*-ideal and simple* $(r, +)$*-ideal.*

Note that $\mathcal{L}_p$, $p \le 1$ is an example of a (p, I)-ideal metric, while (p, I)-ideal nontrivial metrics with $r > 1$ do not exist according to (3.3.2). In [69] is shown that metrics that are simple $(r, \bigvee)$-ideal and simple $(r, +)$-ideal for $r > 1$ are also trivial (and hence the metrics that are compound $(r, \bigvee)$-ideal and simple $(r, +)$-ideal).

Theorem 3.3.10 *Let* $r > 1$, *the simple probability metric* μ *be* (r, III)*-ideal on* $\mathcal{X}(R)$, *and let* μ *satisfy the following regularity conditions:*

- *If* $X_n \stackrel{d}{\to} a$, $Y_n \stackrel{d}{\to} b$, *where* a *and* b *are real constants, then*

$$\limsup_{n \to \infty} \mu(X_n, Y_n) \ge \mu(a, b);$$

- $\mu(a, b) = 0$ *iff* $a = b$.

Then, for any integrable $X, Y \in \mathcal{X}(R)$, $\mu(X, Y)$ *is either zero or infinity.*

3.3.6   Examples of Probability Metrics

Although there exist dozens of probability metrics, we shall give only those that are met most often in the literature as well as those used in what follows.

The Kolmogorov Metric

The *Kolmogorov metric*, also known as the *sup-* or *uniform metric*, is the most useful metric since many results concerning the CLT are expressed in it. It is defined by

$$\rho(X,Y) := \sup\{|F_X(x) - F_Y(x)| : x \in R\}, \tag{3.3.31}$$

where X, Y are r.v.s and $F_X(x)$, $F_Y(x)$ are their d.f.s.

For $X, Y \in R^\infty$, $\rho(X,Y)$ is defined similarly to (3.3.31), with the supremum taken over all $x \in R^\infty$. Here $F_X(x) := P(\cap_{i=1}^{\infty} X^{(i)} \leq x^{(i)})$.

We already discussed in this section an extension of the Kolmogorov metric, the *weighted Kolmogorov metric*

$$\rho_p(X,Y) := \sup\{M^p(x)|F_X(x) - F_Y(x)| : x \in R^\infty\}, \tag{3.3.32}$$

where $M(x) = \inf_{i \geq 1} |x^{(i)}|$, $x \in R^\infty$, $p > 0$.

The Lévy Metric L

The *Lévy metric* is the second most important metric in probability theory and is defined by

$$\mathbf{L}(X,Y) := \inf\{\varepsilon > 0 : F_X(x - \varepsilon) - \varepsilon \leq F_Y(x) \leq F_X(x + \varepsilon) + \varepsilon, \forall x \in R\}. \tag{3.3.33}$$

The geometric meaning of $\mathbf{L}(X,Y)$ is simple — it is the side of the largest square (with sides parallel to the axes of the coordinate system) inscribable between the graphs of the distribution functions F_X and F_Y.

Here is a list of notable properties of the Lévy metric as well as its relations with some other metrics.

- The Lévy metric metrizes the weak convergence (convergence in distribution) in the space $\mathcal{F}$ of probability laws on a separable metric space U.

- In (3.3.7), (3.3.8), and (3.3.9) are given several relations between $\mathbf{L}$ and $\zeta_{s,\delta}$.

- If X has a bounded density p_X, then

$$\rho(x,y) \leq \left(1 + \sup_{x \in R} p_X(x)\right)\mathbf{L}(x,y). \tag{3.3.34}$$

The Prokhorov Metric

The celebrated *Prokhorov metric* [55] is defined by

$$\pi(X,Y) := \inf\{\epsilon > 0 : \Pi_\epsilon(X,Y) < \epsilon\}, \tag{3.3.35}$$

where $\Pi_\epsilon = \sup\{P\{X \in A\} - P\{Y \in A^\epsilon\} : A \in \mathcal{B}(U)\}$ and A^ϵ is the ϵ-neighborhood of A, $A^\epsilon := \{x : d(x, A) < \epsilon\}$. The Prokhorov metric is a generalization of the Lévy metric and metrizes the weak convergence in the space of d.f.s. It is the minimal of the Ky Fan metric, being therefore a simple metric.

The Ky Fan Metric

The *Ky Fan metric*, defined by

$$\mathbf{K}(X, Y) := \inf\{\varepsilon > 0 : P[d(X, Y) > \varepsilon] \leq \varepsilon\}, \qquad (3.3.36)$$

is compound and metrizes convergence in probability in the space of real random variables.

The L_p Metrics Between Distribution Functions

The well-known functional analysis family of metrics

$$\theta_p(X, Y) := \left[\int_{-\infty}^{\infty} |F_X(t) - F_Y(t)|^p dt \right]^{\frac{1}{p}} \qquad (3.3.37)$$

for $p \geq 1$ and X, Y — random variables with d.f.s F_X, F_Y, is fundamental in the theory of probability metrics, too. For $p = 1$, we obtain the *Kantorovich metric*. One can extend the definition of θ_p when $p = \infty$ by setting $\theta_\infty = \rho$, the Kolmogorov metric. One reason for this extension is the following dual representation for $1 \leq p \leq \infty$

$$\theta_p(X, Y) = \sup_{f \in \mathcal{F}_p} \{|\mathbf{E}f(X) - \mathbf{E}f(Y)|\},$$

where $\mathcal{F}_p$ is the class of all measurable functions f with $\|f\|_q < 1$, $1/p + 1/q = 1$; $\|f\|_q$ is the usual L_q-norm for $1 \leq q < \infty$, and $\operatorname{ess\,sup} |f|$ for $q = \infty$.

Zolotarev's Metric ζ_r

Zolotarev's metric was already discussed in Section 3.4 — it is defined by

$$\zeta_r(X, Y) := \sup\{|E[f(X) - f(Y)]| : |f^{(m)}(x) - f^{(m)}(y)|$$

$$\leq \|x - y\|^\beta \quad \text{a.e. on } U\}, \qquad (3.3.38)$$

where the supremum is taken over all continuous and bounded functions $f : U \to R$, $m = 0, 1, 2, \ldots$, and $\beta \in (0, 1]$ is such that $m + \beta = r$; $f^{(m)}$ denotes the mth Fréchet derivative of f (cf. [96], [97]). ζ_r is an ideal metric of order r and will be used to obtain a CLT for independent, nonidentically distributed r.v.s.

The p-Average Compound Metric

Let (U, d) be a Polish space. Let $p > 0$ and $p' = \min(1, 1/p)$. Then

$$\mathcal{L}_p(X, Y) := \left[\int_{U^2} d^p(x, y) P_{XY}(dx, dy) \right]^{p'} = [Ed^p(X, Y)]^{p'} \qquad (3.3.39)$$

is a compound metric in $\mathcal{X}(U)$.

$\mathcal{L}_p$ is a compound $\min(1, p)$-ideal metric.

For $p = 0$ and $p = \infty$,

$$\mathcal{L}_0(X, Y) := P(X \neq Y),$$

$$\mathcal{L}_\infty(X, Y) := \text{ess sup } d(X, Y) = \inf\{\varepsilon > 0 : P(d(X, Y) > \varepsilon) = 0\},$$

correspondingly.

We shall make use of this metric in Section 3.5, where $U := C^4[0, T]$ will be the set of all continuous random processes on the interval $[0, T]$ with $d(X, Y) := (\sum_{k=1}^4 \sup_{t \leq T} |X_k(t) - Y_k(t)|)$, $p > 1$ and

$$L_{p,T}^*(X, Y) := \mathcal{L}_p(X, Y) = \left\{ E\left(\sum_{k=1}^4 \sup_{t \leq T} |X_k(t) - Y_k(t)| \right)^p \right\}^{\frac{1}{p}} \qquad (3.3.40)$$

for $X, Y \in \mathcal{X}(C^4[0, T])$.

The minimal of the p-average metric is

$$l_{p,T}^*(m_1, m_2) := \inf \left\{ L_{p,T}^*(X, Y) : X \overset{d}{=} m_1, Y \overset{d}{=} m_2 \right\} \qquad (3.3.41)$$

for $m_1, m_2 \in M(C_T^4)$, and will be used in Section 3.5 too.

The Engineer's Metric

Very popular among engineers is the metric

$$\mathbf{EN}(X, Y) := |E(X) - E(Y)|, \qquad (3.3.42)$$

which does not go beyond the first moments of the r.v.s X and Y, and is obviously quite restrictive since the first moment may not exist in the first place. Secondly, more important is the closeness of the distribution functions F_X and F_Y.

A generalization of the engineer's metric, for $p \geq 1$ is

$$\mathbf{MOM}_p(X, Y) := |m^p(X) - m^p(Y)|, \qquad (3.3.43)$$

where X and Y are r.v.s with finite pth moment and $m^p(X) := \{E|X|^p\}^{1/p}$.

The following equivalence relates the metrics $\mathcal{L}_p$, $\mathbf{K}$, and $\mathbf{MOM}$. Let $X_0, X_1, X_2, \ldots$ have finite pth moment for $p \geq 1$; then, as $n \to \infty$,

$$\mathcal{L}^p(X_n, X_0) \to 0 \Leftrightarrow \{\mathbf{K}(X_n, X_0) \to 0\} \& \{\mathbf{MOM}_p(X_n, X_0) \to 0\}. \qquad (3.3.44)$$

3.4 Spread of AIDS Among Drug Users and Probability Metrics

The method of probability metrics has been recently applied to problems from AIDS modeling, and the results discussed in this and the next sections demonstrate the potential of the method in that area. Here we shall follow the exposition in [29] and [20], while in the next section — that from [21].

3.4.1 Modeling Group Interactions

Consider a group of n IVDUs at the time t of its inception. We shall think of this group as the initial one, relative to the groups to which it gives rise. In order to simplify the explanation that follows, we shall consider its inception time to be zero, i.e., $t = 0$, although it would be more precise to denote the inception time by some t_0 since the group dynamics, in fact, is event–driven, where the events are the group needle-sharing sessions. Some of the individuals in the group may share needles, some members are infectives (infected with HIV), and others are susceptibles. By the next needle-sharing session, some members B_n, of whom R_n are infected, leave the group. Then the remaining members of the group may split into two subgroups, of size I_n and $n - B_n - I_n$, respectively. Finally, each of the two subgroups are joined by other individuals or perhaps whole groups of them. We let P_n denote the number of individuals joining the partition of size I_n, while Q_n is the number of those joining the other part $n - B_n - I_n$. Let $n_1 = I_n + P_n$ where n_1 may be greater or smaller than n, be the (random) size of a "new" group subject to the same dynamics until the next needle-sharing session, and let n_2 denote the size of the other branch of the bifurcation. At the end of these operations, which together constitute an epoch (or time period) of our discrete-time epidemic, we have two new groups. We now give a more detailed picture of the above description. Suppose that the dynamics of group formation as the group of n IVDUs evolves in (discrete) time $t = 0, 1, 2, \ldots$, is as follows.

(1) Starting at $t = 0$, a group of n IVDUs will "produce" a total of Y_n infectives at $t \to \infty$, where $t = 0, 1, 2, \ldots$, are the instants at which the group holds needle-sharing sessions. Here, by infectives produced by the group we mean those and only those IVDUs who at some point are removed from the group and are diagnosed as infectives (see below).

(2) By the time of the next needle-sharing session, a number B_n of IVDUs is removed from the group, where $B_n \overset{d}{\sim}$ Binomial $(n, b), 0 < b < 1$. These IVDUs consist of:

  (a) individuals who developed AIDS and were hospitalized or died – we denote their number by $R_n \equiv R_{n,B_n}$, where $R_n \overset{d}{\sim}$ Binomial

$(B_n, a), 0 < a < 1;$

(b) the remaining $(B_n - R_n)$ individuals who left the group without being infected.

(3) By the time of the next session the group, now of size $(n - B_n)$, has split into two subgroups: each member of the initial group migrates with probability $p \in (0, 1)$ to a "new" group, and remains in the "old" group with probability $(1 - p)$. We denote by $I_n \equiv I_{n,B_n}$ the number of IVDUs in the "new" group, and hence $I_n \overset{d}{\sim}$ Binomial $(n - B_n, p)$. If $B_n = n$, we set $I_{n,n} = 0$. Clearly, the size of the "old" group after the I_n departures is $(n - B_n - I_n)$.

(4) A number P_n of new IVDUs join the "new" group, while Q_n new IV-DUs join the "old" group, respectively, just before their next needle-sharing sessions. Here we may, for example, assume that P_n (respectively, Q_n) is the sum of n i.i.d. r.v.s with mean $\lambda_P > 0$ (respectively, $\lambda_Q > 0$) each.

Note that as time evolves, the variables corresponding to B_n, R_n, etc., do not depend directly on the group size at time zero, but rather on the current subgroup sizes.

The proposed dynamics of the group of n IVDUs leads to the following stochastic recursive equation for Y_n as $t \to \infty$:

$$Y_n \overset{d}{=} R_n + Y_{I_n + P_n} + \tilde{Y}_{n - B_n - I_n + Q_n}, \qquad (3.4.1)$$

where " $\overset{d}{=}$ " means equality in distribution. It indicates that if a group of n individuals split into two independent groups of sizes $I_n + P_n$ and $n - B_n - I_n + Q_n$, after having divested itself of B_n individuals of whom R_n are infective, then the number of infectives produced by the (parental) group of n IVDUs as $t \to \infty$ must equal the sum of infectives produced by the bifurcation groups as $t \to \infty$, after having accounted for R_n infectives among the B_n removals from the parental group.

Here we make the following set of assumptions:

(i) $\{Y_n\}_{n=0}^{\infty}$ are independent r.v.s, while $\tilde{Y}_n$ is an independent copy of Y_n for all $n = 0, 1, \ldots$, and $\{\tilde{Y}_n\}_{n=1}^{\infty}$ are also independent r.v.s;

(ii) $\{Y_n\}_{n=0}^{\infty}$, $\{\tilde{Y}_n\}_{n=0}^{\infty}$, $\{B_n\}_{n=0}^{\infty}$, $\{R_n\}_{n=0}^{\infty}$, $\{I_n\}_{n=0}^{\infty}$, $\{P_n\}_{n=0}^{\infty}$, $\{Q_n\}_{n=0}^{\infty}$ are mutually independent.

To start the recursion, we assume that $Y_0 \equiv 0$, and Y_1 is an arbitrary r.v. In the simplest case when B_n, P_n and Q_n are set to zero, Y_1 can be viewed as a Bernoulli r.v. with probability $c \in (0, 1)$ of success corresponding to the case when an IVDU is, in fact, HIV-infected.

In what follows, we shall investigate the limiting behavior of Y_n as $n \to \infty$, under the regularity conditions given below.

Condition A: Suppose the following limit relationships hold a.s. and in the L_2-sense:

(i) $\frac{B_n}{n} \to b \in (0,1)$ $\qquad\qquad$ as $n \to \infty$;

(ii) if $R_{n,\beta}$ is the number of infectives leaving the group among the $B_n = \beta$ removals, then for $\beta \in \{0,1,\ldots,n\}$,

$$\frac{R_{n,\beta}}{\beta} \to a \in (0,1) \text{ as } n \to \infty \text{ and } \beta \to \infty;$$

(iii) for $\beta \in \{0,1,\ldots,n\}$,

$$\frac{I_{n,\beta}}{n-\beta} \to p \in (0,1) \text{ as } \beta \to \infty \text{ and } n - \beta \to \infty;$$

(iv) $\frac{P_n}{n} \to \lambda_P \geq 0$, $\frac{Q_n}{n} \to \lambda_Q \geq 0$ as $n \to \infty$, and $\lambda_P + \lambda_Q = b$.

Condition B: Let $L_n = \frac{Y_n - y_n}{\sqrt{n}}$, where $y_n = EY_n$. Then we suppose that

$$\sup_{n>0} E\{|L_n|^r\} < \infty, \text{ for some } r \in (2,3].$$

Condition C: Assume that

$$\sigma_n^2 := Var(L_n) \to \sigma^2 = \text{const} > 0 \text{ as } n \to \infty.$$

Remarks:

(a) Condition A holds when, for example,

$B_n \stackrel{d}{\sim} \text{Binomial}(n,b)$, $\quad I_{n,\beta} \stackrel{d}{\sim} \text{Binomial}(n-\beta,p)$, $\quad P_n \stackrel{d}{\sim} \text{Poisson}(npb)$,
$Q_n \stackrel{d}{\sim} \text{Poisson}(n(1-p)b)$.

(b) The equation

$$CR := \frac{b}{\lambda_p + \lambda_Q} = 1$$

in Condition A, part (iv), corresponds to what we shall call the *critical case* and will be treated in Section 4. The cases $CR > 1$ and $CR < 1$ will be considered in Section 6.

(c) One would expect conditions B and C to be checked for any particular choice of the distributions of $Y_1, B_n, R_n, , I_n, P_n$, and Q_n. Since this task is impossible analytically, one should in general rely on simulation and numerical analysis. Conditions B and C are typically imposed in the cases when an asymptotically normal or stable approximation for Y_n is to be obtained.

3.4.2 The Limit Theorem and Its Proof

Theorem 3.4.1 (Central Limit Theorem (CLT) for independent nonidentically distributed r.v.s.)

Under regularity conditions A, B, and C, the distribution of L_n is asymptotically normal with mean zero and variance σ^2.

Proof: Consider the normalized sequence

$$L_n = \frac{Y_n - y_n}{\sqrt{n}}, \tag{3.4.2}$$

where $y_n = EY_n$. Then, from

$$Y_n = \sqrt{n}L_n + y_n, \tag{3.4.3}$$

we have the following recursion for L_n :

$$\sqrt{n}L_n + y_n \stackrel{d}{=} R_{n,B_n} + \sqrt{I_{n,B_n} + P_n} \cdot L_{I_{n,B_n}+P_n} + EY_{I_{n,B_n}+P_n}$$
$$+ \sqrt{n - B_n - I_{n,B_n} + Q_n} \cdot \tilde{L}_{n-B_n-I_{n,B_n}+Q_n}$$
$$+ E\tilde{Y}_{n-B_n-I_{n,B_n}+Q_n}. \tag{3.4.4}$$

From this we obtain the recursion

$$L_n \stackrel{d}{=} \sqrt{\frac{I_{n,B_n} + P_n}{n}} \cdot L_{I_{n,B_n}+P_n}$$
$$+ \sqrt{\frac{n - B_n - I_{n,B_n} + Q_n}{n}} \cdot \tilde{L}_{n-B_n-I_{n,B_n}+Q_n}$$
$$+ C_n(B_n, R_{n,B_n}, I_{n,B_n}, P_n, Q_n), \tag{3.4.5}$$

where

$$C_n(B_n, R_{n,B_n}, I_{n,B_n}, P_n, Q_n)$$
$$= \frac{1}{\sqrt{n}} \left(R_{n,B_n} - y_n + EY_{I_{n,B_n}+P_n} + E\tilde{Y}_{n-B_n-I_{n,B_n}+Q_n} \right)$$
$$= \frac{1}{\sqrt{n}} \left(R_{n,B_n} - ER_{n,B_n} \right).$$

We will show later that

$$C_n(B_n, R_{n,B_n}, I_{n,B_n}, P_n, Q_n) \to 0 \tag{3.4.6}$$

in probability as $n \to \infty$. To make the idea of the proof clear, assume for the moment that (3.4.6) holds and, furthermore,

$$L_n \to L \text{ in distribution.}$$

Consider then the limit behavior of the size $(I_{n,B_n} + P_n)$ of the "new" group:

$$
\begin{aligned}
\lim_{n\to\infty} \frac{I_{n,B_n} + P_n}{n} &= \lim_{n\to\infty} \frac{I_{n,B_n}}{n} + \lim_{n\to\infty} \frac{P_n}{n} \\
&= \lim_{n\to\infty} \frac{I_{n,B_n}}{n - B_n} \frac{n - B_n}{n} + \lambda_P \\
&= p(1 - b) + \lambda_P \\
&=: p^*
\end{aligned}
\tag{3.4.7}
$$

a.s. and in the L_2-sense. Similarly, for the size of the "old" group we have

$$
\begin{aligned}
\lim_{n\to\infty} \frac{n - B_n - I_{n,B_n} + Q_n}{n} &= (1 - p)(1 - b) + \lambda_Q \\
&= (1 - p)(1 - b) + b - \lambda_P \\
&= 1 - p^*,
\end{aligned}
\tag{3.4.8}
$$

a.s. and in the L_2-sense. Then (3.4.5), (3.4.6), (3.4.7) and (3.4.8) yield

$$
L \stackrel{d}{=} \sqrt{p^*}\, L + \sqrt{1 - p^*}\, \tilde{L},
\tag{3.4.9}
$$

where $\tilde{L}$ is an independent copy of L. The limiting fixed point equation (3.4.9) immediately yields [see, e.g., Breiman (1968)] that the r.v. L is normally distributed with mean zero and variance $\sigma^2 \geq 0$.

We shall now attempt to prove (3.4.6). First we will show that if N_n are independent normal random variables with mean zero and variance $\sigma_n^2 := \mathrm{Var}(L_n) \geq 0$ (see Condition C), then L_n, N_n and N_n^*, defined as

$$
\begin{aligned}
N_n^* \stackrel{d}{=} &\sqrt{\frac{I_{n,B_n} + P_n}{n}}\, N_{I_{n,B_n} + P_n} \\
&+ \sqrt{\frac{n - B_n - I_{n,B_n} + Q_n}{n}}\, N_{n - B_n - I_{n,B_n} + Q_n} \\
&+ C_n(B_n, R_{n,B_n}, I_{n,B_n} P_n, Q_n),
\end{aligned}
\tag{3.4.10}
$$

are three merging sequences whose common limiting distribution is normal with mean zero and variance σ^2. Note that showing L_n and N_n to have the same limiting distribution proves the theorem.

To analyze the closeness between L_n, N_n, and N_n^*, we need a suitable metric in the space of distribution functions. A good candidate for such a metric is Zolotarev's metric $\zeta_{s,\delta}$.

Since our goal is to show that L_n, N_n, and N_n^* merge in distribution to a $N(0, \sigma^2)$ r.v., let us first evaluate the $\zeta_{s,\delta}$-distance between N_n^* and L_n^*, where L_n^* *represents the right-hand side of the recursive equation* (3.4.5) for

L_n. Then, as a second step, we shall bound $\zeta_{s,\delta}(N_n^*, L_n)$, and then finally show that

$$\limsup_{n\to\infty} \zeta_{s,\delta}(L_n, N_n) = 0. \tag{3.4.11}$$

Here, we begin by finding bounds for $\zeta_{s,\delta}(L_n^*, N_n^*)$ using the "ideality" of order r of the metric $\zeta_{s,\delta}$. By the definitions of L_n^* and N_n^* and by inequality (3.3.1),

$$\zeta_{s,\delta}(L_n^*, N_n^*) \leq \sum_{\beta,\rho,k,l,m} \Pr\{B_n = \beta, R_n = \rho, I_n = k, P_n = 1, Q_n = m\}$$

$$\times \left[\left(\frac{k+1}{n}\right)^{\frac{r}{2}} \zeta_{s,\delta}(L_{k+1}, N_{k+1}) \tag{3.4.12} \right.$$

$$\left. + \left(\frac{n-\beta-k+m}{n}\right)^{\frac{r}{2}} \zeta_{s,\delta}(L_{n-\beta-k+m}, N_{n-\beta-k+m}) \right].$$

Now set

$$S := \sup_{n>0} \zeta_{s,\delta}(L_n, N_n) \leq \infty.$$

Then the above estimates yield the following bound:

$$\zeta_{s,\delta}(N_n^*, L_n^*) \leq S \cdot E\left(\frac{I_{n,B_n} + P_n}{n}\right)^{\frac{r}{2}}$$

$$+ S \cdot E\left(\frac{n - B_n - I_{n,B_n} + Q_n}{n}\right)^{\frac{r}{2}}. \tag{3.4.13}$$

By the assumptions of the theorem (Condition A), and since $E|P_n|^{r/2} < \infty$,

$$\lim_{n\to\infty} E\left\{\frac{I_{n,B_n} + P_n}{n - B_n}\right\}^{\frac{r}{2}} = \lim_{n\to\infty} E\left\{\frac{I_{n,B_n}}{n - B_n} \cdot \frac{n - B_n}{n} + \frac{P_n}{n}\right\}^{\frac{r}{2}}$$

$$= \{p(1-b) + \lambda_P\}^{\frac{r}{2}}$$

$$= (p^*)^{\frac{r}{2}}. \tag{3.4.14}$$

Similarly,

$$\lim_{n\to\infty} E\left\{\frac{n - B_n - I_{n,B_n} + Q_n}{n}\right\}^{\frac{r}{2}} = (1 - p^*)^{\frac{r}{2}}. \tag{3.4.15}$$

Summarizing, the recursion (3.4.5), $L_n \overset{d}{=} L_n^*$, yields

$$\limsup \zeta_{s,\delta}(L_n, N_n^*) = \limsup \zeta_{s,\delta}(L_n^*, N_n^*)$$

$$\leq S\left[(p^*)^{\frac{r}{2}} + (1 - p^*)^{\frac{r}{2}}\right], \tag{3.4.16}$$

implying that $\zeta_{s,\delta}(L_n, N_n^*)$ is finite provided that $S < \infty$. To check that S is finite, we use the universal bounds for $\zeta_{s,\delta}$ by the r-th absolute moment:

$$S = \sup_{n>0} \zeta_{s,\delta}(L_n, N_n) \le c_r \sup_{n>0}(E|L_n|^r + E|N_n|^r) < \infty,$$

where c_r is an absolute constant depending on r only (see [47]), where the fact that $EL_n = EN_n = 0, EL_n^2 = EN_n^2 = \sigma_n^2$, and $2 < r \le 3$ is used.

Our next step is to compare N_n^* and N_n in terms of $\zeta_{s,\delta}$. Note that the finiteness of $\zeta_{s,\delta}(N_n^*, L_n)$ for $r > 2$ implies $EN_n^* = EL_n = 0$ and $\text{Var}N_n^* = \text{Var}L_n \equiv \sigma_n^2$ [see [62]]. Next, let N_0 be a standard normal r.v. independent of all r.v.s that we have already introduced. Then for N_n^* we have

$$N_n^* \overset{d}{=} N_0 \left(\frac{I_{n,B_n} + P_n}{n} \cdot \sigma_{I_{n,B_n}+P_n}^2 \right.$$

$$\left. + \frac{n - B_n - I_{n,B_n} + Q_n}{n} \cdot \sigma_{n-B_n-I_{n,B_n}+Q_n}^2 \right)^{\frac{1}{2}} \quad (3.4.17)$$

$$+ C_n(B_n, R_n, I_n, P_n, Q_n).$$

Since $\sigma_n \to \sigma$, $I_{n,B_n}/n \to p$, $P_n/n \to \lambda_P$, $B_n/n \to b$, $Q_n/n \to \lambda_Q$ a.s. and in the L_2-sense, then as $n \to \infty$ the expression in the brackets on the right-hand side of (3.4.17), *which we denote by* δ_n^2, converges to

$$[p(1-b) + \lambda_P]\sigma^2 + [(1-p)(1-b) + \lambda_Q]\sigma^2 = \sigma^2 \quad (3.4.18)$$

a.s. and in the L_2-sense. Now, since

$$N_n^* = \delta_n N_0 + C_n(B_n, R_n, I_n, P_n, Q_n)$$

has the same mean and variance as L_n, we get

$$\sigma_n^2 \equiv \text{Var}L_n \equiv \text{Var}N_n^* = E[\delta_n N_0 + C_n(B_n, R_n, I_n, P_n, Q_n)]^2$$
$$= E\delta_n^2 + EC_n^2(B_n, R_n, I_n, P_n, Q_n),$$

and hence $\sigma_n \to \sigma, \delta_n^2 \to \sigma^2$ yield

$$EC_n^2(B_n, R_n, I_n, P_n, Q_n) \to 0 \quad \text{as } n \to \infty.$$

The L_2-merging of the sequences N_n^* and $N_n \overset{d}{=} \delta_n N_0$ is therefore established. In particular, we have that

$$\zeta_{s,\delta}(N_n^*, N_n) \to 0 \quad \text{as } n \to \infty, \quad (3.4.19)$$

since $E|N_n^* - N_n|^2 \to 0$ implies (3.4.19); see [62].

Going back to the initial problem of showing that $\zeta_{s,\delta}(L_n, N_n) \to 0$, we shall make use of the triangle inequality:

$$\zeta_{s,\delta}(L_n, N_n) \le \zeta_{s,\delta}(L_n, N_n^*) + \zeta_{s,\delta}(N_n^*, N_n), \quad (3.4.20)$$

as well as the fact that $\zeta_{s,\delta}(N_n^*, N_n) \to 0$.

To check that the first term on the right-hand side of inequality (3.4.20) vanishes as $n \to \infty$, choose an arbitrary $\varepsilon > 0$ and fix a positive integer n_ε such that, for all $k \geq n_\varepsilon$,

$$S_k := \zeta_{s,\delta}(L_k, N_k) \leq \bar{S} + \varepsilon,$$

where

$$\bar{S} = \limsup_k \zeta_{s,\delta}(L_k, N_k).$$

Now using the ideality of order r of $\zeta_{s,\delta}$ we have, for $n > n_\varepsilon$ and

$$A := \{(\beta, \rho, k, l, m) : 0 \leq \beta \leq n, 0 \leq \rho \leq \beta, 0 \leq k \leq n - \beta, 0 \leq l, 0 \leq m\},$$
$$A_{n,\varepsilon} := \{(\beta, \rho, k, l, m) \in A : (k + l \leq n_\varepsilon - 1) \text{ or } (n - \beta - k + m \leq n_\varepsilon - 1)\},$$

the following bound for $\zeta_{s,\delta}(L_n, N_n^*)$:

$$
\begin{aligned}
\zeta_{s,\delta}(L_n, N_n^*) \leq &\sum_{A_{n,\varepsilon}} P\{(B_n, R_n, I_n, P_n, Q_n) = (\beta, \rho, k, l, m)\} \\
&\times \sup_{A_{n,\varepsilon}}\{S_{k+l} + S_{n-\beta-k+m}\} \\
&\times \left\{ E\left(\frac{k+l}{n}\right)^{\frac{r}{2}} + E\left(\frac{n-\beta-k+m}{n}\right)^{\frac{r}{2}} \right\} \\
&+ \sum_{A \backslash A_{n,\varepsilon}} P\{(B_n, R_n, I_n, P_n, Q_n) = (\beta, \rho, k, l, m)\} \\
&\times \left\{ E\left(\frac{k+l}{n}\right) + E\left(\frac{n-\beta-k+m}{n}\right)^{\frac{r}{2}} \right\} \times (\bar{S} + \varepsilon).
\end{aligned}
$$

Now recall that $S = \sup_n S_n < \infty$, and note that

$$
\sum_{A_{n,\varepsilon}} P\{(B_n, R_n, I_n, P_n, Q_n) = (\beta, \rho, k, l, m)\}
$$
$$
\times \left\{ E\left(\frac{k+l}{n}\right)^{\frac{r}{2}} + E\left(\frac{n-\beta-k+m}{n}\right)^{\frac{r}{2}} \right\} \to 0
$$

as $n \to \infty$. Then

$$\limsup_n \zeta_{s,\delta}(L_n, N_n^*) \leq 2S \cdot 0 + \left[(p^*)^{\frac{r}{2}} + (1 - p^*)^{\frac{r}{2}}\right](\bar{S} + \varepsilon),$$

where p^* is defined in (3.4.7), and the triangle inequality yields

$$
\begin{aligned}
\bar{S} &\leq \limsup_n \zeta_{s,\delta}(L_n, N_n^*) + \limsup_n \zeta_{s,\delta}(N_n^*, N_n) \\
&\leq \left[(p^*)^{\frac{r}{2}} + (1 - p)^{\frac{r}{2}}\right](\bar{S} + \varepsilon).
\end{aligned}
$$

Since for $r/2 > 1$, $\Delta_p := (p^*)^{r/2} + (1 - p^*)^{r/2} < 1$, and $\varepsilon > 0$ was chosen arbitrarily, we have the contraction

$$0 \leq (1 - \Delta_p)\bar{S} < \varepsilon,$$

giving $\bar{S} := \limsup_n \zeta_{s,\delta}(L_n, N_n) = 0$. Finally, we invoke the fact (3.3.7). Since $\mathbf{L}$ metrizes the weak convergence, we have that L_n inherits the limiting property of N_n, i.e., L_n is asymptotically normal with mean zero and variance σ^2. The proof of the theorem is now complete.

3.5 The Spread of AIDS Among Interactive Transmission Groups

One of the most important questions in the spread of an epidemic is the prediction of its future behavior at time t, given its initial state at time $t = 0$. There is a rich mathematical literature on this subject (**cf.** e.g., Bailey [3] and Anderson and May [2]), where both deterministic and stochastic models are considered.

3.5.1 Formulation of the Model

The four main transmission groups in the AIDS epidemic are as follows.

Transmission group 1: Homosexual/bisexual men, infected by homosexual contact with HIV infectives.

Transmission group 2: Blood transfusion recipients, infected by donors with HIV.

Transmission group 3: Intravenous drug users (IVDUs) sharing HIV - infected needles.

Transmission group 4: Persons infected by heterosexual contact with HIV infectives.

In our model for the spread of AIDS, we consider N communities each having a nonempty transmission group (at least one). Let $X^N(i, k, t)$ represent the size of the HIV+ population at time t in the ith community ($1 \leq i \leq N$) and intransmission group k ($1 \leq k \leq 4$) (briefly the (i, k) transmission group). We shall omit the superscript N when this does not cause any confusion.

The change in the size of the HIV+ in the (i, k) transmission group can be due to some of the following.

- Individuals newly infected by the HIV+ members in the (j, l) transmission group ($1 \leq j \leq N, 1 \leq l \leq 4$).

- Immigrants from the (i, k) transmission group.

- Removals from the (i, k) transmission group due to death, immigration, etc.

Hence the $4N$ SDE governing the system are

$$dX^N(i, k, t) = dW(i, k, t)$$
$$+ \left\{ \frac{1}{N} \sum_{j=1}^{N} \left[\sum_{l=1}^{4} b_{k,l}(X^N(i, k, t), X^N(j, l, t)) \right]^p \right\}^{\frac{1}{p}} dt$$
$$X^N(i, k, 0) = X(i, k), \quad i = 1, ..., N, \ k = 1, ..., 4. \tag{3.5.1}$$

Here

- $b_{k,l}(., .)$ stands for the net rate of growth of the kth transmission group due to the interaction with the lth transmission group, $b_{k,l} \geq 0$.

- $dW(i, k, t)$ is the net rate of the new infectives and removals into the (i, k) group.

- $(W(i, k, t), X(i, k))$ are independent and identically distributed for all i.

We shall establish that each of the N interacting diffusions $(X^N(i, k, t); 1 \leq k \leq 4, t \in [0, T])$ in (3.5.1) has, as $N \to \infty$, a natural limit $(\overline{X}_{i,k}(t); 1 \leq k \leq 4, 0 \leq t \leq T$, where $(\overline{X}_{i,k}, 1 \leq k \leq 4)$ are N independent copies of the solution of the following Liouville-type nonlinear stochastic differential equation:

$$\begin{cases} dX_k(t) = dW_k(t) + \{ \int [\sum_{l=1}^{4} b(X_k(t), y_l)]^p u_t(dy) \}^{\frac{1}{p}} dt, \\ X_k(0) = X_k, \end{cases} \tag{3.5.2}$$

with $W_k \overset{d}{=} W_{1,k}$, $X_k \overset{d}{=} X_{1,k}$ and $u_t = P^{X(t)}$ is the measure generated by $X(t) := \{X_l(t), 1 \leq l \leq 4\}$.

3.5.2 Analysis of the Model

The first goal will be to prove that (3.5.2) has indeed a unique weak and strong solution. We will use the p-average metrics $L_{p,T}^*$ and $l_{p,T}^*$.

$$L_{p,T}^*(X, Y) := \mathcal{L}_p(X, Y) = \left\{ E \left(\sum_{k=1}^{4} \sup_{t \leq T} |X_k(t) - Y_k(t)| \right)^p \right\}^{\frac{1}{p}} \tag{3.5.1}$$

for $X, Y \in \mathcal{X}(C^4[0, T])$.

$$l^*_{p,T}(m_1, m_2) := \inf \left\{ L^*_{p,T}(X, Y) : X \overset{\mathrm{d}}{=} m_1, Y \overset{\mathrm{d}}{=} m_2 \right\} \tag{3.5.2}$$

for $m_1, m_2 \in M(C^4_T)$.

Denote by $M(C^4_T)$ the set of all probability measures generated by the C^4_T-valued random processes.

Let (B_t) be a R^4-valued process on C^4_T with finite pth absolute moment i.e.,

$$E\left(\sum_{k=1}^{4} \sup_{t \leq T} |B^k_t| \right)^p < \infty. \tag{3.5.3}$$

Denote by m_0 the law of (B_t) and define for $m_0 \in M(C^4_T)$

$$M_p(C^4_T, m_0) := \{m_1 \in M(C^4_T), l^*_{p,T}(m_0, m_1) < \infty\}.$$

For $m \in M_p(C^4_T, m_0)$ consider the Liouville-type equation

$$X^k_t = B^k_t + \int_0^t \left\{ \int \left[\sum_{l=1}^{4} b_{k,l}(X^k_s, y^l_s) \right]^p dm(y) \right\}^{\frac{1}{p}} ds. \tag{3.5.4}$$

Let $b_{k,l} \geq 0$ be a Lipschitz function in x:

$$\begin{aligned} &|b_{k,l}(x_1, y) - b_{k,l}(x_2, y)| \leq c|x_1 - x_2|, \\ &k, l = 1, ..., 4. \end{aligned} \tag{3.5.5}$$

Lemma 3.5.1 *Assume (3.5.5) and let*

$$\int_0^T \left\{ \int \left[\sum_{l=1}^{4} b_{k,l}(0, y^l_s) \right]^p dm(y) \right\}^{\frac{1}{p}} ds < \infty, \ k = 1, ...4. \tag{3.5.6}$$

Then
a) Equation (3.5.4) has a unique strong solution X for all $t \leq T$;
b) $\Phi(m) \in M_p(C^4_t, m_0)$, given $\Phi(m)$, is the law of X:

$$\Phi : M_p(C^4_T, m_0) \to M_p(C^4_T, m_0).$$

Proof: Let X be a process in C^4_T with distribution in $M_p(C^4_T, m_0)$; we denote this by $X \in \mathcal{X}_p(C^4_T, m_0)$, and define

$$(SX)^k_t := B^k_t + \int_0^t \left\{ \int \left[\sum_{l=1}^{4} b_{k,l}(X^k_s, y^l_s) \right]^p dm(y) \right\}^{\frac{1}{p}} ds.$$

Then, for $Y \in \mathcal{X}_p(C_T^4, m_0)$,

$$|(SX)_t^k - (SY)_t^k| \leq \int_0^t ds \left\{ \int \left| \left[\sum_{l=1}^4 b_{k,l}(X_s^k, y_s^l) \right] \right. \right.$$
$$\left. \left. - \left[\sum_{l=1}^4 b_{k,l}(Y_s^k, y_s^l) \right] \right|^p dm(y) \right\}^{\frac{1}{p}}$$
$$\leq \int_0^t ds \left\{ \int \left(4c|X_s^k - Y_s^k| \right)^p dm(y) \right\}^{\frac{1}{p}}$$
$$= 4c \int_0^t |X_s^k - Y_s^k| ds.$$

This implies

$$\sup_{s \leq t} |(SX)_s^k - (SY)_s^k| \leq 4c \int_0^t |X_s^k - Y_s^k| ds \leq 4c \int_0^t \sup_{u \leq s} |X_u^k - Y_u^k| ds.$$

Let us now estimate the distance between SX and SY:

$$L_{p,t}^*(SX, SY) \leq \left\{ E \left(\sum_{k=1}^4 4c \int_0^t \sup_{u \leq s} |X_u^k - Y_u^k| ds \right)^p \right\}^{\frac{1}{p}}$$
$$\leq 4c \int_0^t \left\{ E \left(\sum_{k=1}^4 \sup_{u \leq s} |X_u^k - Y_u^k| \right)^p \right\}^{\frac{1}{p}} ds = 4c \int_0^t L_{p,s}^*(X, Y) ds.$$

Define recursively $X^n := SX^{n-1}$, where $X^0 := B$. Then, by induction, for all $t \leq T$,

$$L_{p,t}^*(X^n, X^{n-1}) \leq \frac{(4ct)^n}{n!} L_{p,t}^*(X^1, X^0).$$

That means the operator S has a contraction property w.r.t. the metric $L_{p,T}^*$.

Also from (3.5.3) and (3.5.6) $L_{p,T}^*(X^0, X^1)$ is finite.
Consequently

$$\sum_{n=1}^\infty L_{p,T}^*(X^n, X^{n-1}) \leq e^{4cT} L_{p,T}^*(X^1, X^0) < \infty,$$

so that

$$\sum_{n=1}^\infty \left\{ E \left(\sum_{k=1}^4 \sup_{s \leq T} |X_s^{n,k} - Y_s^{n,k}| \right)^p \right\}^{\frac{1}{p}}$$

is finite.

Thus X^n converges to some process X a.s., and uniformly on bounded intervals. Hence obviously X is a.s. continuous and has finite pth moment, and X is the fixed point of S.

Also, $L_{p,T}^*(X, B) < \infty$, so $\Phi(m) = P^X \in M_p(C_T^4, m_0)$.

Lemma 3.5.2 *Under the condition*

$$|b_{k,l}(x_1, y_1) - b_{k,l}(x_2, y_2)| \le c\{|x_1 - x_2| + |y_1 - y_2|\} \qquad (3.5.7)$$

and the assumptions of Lemma 1, we have that for $t \le T$ and $m_1, m_2 \in M_p(C_T^4, m_0)$

$$l_{p,t}^*(\Phi(m_1), \Phi(m_2)) \le 4ce^{4ct} \int_0^t l_{p,u}^*(m_1, m_2) du. \qquad (3.5.8)$$

holds.

Proof: Let $X^{i,k}$ be the strong solution of the SDE.

$$X_t^{i,k} = B_t^k + \int_0^t \left\{ \int \left(\sum_{l=1}^4 b_{k,l}(X_s^{i,k}, y_s^l) \right)^p dm(y) \right\}^{\frac{1}{p}} ds,$$

and let $m \in M(m_1, m_2)$ be the class of probability measures on $C_T^4 \times C_T^4$ with marginals m_1 and m_2. Then

$$\sup_{s \le t} |X_s^{1,k} - X_s^{2,k}| \le 4c \int_0^t |X_s^{1,k} - X_s^{2,k}| ds$$

$$+ c \int_0^t \left\{ \int \left(\sum_{l=1}^4 |y_s^{1,l} - y_s^{2,l}| \right)^p dm(y^1, y^2) \right\}^{\frac{1}{p}} ds.$$

Minimizing the right-hand side with respect to m, we obtain

$$\sup_{s \le t} |X_s^{1,k} - X_s^{2,k}| \le 4c \int_0^t \sup_{u \le s} |X_u^{1,k} - X_u^{2,k}| ds + c \int_0^t l_{p,s}^*(m_1, m_2) ds.$$

That means

$$L_{p,t}^*(X^1, X^2) \le 4c \int_0^t L_{p,s}^*(X^1, X^2) ds + c \int_0^t l_{p,s}^*(m_1, m_2) ds.$$

But $l_{p,t}^*$ is a minimal metric w.r.t. $L_{p,t}^*$. Hence

$$l_{p,t}^*(\Phi(m_1), \Phi(m_2)) \le 4c \int_0^t l_{p,s}^*(\Phi(m_1), \Phi(m_2)) ds + c \int_0^t l_{p,s}^*(m_1, m_2) ds.$$

Consequently, by the Gronwall Lemma we obtain

$$l_{p,t}^*(\Phi(m_1), \Phi(m_2)) \le 4ce^{4ct} \int_0^t l_{p,s}^*(m_1, m_2) ds.$$

Theorem 3.5.1 *Under condition (3.5.7) and*

$$\int_0^T \left\{ \int \left(\sum_{l=1}^4 b_{k,l}(0, y_s^l) \right)^p dm_0(y) \right\}^{\frac{1}{p}} ds < \infty,$$

the system

$$X_t^k = B_t^k + \int_0^t \left\{ \int \left(\sum_{l=1}^4 b_{k,l}(X_s^k, y_s^l) \right)^p u_t(dy) \right\}^{\frac{1}{p}} ds, \quad k = 1, ..., 4, \quad (3.5.9)$$

has a unique weak and strong solution in $\mathcal{X}(C_T^4, m_0)$*, where* m_0 *is the distribution of* B_t*.*

Proof: From Lemma 2, writing $c_T = 4ce^{4cT}$, we obtain

$$l_{p,t}^*(\Phi^{k+1}(m_0), \Phi^k(m_0)) \le c_T^k \frac{T^k}{k!} l_{p,T}^*(\Phi(m_0), m_0) < \infty$$

Consequently, $(\Phi^k(m_0))$ is a Cauchy sequence in $(C_T^4, l_{p,T}^*)$ and converges to a fixed point of Φ, say $m \in M_p(C_T^4, m_0)$. Then the strong solution of (3.5.5) is the unique strong solution of (3.5.9).

Next we investigate the system of interacting equations in (3.5.1).

Theorem 3.5.2 *Let the* $b_{k,l}$ *satisfy the Lipschitz condition (3.5.8), and suppose that*

$$\int \left[\sum_{l=1}^4 b_{k,l}(\overline{X}_s^1, y_s^l) \right]^{2p} u_s(dy) < \infty \text{ a.s., } k = 1, ..., 4.$$

Then

$$\sup_N \sqrt{N} L_{p,T}^*(X^{i,N}, \overline{X}^{i,N}) < \infty, \quad (3.5.10)$$

where $\overline{X}^{i,N} = (\overline{X}^{i,k,N}, k = 1, ..., 4)$ *are independent copies of the solution of the nonlinear Equation (3.5.2).*

Proof: For notational convenience, we shall use $X_t^{i,k}$ instead of $X^N(i, k, t)$. Since the $b_{k,l}$ satisfy the Lipschitz condition, then

$$|X_t^{i,k} - \overline{X}_t^{i,k}| \le \int_0^t 4c |X_s^{i,k} - \overline{X}_s^{i,k}| ds$$

$$+ \frac{c}{N^{\frac{1}{p}}} \int_0^t \left\{ \sum_{j=1}^N \left[\sum_{l=1}^4 |X_s^{j,l} - \overline{X}_s^{j,l}| \right]^p \right\}^{\frac{1}{p}} ds$$

$$+ \int_0^t \left| \left\{ \frac{1}{N} \sum_{j=1}^N \left[\sum_{l=1}^4 b_{l,k}(\overline{X}_s^{i,k}, \overline{X}_s^{j,l}) \right]^p \right\}^{\frac{1}{p}} \right|$$

$$-\left\{\int\Big[\sum_{l=1}^{4}b_{l,k}(\overline{X}_s^{i,k},y_s^l)\Big]^p u_s(dy)\right\}^{\frac{1}{p}}\Bigg|ds.$$

The above and the Hölder inequality imply

$$L_{p,T}^*(X^i,\overline{X}^i)\leq 4c\int_0^T L_{p,s}^*(X^i,\overline{X}^i)ds$$

$$+4\frac{c}{N^{\frac{1}{p}}}\int_0^T\left\{\sum_{j=1}^{N}\Big[L_{p,s}^*(X^j,\overline{X}^j)\Big]^p\right\}^{\frac{1}{p}}ds$$

$$+\sum_{k=1}^{4}\int_0^T\left\{E\Big|\frac{1}{N}\sum_{j=1}^{N}\Big[\sum_{l=1}^{4}b_{l,k}(\overline{X}_s^{i,k},\overline{X}_s^{j,l})\Big]^p\right\}^{\frac{1}{p}}$$

$$-\left\{\int\Big[\sum_{l=1}^{4}b_{l,k}(\overline{X}_s^{i,k},y_s^l)\Big]^p u_s(dy)\right\}^{\frac{1}{p}}\Big|^p\right\}^{\frac{1}{p}}ds.$$

Summing up over i and using the symmetry, we find

$$NL_{p,T}^*(X^1,\overline{X}^1)=\sum_{i=1}^{N}L_{p,T}^*(X^i,\overline{X}^i).$$

Thus

$$L_{p,T}^*(X^1,\overline{X}^1)\leq 8c\int_0^T L_{p,s}^*(X^1,\overline{X}^1)ds$$

$$+\sum_{k=1}^{4}\int_0^T\left\{E\Big|\Big\{\frac{1}{N}\sum_{j=1}^{N}\Big[\sum_{l=1}^{4}b_{l,k}(\overline{X}_s^{i,k},\overline{X}_s^{j,l})\Big]^p\Big\}^{\frac{1}{p}}\right.$$

$$-\left\{\int\Big[\sum_{l=1}^{4}b_{l,k}(\overline{X}_s^{i,k},y_s^l)\Big]^p u_s(dy)\Big\}\frac{1}{p}\Big|^p\right\}^{\frac{1}{p}}ds.$$

By the Gronwall Lemma

$$L_{p,T}^*(X^i,\overline{X}^i)\leq 8ce^{8cT}\int_0^T\sum_{k=1}^{4}\left\{E\Big|\Big\{\frac{1}{N}\sum_{j=1}^{N}\Big[\sum_{l=1}^{4}b_{l,k}(\overline{X}_s^{i,k},\overline{X}_s^{j,l})\Big]^p\right\}^{\frac{1}{p}}$$

$$-\left\{\int\Big[\sum_{l=1}^{4}b_{l,k}(\overline{X}_s^{i,k},y_s^l)\Big]^p u_s(dy)\right\}^{\frac{1}{p}}\Big|^p\right\}^{\frac{1}{p}}ds.$$

From our assumption about the $b_{k,l}$ and the Central Limit Theorem, we conclude that the result (3.5.10) holds.

The previous result shows us that when the initial distribution of the HIV+ individuals in the transmission group is independent of all the others, the interaction in (3.5.1) soon destroys that independence. But, for a given n, when the number of communities N becomes large, Theorem 3.5.2 implies that the distributions become approximately independent again, so that independence still spreads. This is a type of a propagation of chaos result.

As result of the complexity of (3.5.2), it is difficult to characterize the limiting distribution even in simple cases, cf. [10].

Bibliography

[1] G. Anastassiou and S.T. Rachev (1992), Moment Problems and Stability of Queueing Models, *Comp. Math. Appl.*, **24**(8/9), 229-246.

[2] R.M. Anderson and R.M. May (1992), *Infectious Diseases of Humans*, Oxford University Press, Oxford.

[3] N.T.J. Bailey (1975), *The Mathematical Theory of Infectious Diseases and Its Applications*, Griffin, London.

[4] N.T.J. Bailey (1994), Core-Group Dynamics and Public Health Action, *Modeling the AIDS Epidemic*, Raven Press, New York.

[5] J. Beirlant and S.T. Rachev (1987), The Problem of Stability in Insurance Mathematics, *Insur., Math.: Math. & Econ.*, **6**, 179-188.

[6] M. Balinski, B. Athanasopoulos, and S.T. Rachev (1993), Some Developments on the Theory of Rounding Proportions, *Bull. ISI, 49th Session, Firenze, Book 1*, 71-72.

[7] M. Balinski and S.T. Rachev (1993), Rounding Proportions: Rules of Rounding, *Numer. Funct. Anal. Optim.*, **14**, 475-501.

[8] A.D. Barbour (1974), On a Functional Central Limit Theorem for Markov Population Processes, *Adv. Appl. Probab.* **6**, 21-39.

[9] T. Chen (1980) Inequalities for Distributions with Given Marginals, *Ann. Probab.* **8**, 814-827.

[10] T.S. Chiang, G. Kallianpur and P. Sundar (1991) Propagation of Chaos and the McKean-Vlasov Equation in Duals of Nuclear Spaces, *Appl. Math. Optim.*, **24**, 55-83.

[11] J.A. Cuesta-Albertos, C. Matran, S.T. Rachev, and L. Rüschendorf (1996), Mass Transportation Problems in Probability Theory, *Math. Sciientist* **21**, 34-72.

[12] P. Diaconis and D. Freedman (1987), A Dozen de Finnety-Style Results in Search of a Theory, *Ann. Inst. Henri Poincare*, **23**, 397-423.

[13] B.N. Dimitrov, S.T. Rachev, and A.Yu. Yakovlev (1985), Maximum Likelihood Estimation of the Mortality Rate Function, *Biomed. J.*, **27**, 317-326.

[14] R.L. Dobrushin (1970), Prescribing a System of Random Variables by Conditional Distributions, *Theory Probab. Appl.* **15**, 458-486.

[15] R.M. Dudley (1968), Distances of Probability Measures and Random Variables, *Ann. Math. Stat.*, **39**, 1563-1572.

[16] N. Dunford and J. Schwartz (1988), *Linear Operators*, Vol. 1, John Wiley & Sons, New York.

[17] P. Feldman, S.T. Rachev, and L. Lüschendorf (1994), Limit Theorems for Recursive Algorithms, *J. Comp. Appl. Maths.*, **56**, 169-182.

[18] P. Feldman, S.T. Rachev, and L. Lüschendorf (1995), Limit Distribution of the Collision Resolution Interval, *Stat. Neder.* (to appear).

[19] R. Fortét and E. Mourier (1953), Convergence de la répartition empirique vers la répétition theoretique, *Ann. Sci. Ecole Norm.*, **70**, 267-285.

[20] J. Gani, G. Haynatzki, S.T. Rachev, and S. Yakowitz (1995) Steady-State Model for the Spread of HIV Among Drug Users, Techn. Report, Dept. of Stats & Appl. Probab., University of California, Santa Barbara.

[21] J. Gani, V. Haynatzka, and S.T. Rachev (1996), The Spread of AIDS Among Interactive Transmission Groups, Techn. Report, Dept. of Stats & Appl. Probab., University of California, Santa Barbara.

[22] M. Gelbrich (1990), On a Formula for the L^2-Wasserstein Metric Between Measures on Euclidean and Hilbert Spaces, *Math. Nachr.*, **147**, 185-203.

[23] M. Gelbrich and S.T. Rachev (1995), Discretization for Stochastic Differential Equations, L^2-Wasserstein Metrics, and Econometric Models, Proceedings: *Distributions with Given Marginals*, IMS (to appear).

[24] L. de Haan (1984), A Spectral Representation for Max-Stable Processes, *Ann. Probab.* **12**, 1194-1204.

[25] L. de Haan and S.T. Rachev (1989), Estimates of the Rate of Convergence for Max-Stable Processes, *Ann. Probab.* **17**(2), 651-677.

[26] L.G. Hanin, S.T. Rachev, and A.Yu. Yakovlev (1993), On Optimal
     Control of Cancer Radiotherapy for Nonhomogeneous Cell Populations,
     Adv. Appl. Probab. **25**, 1-23.

[27] L.G. Hanin and S.T. Rachev (1994), Mass-Transshipment Problems
     and Ideal Metrics, *J. Comp. Appl. Math.*, **56**, 183-196.

[28] L.G. Hanin and S.T. Rachev (1995), An Extension of the Kantorovich-
     Rubinstein Mass-Transshipment Problem, *Numer. Funct. Anal. Optim.*
     16, 701-735.

[29] G.R. Haynatzki (1995), *Stochastic Methods for Birth, Death and Epi-
     demic Processes*, Ph. D. dissertation, Dept. of Stats & Appl. Prob.,
     University of California, Santa Barbara.

[30] W. Hoeffding (1950), Masstabvariate Korrelationstheorie, *Sehr. Math.
     Inst. Univ. Berlin*, **5**, 181-233.

[31] Zv. Ignatov and S.T. Rachev (1986), Minimality of Ideal Probabilistic
     Metrics, *J. Soviet Math.*, **32**(6), 595-608.

[32] N.O. Kadyrova, S.T. Rachev, and A.Yu. Yakovlev (1989), Maximum
     Likelihood Estimation of the Bimodal Failure Rate of Censored and
     Tied Observations, *Statistics*, **20**, 135-140.

[33] V.V. Kalashnikov and S.T. Rachev (1986), Characterization of Queue-
     ing Models and Their Stability, **In:** *Probability Theory and Mathemat-
     ical Statistics* (Eds. Prokhorov *et al.*), Vol. **2**, UNU, Science Press, New
     York, 37-53.

[34] V.V. Kalashnikov and S.T. Rachev (1990), *Mathematical Methods for
     Construction of Queueing Models*, Wadsworth & Brooks/Cole, Pacific
     Grove, California.

[35] L.V. Kantorovich (1942) On the Transfer of Masses, *Dokl. Akad. Nauk
     SSSR*, **37**, 7-8.

[36] L.V. Kantorovich and G.P. Akilov (1984), *Functional Analysis*, Nauka,
     Moscow (in Russian).

[37] H.G. Kellerer (1984) Duality Theorems for Marginal Problems, *Z.
     Wahrsch. Verw. Geb.*, **67**, 399-432.

[38] L.B. Klebanov, S.T. Rachev, and A.Yu. Yakovlev (1993), A Stochas-
     tic Model of Radiation Carcinogenesis: Latent Time Distibutions and
     Their Properties, *Math. Biosci.*, **113**, 51-75.

[39] L.B. Klebanov, S.T. Rachev, and A.Yu. Yakovlev (1993) On the Parametric Estimation of Survival Functions, *Stat. Decisions*, Suppl. Issue 3, 83-102.

[40] L.B. Klebanov and S.T. Rachev (1994), Proximity of Probability Measures with Common Marginals in a Finite Number of Directions, Proceedings: *Distributions with Given Marginals*, IMS (to appear).

[41] L.B. Klebanov and S.T. Rachev (1995), The Method of Moments in Computer Tomography, *Math. Sci.*, **20**, 1-14.

[42] M. Knott and C.S. Smith (1984), On the Optimal Mapping of Distributions, *J. Optim. Theory Appl.*, **43**, 39-49.

[43] T.G. Kurtz (1976), Limit Theorems and Diffusion Approximations for Density Dependent Markov Chains, *Math. Programm. St.*, **5**, 67-78.

[44] V.L. Levin and S.T. Rachev (1990), General Monge-Kantorovich Problem and Its Applications in Measure Theory and Mathematical Economics, **In:** *Functional Analysis, Optimization and Mathematical Economics* (A collection of papers dedicated to the memory of L.V. Kantorovich), (Ed. L.J. Leifman), Oxford University Press, New York.

[45] V.L. Levin (1990), New Duality Theorems for Marginal Problems with Some Application in Statistics, *Lecture Notes in Mathematics*, Springer-Verlag, New York, **1412**, 137-171.

[46] M. Maejima (1983), On a Class of Self-Similar Processes with Stationary Increments, *Z. Wahrsch.*, **62**, 235-245.

[47] M. Maejima and S.T. Rachev (1987), An Ideal Metric and the Rate of Convergence to a Self-Similar Process, *Ann. Probab.*, **14**, 513-525.

[48] M. Maejima and S.T. Rachev (1996), Rates of Convergence in the Operator-Stable Limit Theorem, *J. Theor. Probab.*, **9**, 37-85.

[49] S. Mittnik and S.T. Rachev (1989), Stable Distributions for Asset Returns, *Appl. Math. Lett.*, **2/3**, 301-304.

[50] S. Mittnik and S.T. Rachev (1993), Modeling Asset Returns with Alternative Stable Laws, *Econometric Rev.*, **12**(3), 261-330.

[51] S. Mittnik and S.T. Rachev (1995), *Modeling Financial Assets with Alternative Stable Models*, Wiley Series in Financial Economics and Quantitative Analysis, John Wiley & Sons, New York.

[52] I. Olkin and S.T. Rachev (1990), Marginal Problems with Additional Constraints, Tech. Report 270, Dept. of Stats, Stanford Universit, Stanford, CA.

[53] I. Olkin and S.T. Rachev (1993), Maximum Submatrix Traces for Positive Definite Matrices, *SIAM J. Matrix Anal. Appl.*, **14**, 390-397.

[54] E.Omey and S.T. Rachev (1988), On the Rate of Convergence in Extreme Value Theory, *Theory Probab. Appl.*, **33**, 560-565.

[55] Yu.V. Prokhorov (1956), Convergence of Random Processes and Limit Theorems in Probability Theory, *Theory Probab. Appl.*, **1**, 157-214.

[56] S.T. Rachev (1978), Hausdorff Metric Structures of the Space of Probability Measures, *Zap. Nauchn. Sem. LOMI* **87**, 87-103. (In Russian). (Engl. transl. (1981) *J. Soviet. Math.*, **17**, 2218-2232.)

[57] S.T. Rachev (1982), Minimal Metrics in the Minimal Variables Space, *Publ. Inst. Stat. Univ. Paris.*, **XXVII**, 27-47.

[58] S.T. Rachev (1982), Minimal Metrics in the Random Variables Space, *Probability and Statistical Inference*, Proceedings of the 2nd Pannonian Symposium (Eds. W. Grossmann, et al.), D. Reidel, Dordrecht, 319-327.

[59] S.T. Rachev (1984), On a Class of Minimal Functions on a Space of Probability Measures, *Theory Probab. Appl.*, **29**, 41-49.

[60] S.T. Rachev (1984), The Monge-Kantorovich Mass-Transfer Problem and Its Stochastic Applications, *Theory Probab. Appl.*, **29**, 647-676.

[61] S.T. Rachev (1989), The Problem of Stability in Queuing Theory, *Queuing Sys. Theory Appl.*, **4**, 287-318.

[62] S.T. Rachev (1991), *Probability Metrics and Stability of Stochastic Models*, John Wiley & Sons, New York.

[63] S.T. Rachev (1991), Mass-Transshipment Problems and Ideal Metrics, *Numer. Funct. Anal. Optim.*, **12**, 563-573.

[64] S.T. Rachev (1992), Moment Problems and Stability of Queueing Models, *Comp. Math. Appl.*, **24**(8/9), 229-246.

[65] S.T. Rachev (1993), Rate of Convergence of Maxima of Random Arrays with Applications to Stock Returns, *Stat. & Decisions*, **11**, 279-288.

[66] S.T. Rachev, B. Dimitrov, and Z. Khalil (1991), A Probabilistic Approach to Optimal Quality Usage, *Numer. Funct. Anal. Optim.*, **12**(5-6), 563-573.

[67] S.T. Rachev and L. Rüschendorf (1991), Approximate Independence of Distributions of Spheres and Their Stability, *Ann. Probab.*, **19**, 1311-1337.

[68] S.T. Rachev and L. Rüschendorf (1991), Recent Results in the Theory of Probability Metrics, *Stat. Decisions*, **9**, 327-373.

[69] S.T. Rachev and L. Rüschendorf (1992), Rate of Convergence for Sums and Maxima and Doubly Ideal Metrics, *Theory Probab. Appl.*, **37**(2), 276-289.

[70] S.T. Rachev and L. Rüschendorf (1993), On the Rate of Convergence in the CLT with Respect to the Kantorovich Metric, Proceedings: *9th Confer. on Probability in Banach Spaces*, (Eds. J. Kuelbs, M. Marcus and J. Hoffman-Jorgensen).

[71] S.T. Rachev and L. Rüschendorf (1994), Propagation of Chaos and Contraction of Stochastic Mappings, *Sib. Adv. Math.*, **4**, 114-150.

[72] S.T. Rachev and L. Rüschendorf (1994), Solution of some transportation problems with relaxed or additional constraints, *SIAM J. Control Optim.*, **32**, 673-689.

[73] S.T. Rachev and L. Rüschendorf (1995), Probability Metrics and Recursive Algorithms, *Adv. Appl. Probab.*, **27**, 770-799.

[74] S.T. Rachev and L. Rüschendorf (1996), *Transportation Problems in Probability Theory*, Springer, New York.

[75] S.T. Rachev and G. Samorodnitsky (1995), Limit Laws for a Stochastic Process and Random Recursion Arising in Probabilistic Modeling, *J. Appl. Probab.*, **27**, 185-202.

[76] S.T. Rachev and A. Schief (1992), On L_p Minimal Metrics, *Probab. Math. Stat.*, **13**(2), 311-320.

[77] S.T. Rachev and A. SenGupta (1993), Laplace-Weibull Mixtures for Modeling Price Changes, *Manage. Sci.*, 1029-1038.

[78] S.T. Rachev and R.M. Shortt (1990), Duality Theorems for Kantorovich-Rubinstein and Wasserstein Functionals, *Diss. Math.*, **299**, 647-676.

[79] S.T. Rachev and P. Todorovich (1990), On the Rate of Convergence of Some Functionals of a Stochastic Process, *J. Appl. Probab.*, **28**, 805-814.

[80] S.T. Rachev, C.-F. Wu and A.Yu. Yakovlev (1995), A Bivariate Limiting Distribution of Tumor Latency Time, *Math. Biosci.*, **127**, 127-147.

[81] S.T. Rachev and A.Yu. Yakovlev (1988), Theoretical Bounds for the Tumor Treatment Efficiency, *Syst. Anal. Model Simul.*, **5**(1), 37-42 (in Russian).

[82] S.T. Rachev and A.Yu. Yakovlev (1988), Bounds for Crude Survival Probabilities Within Competing Risks Framework and Statistical Applications, *Stat. Probab. Lett.*, 389-394.

[83] S.T. Rachev and J.E. Yukich (1989), Rates for the CLT via New Ideal Metrics, *Ann. Probab.*, $17(2)$, 775-788.

[84] S.T. Rachev and J.E. Yukich (1989), Smoothing Metrics for Measures on Groups, *Ann. Inst. Henri Poincare*, **25**, 429-441.

[85] S.T. Rachev and J.E. Yukich (1991), Rates of Convergence of α-stable Random Motions, *J. Theor. Probab.*, **4**, 333-352.

[86] D. Ramachandran and L. Rüschendorf (1995), A general Duality Theorem for Marginal Problems, *Probab. Theory Relat. Fields*, **101**, 311-319.

[87] L. Rüschendorf (1981), Sharpness of Fréchet Bounds, *Z. Wahrsch. Werm. Geb.*, **57**, 293-302.

[88] L. Rüschendorf (1983), Solution of a Statistical Optimization Problem, *Metrika*, **30**, 55-62.

[89] L. Rüschendorf (1991), Bounds for Distributions with Multivariate Marginals, Proceedings: *Stochastic Order and Decision Under Risk* (Eds., K. Mosler and M. Scarsini), IMS Lecture Notes **19**, 285-310.

[90] L. Rüschendorf (1991), Fréchet Bounds and Their Applications, **In:** *Advances in Probability Measures with Given Marginals*, (Eds., G. Dall'Aglio, S. Kotz, and G. Salinetti), 151-188.

[91] L. Rüschendorf and S.T. Rachev (1990), A Characterization of Random Variables with Minimum L^2-Distance, *J. Multivariate Anal.*, **32**, 48-54.

[92] A.V. Skorohod (1956), Limit Theorems for Stochastic Processes, *Theory Probab. Appl.*, **1**, 261-290.

[93] V. Strassen (1965), The Existence of Probability Measures with Given Marginals, *Ann. Math. Stat.*, **36**, 423-439.

[94] A.S. Sznitman (1989), Propagation of Chaos, **In:** Ecole d'Eté de Probabilities de Saint-Flour, *Lecture Notes in Mathematics*, **1464**, Springer-Verlag, Heidelberg, 165-251.

[95] V.M. Zolotarev (1975), On the Continuity of Stochastic Sequences Generated by Recurrent Procedures, *Theory Probab. Appl.*, **20**, 819-832.

[96] V.M. Zolotarev (1976), Metric Distances in Spaces of Random Variables and Their Applications, *Math. USSR Sb.*, **30**(3), 373-401.

[97] V.M. Zolotarev (1976), Approximation of Distributions of Sums of Independent Random Variables with Values in Infinite-Dimensional Spaces, *Theory Probab. Appl.*, **21**, 721-737.

[98] V.M. Zolotarev (1977), Ideal Metrics in the Problem of Approximating Distributions of Sums of Independent Random Variables, *Theory Probab. Appl.*, **22**, 433-439.

[99] V.M. Zolotarev (1979), Ideal Metrics in the Problems of Probability Theory and Mathematical Statistics, *Aust. J. Stat.* **21**, 193-208.

[100] V.M. Zolotarev (1983), Probability Metrics, *Theory Probab. Appl.*, **28**, 278-302.

[101] V.M. Zolotarev (1986), *The Modern Theory of Summation of Independent Random Variables*, Nauka, Moscow (in Russian).

Chapter 4

Higher Order Stochastic Differential Equations

M. M. Rao

4.1. INTRODUCTION.

A classical application in physics on the motion of a free particle starts with the Langevin equation

$$\frac{du}{dt} + \beta u = \varepsilon(t) \tag{1}$$

where u is the velocity of the particle, βu is the dynamical friction exerted on it, and $\varepsilon(t)$ is the random fluctuation, usually taken to be white noise. This may be generalized, in the presence of an external force, by addition of the acceleration term $a(t)$ produced by the former, so that one has

$$\frac{du}{dt} + \beta u = \varepsilon(t) + a(t). \tag{2}$$

It can be illustrated also by a one-dimensional harmonic oscillator with the circular frequency ω and the distance traversed x, so that

$$\frac{d^2 x}{dt^2} + \beta \frac{dx}{dt} + \omega^2 x = \varepsilon(t). \tag{3}$$

Since in these equations $\{\varepsilon(t), t \geq 0\}$ is a white noise which typically is interpreted as a "derivative" of the irregular Brownian motion, these equations cannot be defined and solved by the classical methods. Instead, one has to devise and present a probabilistic solution, involving a new meaning of the results. This leads to the theory of stochastic differential equations. A nice description of these equations and of their fundamental role in problems of physics and astronomy is discussed in Chandrasekhar (1943). A rigorous interpretation of such problems is nontrivial, and a proper study must involve an application of stochastic analysis, as outlined and emphasized by Doob (1942). Hence the theory of stochastic differential equations (SDEs)becomes a part of probability theory.

Observing that the Equations (1) to (3) are of first and second order linear differential equations, one begins with the existence and uniqueness of their solutions and then studies the (probabilistic) behavior of their paths or of their distributional properties.

Motivated by these and several other applications, one considers a more general class of problems involving (not necessarily linear) equations subsuming (1) and (2) which may be formally expressed as:

$$\frac{dX_t}{dt} + a(t, X_t) = b(t, X_t)Y_t, \tag{4}$$

or symbolically, setting $Y_t dt = dZ_t$, as

$$dX_t + a(t, X_t)dt = b(t, X_t)dZ_t, \tag{5}$$

or in the integrated form as

$$X_t = X_{t_0} - \int_{t_0}^{t} a(s, X_s)ds + \int_{t_0}^{t} b(s, X_s)dZ_s, \tag{6}$$

where $a(\cdot, \cdot)$ and $b(\cdot, \cdot)$ satisfy certain growth conditions. But now the last integral in (6) will have to be defined suitably for a large class of processes containing Brownian motion as a (key) special case. In a similar manner the second order Equation (3) admits a (not necessarily linear) generalization which may again be stated symbolically as:

$$d\dot{X}_t + \alpha(t, X_t, \dot{X}_t)dt = \beta(t, X_t, \dot{X}_t)dZ_t, \tag{7}$$

or in integrated form as

$$X_t = A + B(t - t_0) - \int_{t_0}^{t} \left[\int_{t_0}^{s} \alpha(u, X_u, \dot{X}_u)du \right. $$
$$\left. + \int_{t_0}^{s} \beta(u, X_u, \dot{X}_u)dZ_u \right]ds, \tag{8}$$

where $X_{t_0} = A$, $\dot{X}_{t_0} = B$ is the initial data, and $\dot{X}_t = \frac{dX_t}{dt}$.

The last two equations already raise an additional problem of defining the $\{\dot{X}_t, t \geq t_0\}$ (i.e., the derived) process along with the concepts of stochastic integrals. These two points will be discussed in the next section and the results used in the rest of the work. Here a general boundedness principle will play a fundamental role. Moreover, one can consider various forms of the linear equations themselves, naturally patterned after the classical deterministic studies. This aspect is the theme of Section 3 where the existence and uniqueness of solutions of these equations are considered.

Starting with Section 4 the work concentrates on higher (mostly second) order nonlinear nonconstant coefficient stochastic differential equations, and then their infinitesimal characteristics are treated in Section 5. Specialized analysis and the associated operator theoretical representations will be detailed when the "noise" process $\{Z_t, t \geq t_0\}$ is Brownian motion. This forms the content of Sections 6 and 7. The sample path properties of the solution processes for large times and the corresponding discussion on stochastic flows continues in the next two sections. The final section briefly considers some multiparameter analogs together with a discussion of the stochastic PDEs from this point of view. It should also be noted that the treatment in Sections 5 to 7, as well as the ensuing work, contains the details of an outline given in the author's recent book (Rao (1995), pp.510–525) on this topic.

One of the key points seen in this study is a certain degeneracy for the higher order equations of type (7) or (8). Indeed, the generators of the associated evolution or semi-group operator families lead to studies of degenerate elliptic or parabolic equations and that difficulty cannot be eliminated. This contrasts with the solutions of the first order problems of (5) or (6). These matters are made explicit in what follows, and a number of new and nontrivial questions awaiting solutions in the study of such higher order equations are also pointed out.

4.2. STOCHASTIC INTEGRALS AND BOUNDEDNESS PRINCIPLES.

Let $X : T \to L^p(P)$ be a mapping of $T = [t_0, b)$ into the function space $L^p(P)$, on a probability triple (Ω, Σ, P) assumed complete for convenience. If $f : T \times \Omega \to L^\infty(P)$ is a simple function, with a standard representation as $f = \sum_{i=1}^n \alpha_i \chi_{A_i}$, $A_i = [t_i, t_{i+1})$ where $t_0 < t_1 < \cdots < t_{n+1} \leq b$ and $\alpha_i \in L^\infty(P)$, then define $(\tau_X(f) = \tau(f))$

$$\tau(f) = \int_{t_0}^b f dX = \sum_{i=0}^n \alpha_i (X(t_{i+1}) - X(t_i)). \tag{1}$$

The definition of $\tau(f)$ is unambiguous, and $\tau(\cdot)$ is finitely additive on such simple functions, denoted by $\mathcal{S}$. If λ is a σ-finite measure on the Borelian space $(T, \mathcal{T})$, then it is desirable to extend τ continuously onto the space $L^p(\lambda \otimes P)$ based on $(T \times \Omega, \mathcal{T} \otimes \Sigma)$ so that X determines a countably additive $L^p(P)$-valued (or stochastic) measure. Although $\mathcal{S}$ is dense in this space for $1 \leq p < \infty$, such an extension is usually not possible even when the f are deterministic in (1). A good (or "optimal") condition on X is obtainable by a generalized boundedness principle, presented below. This was originally formulated by Bochner (1955) when f is nonstochastic and X is a stable process. The concept is first stated and its specializations and motivation are then discussed to explain its nature.

1. Definition. Let $\varphi_i : \mathbb{R} \to \mathbb{R}^+, i = 1, 2$, be nondecreasing functions, $\varphi_i(0) = 0, \varphi_i(-x) = \varphi_i(x)$, and $X : T \times \Omega \to \mathbb{R}$ be a (jointly) measurable function on $(T \times \Omega, \mathcal{T} \otimes \Sigma, \lambda \otimes P)$, in the above notation. If $\mathcal{O} \subset \mathcal{T} \otimes \Sigma$ is a σ-subalgebra, then X is said to be L^{φ_1, φ_2}-*bounded relative to* $\mathcal{O}$ and φ_i provided there exists a σ-finite measure $\alpha : \mathcal{O} \to \bar{\mathbb{R}}^+$ and a constant $K(= K_{\alpha, \varphi_1, \varphi_2} > 0)$ such that for any $\mathcal{O}$-simple function f the image $\tau(f)$ of (1) satisfies the inequality:

$$E(\varphi_2(\tau(f))) \le K \int_{\Omega'} \varphi_1(f) d\alpha, \qquad (2)$$

where $\Omega' = T \times \Omega$, and E is the expectation symbol on (Ω, Σ, P).

This reduces to Bochner's original formulation if one takes $\varphi_1(x) = |x|^\rho, \varphi_2(x) = |x|^p, \rho > 0, p \ge 1, \alpha = \lambda \otimes P$ with λ as the Lebesgue measure, and $\mathcal{O} = \mathcal{T} \otimes \{\emptyset, \Omega\}$. The latter is then termed the $L^{\rho, p}$-*boundedness*; and the most commonly used case is the $L^{2,2}$-*boundedness* with $p = \rho = 2$, for a σ-finite α. Indeed the results established with the $L^{2,2}$-boundedness principle are extended to the general case using various well-known devices and hypotheses. For instance, let X be the standard Brownian motion, f a (deterministic) simple function (so the $\alpha_i = a_i$ are constants in (1)), and $\lambda =$ Leb. measure. Then using the independence of increments of X in (1), one finds

$$E(|\tau(f)|^2) = \sum_{i=1}^n a_i^2 E(X(t_{i+1}) - X(t_i))^2 + 0 = \int_{\Omega'} |f|^2 d\alpha, \qquad (3)$$

so that (2) holds with $\varphi_1(x) = \varphi_2(x) = |x|^2, K = 1$ and $\alpha = \lambda \otimes P$. This gives the classical Wiener integral, and the $L^{2,2}$-boundedness principle is at work. If X is a stable process of exponent $\rho\,(0 < \rho \le 2)$ and $p \ge 1$, then the $L^{\rho, p}$-bondedness inequality (2) holds in a similar manner, as Bochner himself has shown. It can also be verified, after a nontrivial computation, that X satisfies an $L^{2, p}$-boundedness principle $(1 \le p \le 2)$ if it has orthogonal increments with λ a σ-finite measure on $\mathcal{T}$ so that $\alpha = \lambda \otimes P$ is a more general measure. This includes the Cramér-Kolmogorov-Karhunen-Loève-type integrals. Since the Brownian motion does not have finite variation in any nondegenerate open interval, it follows from the above example that an L^{φ_1, φ_2}-bounded X is usually of unbounded variation with probability 1.

In implementing this principle, two key problems to solve are in finding a dominating measure α, and a function φ_1, given φ_2 and $\mathcal{O}$. For many concrete cases, such as X being a semimartingale, a σ-finite α on $\mathcal{O}$ and a φ_1 can be found without much difficulty. Before illustrating this, it will be useful to recall the concept of a stochastic integrator which incorporates a form of the dominated convergence statement (for these integrals) in its formulation.

2. Definition. Let $X = \{X_t, t \in T\}, \mathcal{O}$ and φ be as in Definition 1. If $\mathcal{S}(\Omega', \mathcal{O})$ is the set of $\mathcal{O}$-simple functions on $\Omega' = T \times \Omega$, then X is called a *stochastic integrator* on $\mathcal{S}(\Omega', \mathcal{O})$ if the following two conditions hold:
  (i) $\{E(\varphi(\tau(f))) : f \in \mathcal{S}(\Omega', \mathcal{O}), \|f\|_\infty \leq 1\}$ is bounded,
  (ii) $f_n \in \mathcal{S}(\Omega', \mathcal{O}), |f_n| \downarrow 0, a.e. \Rightarrow \lim_n \tau(f_n) = 0$ in probability.
This definition is not easy to apply, but is important for the stochastic integration. The next result shows that if X is L^{φ_1, φ_2}-bounded, then it is a stochastic integrator. Since the boundedness principle is in comparison operational, it enables us to devise relatively easy procedures for stochastic integration. Under some conditions (with φ_i as convex) it can be shown that a stochastic integrator (as in the preceding definition) necessarily satisfies a boundedness principle so that the latter is essentially optimal in a well-defined sense. Thus the desired result is given by:

3. Theorem. *Let $X = \{X_t, t \in T\}$ be L^{φ_1, φ_2}-bounded relative to a σ-finite α on a σ-subalgebra $\mathcal{O}$ as in Definition 1. Then X is a stochastic integrator in the sense of Definition 2.*

[Thus the stochastic integral $\tau(f)$ is defined for a class of functions f containing $\mathcal{S}(\Omega', \mathcal{O})$ as a subset.]

Proof. The demonstration is straightforward when the problem is translated into one of abstract analysis. For this it is useful to recall the generalized Orlicz spaces based on (Ω, Σ, P). Thus if $\varphi : \mathbb{R} \to \mathbb{R}^+$ is a symmetric increasing function vanishing at the origin, as in the statement, let $\{L^\varphi(P), \|\cdot\|_\varphi\}$ denote the space:

$$L^\varphi(P) = \{f : \Omega \to \mathbb{R}, P - \text{measurable}, \|f\|_\varphi < \infty\}, \qquad (4)$$

where

$$\|f\|_\varphi = \inf\{k > 0 : \int_\Omega \varphi(f/k)dP \leq k\}. \qquad (5)$$

Then one can verify that $\{L^\varphi(P), \|\cdot\|_\varphi\}$ is a complete linear metric (or F-) space and that simple functions are dense if further $\varphi(2x) \leq C\varphi(x), x \geq x_0 \geq 0$, for some constant C, called the Δ_2-condition. (See, e.g., Rao and Ren (1991), Sec. 10.1.) Replacing φ by φ_i, and P by P and α, two such spaces result, and $\mathcal{S}(\Omega', \mathcal{O}) \subset L^{\varphi_1}(\alpha)$. Hence τ is defined between the spaces:

$$\tau : \mathcal{S}(\Omega', \mathcal{O}) \subset L^{\varphi_1}(\alpha) \to L^{\varphi_2}(P),$$

by (1), and is a linear mapping. Moreover, (2) implies that τ is bounded and continuous on $\mathcal{S}(\Omega', \mathcal{O})$ although not necessarily on all of $L^{\varphi_1}(\alpha)$. Since τ is mapping into a complete metric space, it follows by the principle of extension by continuity [and τ, being linear, is uniformly continuous, cf., Dunford and Schwartz (1958), p. 53], that τ has a unique continuous extension to $M^{\varphi_1}(\alpha) = \bar{sp}(\mathcal{S}(\Omega', \mathcal{O})) \subset L^{\varphi_1}(\alpha)$. Further, if $f_n \in \mathcal{S}(\Omega', \mathcal{O}), |f_n| \downarrow 0, a.e.$, then $\varphi_1(f_n) \downarrow 0, a.e.$, and $\varphi_1(f_n) \in L^1(\alpha)$. Hence by the dominated convergence theorem the right side (hence also the left side) of (2) tends to zero. The boundedness of the set in Def. 2(i) is a consequence of the fact that $\{f \in \mathcal{S}(\Omega', \mathcal{O}) : \|f\|_\infty \leq 1\}$ is contained in a ball of $L^{\varphi_1}(\alpha)$. Thus both conditions of Definition 2 are satisfied and X is a stochastic integrator. Note that, by what has been shown, τ is defined on all of $M^{\varphi_1}(\alpha)$ which contains $\mathcal{S}(\Omega', \mathcal{O})$ as a (dense) subspace, and X integrates all functions of $M^{\varphi_1}(\alpha)$ under these conditions, and this is the set referred to in the parenthetical statement. $\square$

Remarks. 1. The above argument is a slight extension to the F-space context of that given in (Rao (1995)). These spaces are normed iff φ_1, φ_2 are moreover convex. In the latter case, if (Ω, Σ, P) is also separable, φ_2 satisfies the Δ_2-condition, and X is a stochastic integrator, it was shown in the above reference, that there exist an α and a convex φ_1 relative to which X is L^{φ_1, φ_2}-bounded on $(\Omega', \mathcal{O})$. But the argument is considerably involved, and that result is not needed for the following applications and so will not be further discussed.

2. In applications one chooses $\mathcal{O}$ conveniently (often as a "predictable" σ-algebra) and specializes the resulting integral to get, for instance, the Itô or Stratonovich types, and studies the problems of the corresponding differential equations. This will be done below. But the point of the boundedness principle is that it is quite general for the existence of each of the classes of the intervening integrals.

To illustrate the above result, consider $X = \{X_t, \mathcal{F}_t, t \geq t_0\}$ where X_t is $\mathcal{F}_t$-measurable (or $\mathcal{F}_t$-adapted) and $\mathcal{F}_t \subset \mathcal{F}_{t'} \subset \Sigma$ for $t \leq t'$. The family $\{\mathcal{F}_t, t \geq t_0\}$ of σ-algebras is a filtration and is termed *standard* if, moreover, $\mathcal{F}_s = \cap_{t>s}\mathcal{F}_t$ for each $s > t_0$ and each $\mathcal{F}_t$ is P-complete. Let $\mathcal{O} = \sigma\{(s,t] \times A : A \in \mathcal{F}_s, t_0 \leq s < t\} \subset \mathcal{T} \otimes \Sigma$, called a *predictable σ-algebra*. Suppose that X is a semimartingale, i.e., $X = V + Z$ where $V = \{V_t, \mathcal{F}_t, t \geq t_0\}$ is an adapted process of pointwise bounded variation on compact intervals of T and $Z = \{Z_t, \mathcal{F}_t, t \geq t_0\}$ is a martingale so that $Z_s = E^{\mathcal{F}_s}(Z_t)$, for each $t \geq s \geq t_0$. Sometimes $t_0 = 0$ is taken for simplicity.

4. Proposition. *Using the preceding notation and assumptions, let $X_t \in L^2(P)$ and X be $L^{2,2}$-bounded. Then for each bounded $\mathcal{O}$-measurable $f : \Omega' \to \mathbb{R}$ ($\mathcal{O}$ being a predictable σ-algebra) the following stochastic integral is* *well defined and*

$$Y_t^f = \int_0^t f(s)dX_s,$$ (6)

with $\{Y_t^f, t \geq 0\}$ as a semimartingale when $Y_t^f \in L^2(P)$.

Proof. Let us first consider the case that f is a simple function such that $f(t, \cdot)$ is $\mathcal{F}_t$-adapted. It can also be taken that, under our current assumptions, X, V, Z are left continuous with right limits, using such a modification if necessary; it is known that the latter exist. Next observe that the square integrable martingale Z is $L^{2,2}$-bounded relative to a σ-finite measure $\mu : \mathcal{O} \to \bar{\mathbb{R}}^+$. Indeed for the elementary integral $\tau_Z(f)$ given by (1), with Z in lieu of X there, one has

$$E(|\tau_Z(f)|^2) = \sum_{i=0}^n a_i^2 E(\chi_{A_i}(Z_{t_{i+1}}^2 - Z_{t_i}^2)) + 0,$$

$$= \sum_{i=0}^n a_i^2 \mu((t_i, t_{i+1}] \times A_i), \text{(say)}$$

$$= \int_{\Omega'} |f|^2 d\mu,$$ (7)

where the cross-product terms in the first line vanish by the martingale property of Z, and μ is a measure associated with the positive (left continuous) submartingale $\{Z_t^2, \mathcal{F}_t, t \geq 0\}$. If Z is a Brownian motion, then it can be seen that $\mu = Leb. \otimes P$, and in the general case it is a Doléans-Dade measure, the existence of which is a well-known result (cf., e.g., Rao (1995), p. 365).

Now consider the general case with X itself. Here one also uses the (pointwise) Lebesgue-Stieltjes integral for V. Thus one has (taking V to be bounded on $\mathbb{R}$ for convenience):

$$E(|\tau(f)|^2) = E(|\tau_Z(f) + \tau_V(f)|^2)$$

$$\leq 2E((\tau_Z(f))^2 + (\tau_V(f))^2)$$

$$\leq 2 \int_{\Omega'} |f|^2 d\mu + E(|V_\infty| \int_{\mathbb{R}^+} |f|^2 d|V_t|), \text{ by (7)},$$

$$= 2 \int_{\Omega'} |f|^2 d\alpha,$$ (8)

where in the last but one line the Jensen inequality is applied to the measure $d|V_t|/|V_\infty|$, and $\alpha : A \mapsto \int_A (d\mu + |V_\infty| dV_t dP), A \in \mathcal{O}$, a σ-finite measure. This shows that X is $L^{2,2}$-bounded relative to α and $\mathcal{O}$, with $K = 2$ in (2). Consequently τ extends uniquely to the closure of simple functions in $L^2(\alpha)$ which is $L^2(\alpha)$ itself. This yields the main assertion. The fact that for bounded f in $L^2(\alpha), Y_t^f \in L^2(P)$ is obvious, and since $\tau(f) = \tau_Z(f) + \tau_V(f)$, after using the decomposition, and since $\tau_Z(f)$ is easily shown to be a (square integrable) martingale (because f is predictable and $\tau_V(f)$ is of bounded variation on compact sets), one concludes the semimartingale property of $\tau(f)$ using the decomposition again. $\square$

With localization and stopping times techniques which are standard tools in such studies, the stochastic integration of the $L^{2,2}$-bounded case obtained above can be extended to all L^{φ_1, φ_2}-bounded processes. This will not be detailed here. With this development at hand, it is now appropriate to state conditions for the stochastic (strong) derivative of a process $X = \{X_t, t \geq 0\}$ to have a separable and measurable modification so that it can be used for a study of differential equations below. In fact this will be needed to give a rigorous meaning of equations such as (7) of Section 1.

5. Proposition. *Let* $X = \{X_t, t \in I = [a, b] \subset \mathbb{R}\}$ *be a real process such that* $X_t \in L^p(P), p \geq 1, t \in I$. *If* X_t *is strongly differentiable for each* $t \in I$ *(one sidedly at a and b) with derivative* $\dot{X}_t \in L^p(P)$ *where* (Ω, Σ, P) *is taken complete, then the derived process* $\{\dot{X}_t \in I\}$ *admits a separable and measurable modification whenever it has no fixed points of discontinuities (but can have moving discontinuity points). Moreover, the derived processes can be identified as a pathwise differentiated process of* $X, a.e.$

Recall that $t_0 \in I$ is termed a fixed discontinuity of X if almost all of its sample paths are not continuous at t_0.

Proof. By hypothesis $\{\dot{X}_t, t \in I\}$ exists and is a process in $L^p(P)$. Since it has no fixed points of discontinuity so that the process is continuous in probability, the well-known general theory (cf., e.g., Rao (1995), Sec. 3.3) implies that the above process has a measurable modification which can be taken to be separable with any dense denumerable set of I as a universal separating set. Taking such a version, one can express X as

$$X_t(\omega) = X_a(\omega) + \int_a^t \dot{X}_s(\omega) ds, \ a \leq t \leq b, \tag{9}$$

for all $\omega \in \Omega - N, P(N) = 0$. It follows from (9) that, by the (Lebesgue) fundamental theorem of calculus, $\dot{X}_t(\omega) = X_t'(\omega), \omega \notin N$ and for all $t \in I$ so that $\dot{X}_t$ can be taken as an $a.e.$ pathwise derivative of X. $\square$

Based on the general principles introduced in this section it is now possible to consider the subject proper, namely the differential systems, starting with the linear problem.

4.3. LINEAR STOCHASTIC DIFFERENTIAL EQUATIONS.

Consider a deterministic linear differential equation which may be expressed in standard form as:

$$dX(t) + \alpha(t)X(t)dt = d\beta(t), \tag{1}$$

where X is a (possibly vector) function to be determined and α, β are suitable matrix or vector functions. It is linear since it is so in X and will be homogeneous if $\beta = 0$. Now suppose that both $\alpha(\cdot)$ and $\beta(\cdot)$ are subject to (random) perturbations so that it can be written as follows:

$$dX(t) + (\alpha_1(t) + \delta_1(t)b_1(t))X(t)dt = (\beta_1(t) + \delta_2(t)b_2(t))dt, \qquad (2)$$

where $\{b_1(t), b_2(t) : t \in I\}$ are the disturbances which could be independent "white noises". Here from (1), $d\beta(t) = \beta_1(t)dt$ so that separating the random parts (2) can be written (with $dB_i(t) = b_i(t)dt$) as:

$$dX(t) + (\alpha_1(t)X(t) - \beta_1(t))dt = \delta_2(t)dB_2(t) - \delta_1(t)X(t)dB_1(t). \qquad (3)$$

Based on this type of equation one can state a general form of (3) that includes the Langevin equation as follows.

1. Definition. A general (inhomogeneous) *linear stochastic differential equation* is one of the form (with $k \geq 2$)

$$dX(t) = (\alpha(t)X(t) + \alpha_0(t))dt + \sum_{i=1}^{k}(\beta_i(t)X(t) + \gamma_i(t))dB_i(t), \qquad (4)$$

where the noise processes $\{B_i(t), t \geq t_0, i = 1, \ldots, k\}$ are $L^{2,2}$-bounded and the $\alpha_i, \beta_i, \gamma_i$ are suitable (nonstochastic) measurable, possibly matrix valued, functions. If $\alpha_0 = 0$ and $\gamma_i = 0$, $i = 1, \ldots, k$, one has a *homogeneous linear equation* (in X).

It may be noted that, if the B_i are deterministic differentiable functions then (4) reduces to (1). On the other hand, if the $\alpha_i, \beta_i, \gamma_i$ are independent of t, $\alpha_0 = 0 = \beta_i$ for all i, and X is (possibly) a vector, one has a standard differential equation describing a stationary flow and (4) takes the form:

$$dX(t) = \alpha X(t)dt + \beta dB(t), \qquad (5)$$

where α, β are suitable matrices. Although (4) or (1') is only a symbolic equation when B is a (nondifferentiable) noise process, the integrated form is well-defined by the work of the preceding section, and this is how these equations are always interpreted.

First consider the linear Equation (5). Under minimal conditions (on α, β, and B) the existence, uniqueness, and a construction of the solution will be given here. More general (nonlinear) cases, with suitable restrictions on these data, will be obtained (for higher order equations) later. Thus let $\beta_i = 0, i = 0, \ldots, k$ in (4), and define a process $\bar{B}$ by setting $d\bar{B}(t) = \sum_{i=1}^{k} \gamma_i(t)dB_i(t)$ which, for given bounded measurable γ_is and the

B_i any $L^{2,2}$-bounded processes, will again be $L^{2,2}$-bounded. Then (4) can be expressed as:

$$\frac{d(X(t) - \bar{B}(t))}{dt} = \alpha(t)(X(t) - \bar{B}(t)) + (\alpha(t)\bar{B}(t) + \alpha_0(t)). \qquad (6)$$

Letting $Y(t) = X(t) - \bar{B}(t)$, $g(t) = \alpha(t)\bar{B}(t) + \alpha_0(t)$, this becomes

$$\frac{dY(t)}{dt} = \alpha(t)Y(t) + g(t) = k(t, Y)(t), \text{ (say)}, \qquad (7)$$

a familiar ordinary differential equation in a Banach space $\mathcal{X}$, which will be $L^2(P)$ here. Consequently (7) can be solved as follows. Set

$$(TY)(t) = h(t) + \int_{t_0}^t \alpha(s)Y(s)ds, \qquad (8)$$

where $h(t) = X(t_0) + \int_{t_0}^t g(s)ds (= X(t_0) + \int_{t_0}^t (\alpha(s)\bar{B}(s) + \alpha_0(s))ds)$, all the integrals being in Bochner's sense and they all exist. Since B_1 is $L^{2,2}$-bounded and $h(\cdot)$ is strongly continuous with $Y(t_0) = h(t_0) = X(t_0)(\bar{B}(t_0) = 0)$ as the initial value, one notes that $k(\cdot, \cdot)$ satisfies a local Lipschitz condition on $[t_0, b] \subset \mathbb{R}^+$. Indeed,

$$\begin{aligned}
\|k(t, Y_1)(t) - k(t, Y_2)(t)\|_2 &= \|\int_{t_0}^t \alpha(s)(Y_1(s) - Y_2(s))ds\|_2 \\
&\le \int_{t_0}^t \|\alpha(s)(X_1(s) - X_2(s))\|_2 ds \\
&\le M \int_{t_0}^t \sup_{s \in [t_0, b]} \|X_1(s) - X_2(s)\|_2 ds \\
&\le M(t - t_0)\|X_1 - X_2\|_2^-, \qquad (9)
\end{aligned}$$

where $\|X_1 - X_2\|_2^- = \sup_{t_0 \le s \le t} \|X_1(s) - X_2(s)\|_2$ and $|\alpha(s)| \le M$ is used. Now observe that one has from the $L^{2,2}$-boundedness of B,

$$\begin{aligned}
\|k(t, Y)\|_2 &\le \|g(t)\|_2 + M\|Y(t)\|_2 \\
&\le M(\|\bar{B}(t)\|_2 + 1) + M(\|X(t)\|_2 + \|\bar{B}(t)\|_2) \\
&\le M(2K_1(t - t_0) + 1) + M\|X(t)\|_2 \\
&\le M_2[1 + (b - t_0)] < \infty
\end{aligned}$$

for some constants K_1, M_2 since $\sup_{t_0 \le t \le b} \|X(t)\|_2 < \infty$. Thus $\|X_i\|_2^- < \infty$ and hence so is $\|Y_i\|_2^-$. It follows from (8) and (9) that

$$\|T(Y_1)(t) - T(Y_2)(t)\|_2 \le M_1(t - t_0)\|Y_1 - Y_2\|_2^-. \qquad (10)$$

Replacing Y_i by TY_i in (10), it becomes

$$\|T^2(Y_1)(t) - T^2(Y_2)(t)\|_2 \le \frac{M_1(t-t_0)^2}{2!}\|Y_1 - Y_2\|_2,$$

and iterating the procedure, one gets:

$$\|(T^n Y_1)(t) - (T^n Y_2)(t)\|_2 \le \frac{(M_1(t-t_0))^n}{n!}\|Y_1 - Y_2\|_2. \qquad (11)$$

Thus T^n is a contraction on the set $\{Y : \|Y - Y_0\|_2 \le a\}$, for large enough n, and by the Banach Contraction Mapping Theorem, T^n has a unique fixed point Y in the above set so that (8) (or (6)) has a unique solution. It will now be shown how Y, hence X, can be obtained in the form of a series.

From (8) one gets by repeated substitution the following:

$$\begin{aligned}
Y(t) =& h(t) + \int_{t_0}^t \alpha(s)h(s)ds + \cdots \\
&+ \int_{t_0}^t \alpha(s_1)ds_1 \int_{t_0}^{s_1} \alpha(s_2)ds_2 \cdots \int_{t_0}^{s_{n-1}} \alpha(s_n)Y(s_n)ds_n \\
=& h(t) + \sum_{n=1}^\infty h_n(t),
\end{aligned}$$

with $h_n(t) = \int_{t_0}^t \alpha(s)h_{n-1}(s)ds$, $h_0(t) = h(t)$, the series converging absolutely because of the estimate (11) on each term. Thus defining the operator function $V(\cdot)$ by the series

$$V(t) = I + \int_{t_0}^t \alpha(s)ds + \sum_{n=2}^\infty \int_{t_0 < t_n < \cdots < t_1 < t} \alpha(t_n)\cdots\alpha(t_1)dt_1 \cdots dt_n \qquad (12)$$

which converges in the operator norm, the solution is given by:

$$Y(t) = X(t) - \bar{B}(t) = V(t)Y(t_0) = V(t)(X(t_0) - \bar{B}(t_0)), \qquad (13)$$

so that ($\bar{B}(t_0) = 0$ being used)

$$X(t) = \sum_{i=1}^k \int_{t_0}^t \gamma_i(s)dB_i(s) + V(t)X(t_0). \qquad (14)$$

This may be summerized as:

2. Proposition. *Suppose that the linear SDE is given by (4) in which $\beta_i = 0$, the α_i, γ_i are bounded measurable functions, and the $B_i, i = 1, \dots, k$*

are $L^{2,2}$-bounded relative to a σ-finite measure on $\mathbb{R}^+ \times \Omega$. Then the equation has a unique solution having the given initial value $X(t_0)$, and the process $X(t)$ given by (14) where the linear operator $V(t)$ is defined by the uniformly convergent series (12) for t in compact intervals, solves the equation.

The same result is also valid if α, γ take values in suitable Banach spaces, and the B_is are $L^{2,2}$-bounded processes defined in such spaces (cf., e.g., Rao (1995), Sec. 6.3). The procedure, to be used in the next section for the (nonlinear) extension, is a modification of the above one. See also Dalecky and Krein (1974, Chapters 3 and 7). But the infinite-dimensional state space case will not be discussed per se in the following work. (The finite matricial version is immediate).

One should observe that this result did not fully use the force of $L^{2,2}$-boundedness of B_i or of $\bar{B}$; it only used that (6) be meaningful. In fact that $\bar{B}$ is (strongly) measurable, and essentially bounded on compact intervals sufficed. Thus for this type of (linear) SDEs, Z need not define a (σ-additive) vector measure. The existence and uniqueness for the initial value problems hold in a quite general context. Thus many different types of stochastic integrals (or differentials) are covered here.

An extension of the above result, applicable to the general linear case stated in Definition 1, can be obtained. The idea now is to convert the given (linear Itô) SDE to one of (Fisk-) Stratonovich or symmetric type for which the classical differentiation rules apply. Although a direct approach without the latter transformation is possible, the conversion in this case makes the result transparent. The integral is recalled as follows.

Let $X_i = \{X_i(t), \mathcal{F}_t, t \in [a, b) \subset \mathbb{R}^+\}, i = 1, 2$ be a pair of $L^{2,2}$-bounded continuous processes relative to a σ-finite measure α, so that the (generalized Itô) integral $\int_a^t X_1(s) dX_2(s)$ is well defined and is again a process of the same type. In the present case the X_i will be $L^2(P)$-semimartingales (cf. Rao (1995), Thm. VI.2.14). The (Fisk-) Stratonovich integral of X_1 relative to X_2, denoted $\int_a^t X_1(s) \circ dX_2(s)$, is by definition:

$$\int_a^t X_1(s) \circ dX_2(s) = \int_a^t X_1(s) dX_2(s) + \frac{1}{2} \int_a^t d\langle X_1, X_2 \rangle(s), \qquad (15)$$

where $\langle \cdot, \cdot \rangle$ is the quadratic (co)variation process (hence of locally bounded variation) determined by X_1 and X_2. It may also be verified that the left side is the limit in probability of the sums $\frac{1}{2} \sum_{i=1}^n [X_1(t_i) + X_1(t_{i-1})][X_2(t_i) - X_2(t_{i-1})]$, as the partitions $\pi_n : a = t_0 < t_1 < \cdots < t_n = b$ are infinitely refined. This will again be (symbolically) written as

$$X_1 \circ dX_2 = X_1 dX_2 + \frac{1}{2} dX_1 \cdot dX_2 \left(= X_1 dX_2 + \frac{1}{2} d\langle X_1, X_2 \rangle\right). \qquad (16)$$

If dS, dA denote the differentials of (local) semi-martingales and (local) bounded variation processes adapted to the given filtration $\{\mathcal{F}(t), t \in [a, b)\}$, then the following properties, first noted by Itô, obtain (cf., e.g., Itô and S. Watanabe (1978), Sec. 2):

$$dS.dS \subset dA; \quad dS.dA = 0; \quad dS.dS.dS = 0;$$

and further

$$X_1 \circ dX_2 = X_1 dX_2 \text{ (if } X_1 \text{ or } X_2 \in A); \quad (X_1 \cdot dX_2) \cdot dX_2 = X_1 dX_2 \cdot dX_2.$$
$$(16')$$

Moreover, the standard differential calculus rules hold for the symmetric or Stratonovich integral while the same is not generally true for the Itô integral. However, most of the problems in stochastic applications are given as the Itô integrals since they often have martingale, Markovian, and other probabilistic properties, whereas the Stratonovich type lacks them. Nevertheless, for computational simplicity one makes the conversion into symmetric integrals when it is convenient. The following result clearly illustrates these points.

3. Proposition. *Suppose that the $L^{2,2}$-bounded processes B_i and the coefficients $\alpha, \alpha_0, \beta_i, \gamma_i$ of (4) are continuous. Then the linear SDE of (4) has a unique solution for any given initial (constant) value $X(t_0)$, and $X(t)$ is explicitly representable as:*

$$X(t) = M(t)^{-1} \{ X(t_0) + \int_{t_0}^t M(s)\alpha_0(s)\, ds - \frac{1}{2} \sum_{i,j=1}^k \int_{t_0}^t M(s) \times$$

$$\beta_i(s)\gamma_i(s) \circ dB_i(s) + \sum_{i=1}^k \int_{t_0}^t M(s)\gamma_i(s) \circ dB_i(s) \}, \qquad (17)$$

where

$$M(t) = \exp\{ -\int_{t_0}^t \alpha(s)\, ds - \sum_{i=1}^k \int_{t_0}^t \beta_i(s)\, dB_i(s)$$

$$+ \frac{1}{2} \sum_{i,j=1}^k \int_{t_0}^t \beta_i(s)\beta_j(s)\, d\langle B_i, B_j \rangle(s) \}. \qquad (18)$$

Remark. The result holds if X is an n-vector, $\alpha, \beta_i, \gamma_i$ are $n \times n$-matrices, α_0 and B_i are n-vectors, so that M is also an $n \times n$-process. For simplicity only the scalar case ($n = 1$) will be considered here. (See, however, (25) below.)

Proof. One method of proof is to substitute (17) into (4) and verify (with the Itô chain rule formula) that the equation is satisfied. The uniqueness part of Proposition 2 then completes the argument. However, an alternative procedure is to derive (17) constructively, as shown here, since it is not simple to guess the answer.

Let us rewrite (4), for convenience, as:

$$dX(t) = (\alpha(t)X(t) + \alpha(t))dt +$$

$$\sum_{i=1}^{k} \beta_i(t)X(t)dB_i(t) + \sum_{i=1}^{k} \gamma_i(t)dB_i(t). \tag{19}$$

Consider the conversion of the Itô differential term above into the Stratonovich form as:

$$\beta_i(t)X(t) \circ dB_i(t) = \beta_i(t)X(t)\,dB_i(t) + \frac{1}{2}d(\beta_i(t)X(t)) \cdot dB_i(t)$$

$$= \beta_i(t)X(t)\,dB_i(t) + \frac{1}{2}[X(t)d\langle\beta_i, B_i\rangle(t)$$

$$+ \beta_i(t) \cdot dX(t)\,dB_i(t)]$$

$$= \beta_i(t)X(t)\,dB_i(t) + \frac{1}{2}\beta_i(t)[0+$$

$$\sum_{j=1}^{k} (\beta_j(t)X(t) + \gamma_j(t))d\langle B_i, B_j\rangle(t)],$$

using $dX(t)$ and the rules of (16'). Substituting this into (19) one gets

$$dX(t) = (\alpha(t)X(t) + \alpha_0(t))dt + \sum_{i=1}^{k} \beta_i(t)X(t) \circ dB_i(t) -$$

$$\sum_{i,j=1}^{k} \beta_i(t)[\beta_j(t)X(t) + \gamma_j(t)]d\langle B_i, B_j\rangle(t)$$

$$= X(t)[\alpha(t)dt + \sum_{i=1}^{k} \beta_i(t) \circ dB_i(t) - \frac{1}{2}\sum_{i,j=1}^{k} \beta_i(t)\beta_j(t)\,d\langle B_i, B_j\rangle(t)]$$

$$+ [\alpha_0(t)dt - \frac{1}{2}\sum_{i,j=1}^{k} \beta_i(t)\gamma_j(t)d\langle B_i, B_j\rangle(t) +$$

$$\sum_{i=1}^{k} \gamma_i(t)\,dB_i(t)]. \tag{20}$$

The homogeneous equation obtained from (20), by setting $\alpha_0 = 0$ and $\gamma_i = 0$, is given by:

$$dX(t) = X(t)[\alpha(t)dt + \sum_{i=1}^{k} \beta_i \circ dB_i(t) - \frac{1}{2}\sum_{i,j=1}^{k} \beta_i(t)\beta_j(t)d\langle B_i, B_j\rangle(t)].$$

Since the classical ODE rules apply to this equation one has the "integrating factor" as:

$$M(t) = \exp\{ - \int_{t_0}^{t} \alpha(s)\,ds - \sum_{i=1}^{k} \int_{t_0}^{t} \beta_i(s) \circ dB_i(s) +$$

$$\frac{1}{2} \sum_{i,j=1}^{k} \int_{t_0}^{t} \beta_i(s)\beta_j(s)\,d\langle B_i, B_j\rangle(s) \}. \tag{21}$$

Multiplying (20) by $M(t)$ of (21), and integrating $d(X(t)M(t))$, after recalling that the Itô and Stratonovich integrals of deterministic integrands coincide, one has:

$$X(t)M(t) = \{X(t_0) + \int_{t_0}^{t} M(s)\alpha_0(s)ds - \frac{1}{2} \sum_{i,j=1}^{k} \times$$

$$\int_{t_0}^{t} M(s)\beta_i(s)\gamma_j(s)\langle B_i, B_j\rangle(s) + \sum_{i=1}^{k} \int_{t_0}^{t} M(s)\gamma_i(s) \circ dB_i(s) \} M^{-1}(t).$$

This establishes (17). $\Box$

A somewhat specialized form of this result (with B_i independent Brownian motions and a different argument), may be found in Wu ((1985), Chapter 2). Since the integrals with $L^{2,2}$-bounded integrators include, when appropriately identified, Skorokhod integrals which are themselves generalizations of Itô's (cf. Rao (1995), pp.531-32), the existence part of the above result applies to these equations also. Letting the initial value be an arbitrary bounded random variable and $\alpha(t), \beta(t)$ be certain other processes (but the B_i still are Brownian motions) an explicit solution of (4) was obtained by Buckdahn (1991). It appears that the entire work can be extended to the $L^{2,2}$-bounded case. However, this will not be considered further in the present study. On the other hand, first order (even vector) stochastic differential equations driven by a continuous noise process (e.g., white noise) can be solved by extending the classical ODE methods. These turn out to coincide with those of Stratonovich differential equations when the white noise is used, as shown by Sussman (1978). He also observed by a counterexample that this special method does not extend to systems of such equations to include vector Brownian motions.

To indicate another aspect of the above type of results, a multidimensional problem will now be discussed briefly. Consider the following system of Stratonovich differential equations:

$$dX^i(t) = \sum_{k=1}^{p}[\sum_{j=1}^{q} L^i_{jk} X^j(t) + a^i_k] \circ dB^k(t)+$$

$$[\sum_{j=1}^{q} L^i_{0j} X^j(t) + a^i_0]dt$$

$$= \sum_{j=1}^{q} X^j(t)[\sum_{k=1}^{p} L^i_{jk} \circ dB^k(t) + L^i_{0j} dt]+$$

$$\sum_{k=1}^{p} a^i_k \circ dB^k(t) + a^i_0 dt, \qquad (22)$$

$$X^i(t_0) = x^i, \; i = 1, \dots, q.$$

Here $L^i_{jk}, a^i_k, \; k = 0, 1, \dots, p, 1 \le i, j \le q$ are constants, and B^k is an $L^{2,2}$-bounded process relative to a σ-finite measure, or it is just a semimartingale. Let $d\tilde{B}^i_j(t) = \sum_{k=1}^{p} L^i_{jk} \circ dB^k(t) + L^i_{0j} \, dt$, $d\bar{B}^i(t) = \sum_{k=1}^{p} a^i_k \circ dB^k(t) + a^i_0 \, dt$. If $\tilde{B} = (\tilde{B}^i_j, 1 \le i, j \le q)$ and $\bar{B} = (\bar{B}^i, 1 \le i \le q)$, the matrix and vector processes respectively of the same type, then (22) can be expressed in a compact form with $X(t) = (X^1(t), \cdots, X^q(t))$, as:

$$dX(t) = X(t) \circ d\tilde{B}(t) + d\bar{B}(t). \qquad (23)$$

Taking $a^k_i = 0, 0 \le k \le q$, so that $\bar{B}(t) = 0$ in (23), one gets the corresponding homogeneous equation:

$$dX(t) = X(t) \circ d\tilde{B}(t). \qquad (24)$$

Since $\det(\tilde{B}(t)) = \exp(\int_{t_0}^{t} tr(d\tilde{B}(s))) \ne 0$ with probability one for any $t \ge t_0$, (24) and then (23) will have a unique solution satisfying $X(t_0) = x_0$, a.e. (cf., e.g., Coddington and Levinson (1955) p.69 and p.74 in the classical case). The solution of (23) is given by

$$X(t) = M(t)^{-1}[x_0 + \int_{t_0}^{t} \exp(\int_{t_0}^{s} d\tilde{B}(u)) \circ d\bar{B}(s)], \qquad (25)$$

which is an n-dimensional Stratonovich analog of (17). Here $M(t)$ is the following matrix process which is the solution of (24):

$$M(t) = \exp\{-\int_{t_0}^{t} d\tilde{B}(s)\}. \qquad (26)$$

To simplify (25) further, one has to expand the exponential involved there. But by definition, $d\tilde{B}$ is defined as a matrix of vector fields which do not necessarily commute so that $e^{\alpha+\beta} \neq e^{\alpha} \cdot e^{\beta}$. Thus one has to bring in the commutators $[\alpha, \beta] = \alpha\beta - \beta\alpha$ and their iterates. To have some control on these one needs to find a suitable extension of the Campbell-Hausdorff formula for the differentials used in the definition of $B_j^i(\cdot)$, namely the vector fields $L_k = \sum_{i=1}^{q}(\sum_{j=1}^{q} L_{jk}^i x^i + a_k^i)\frac{\partial}{\partial x^i}, k = 0, 1, \ldots, p$ which need not commute. Assuming that $L_{jk}^i = 0, i > k$, and considering the Lie algebra generated by $\{L_0, L_1, \ldots, L_p\}$, which in this case turns out to be solvable, (together with the Jacobi identity of these algebras for simplifications) Kunita (1980) has presented a generalized exponential expansion using the Lie brackets $[L_i, L_j], [L_i, [L_j, L_k]], \cdots$ where the brackets vanish after a finite number of iterates due to solvability. This shows what additional techniques from different areas of mathematics have to be utilized (often in extended forms) to obtain detailed information to such questions. Further analysis is needed with these ideas to study higher order equations. This will be illustrated for a constant coefficient nth order case where the situation is relatively simple and direct calculations are possible.

Consider the nth order linear differential operator L_n, given by

$$L_n = a_0 \frac{d^n}{dx^n} + a_1 \frac{d^{n-1}}{dx^{n-1}} + \cdots + a_n,$$

where $a_0, a_1, \ldots, a_n$ are continuous functions on $I = [t_0, b)$ and $a_0(t) \neq 0$ for any $t \in I$, so that $L_n X(t) = \dot{B}_n(t)$, for a continuously differentiable function B_n, is an nth order ODE. Here one can take $a_0(t) = 1$ for convenience, dividing through by it if necessary. Now if $\dot{B}_n$ is replaced by a white noise process, or more generally by the "derivative" of a semimartingale (or an $L^{2,2}$-bounded) process, interpreting $L_n X(t) = \dot{B}_n(t)$ as

$$\int_I \varphi(t) L_n X(t)\, dt = \int_I \varphi(t)\, dB_n(t), \qquad (27)$$

for all compactly supported continuous [or even simple] functions φ on I, the right side being a standard stochastic integral, then the symbolic stochastic differential equation $L_n X(t) = \frac{dB_n(t)}{dt}$ is meaningful. It may be expressed as a (first order) vector equation as follows. Set $\frac{dX^i(t)}{dt^i} = X^{(i+1)}(t), i = 0, 1, \ldots, n-1, [X^{(0)} = X(t)]$, and $\frac{d\check{X}(t)}{dt} = (X^{(0)}, \cdots, X^{(n-1)})^*, dB = (0, \cdots, 0, dB_n(t))^*$, be column vectors, '$*$' denoting transpose here. If the

$n \times n$-martix A is given as

$$
A = \begin{bmatrix}
0 & 1 & 0 & \cdots & 0 & 0 \\
0 & 0 & 1 & \cdots & 0 & 0 \\
0 & 0 & 0 & \cdots & 0 & 0 \\
\vdots & \vdots & \vdots & \ddots & \vdots & \vdots \\
0 & 0 & 0 & \cdots & 0 & 1 \\
-a_n & -a_{n-1} & -a_{n-2} & \cdots & -a_2 & a_1
\end{bmatrix}
$$

then

$$
d\hat{X}(t) = A\hat{X}(t)\, dt + dB(t), \tag{28}
$$

is the symbolic vector equation corresponding to (27). If one takes the initial condition as $\hat{X}(t_0) = C$, then a comparison of (28) with (1') shows that Proposition 2 is applicable here and one deduces that this equation has a unique solution where the (component) derivatives are taken in the mean square sense. It is of interest to present an explicit solution, using the forms of the matrix A and the noise vector B, especially when A is a constant (i.e., $a_i(t)$ are independent of t). In this case, the series for $V(\cdot)$ in (14) simplifies to an exponential operator and one obtains:

$$
\hat{X}(t) = e^{-A(t-t_0)}C + \int_{t_0}^{t} e^{-(t-u)A}\, dB(u), \tag{29}
$$

where $\hat{X}(t_0) = C$, being the (constant) initial value. But for any $h > 0$, (29) can be expressed as:

$$
\hat{X}(t+h) = e^{-Ah}\hat{X}(t) + \int_{t}^{t+h} e^{-A(t+h-u)}\, dB(u), \tag{30}
$$

where the fact that A is independent of t is used. If now the process $\{B_n(t), t \geq t_0\}$ has independent increments, then with the definition of the stochastic integral, the σ-algebra $\mathcal{B}_t$ generated by the process $\{X(s), t_0 \leq s \leq t\}$ for which each of the $X^{(i)}(s)$ is adapted ($i \leq n$) and the right side integral of (30) are independent, because $B_n(t+h) - B_n(t)$ is independent of $B_n(s), s \leq t$. Hence for any Borel set $B \subset \mathbb{R}^n$, and $h > 0$, one has for the conditional probability relation:

$$
P[\hat{X}(t+h) \in B | \mathcal{B}_{t_0}^t] = P[\hat{X}(t+h) \in B | \mathcal{B}_t^t], \ \forall t \geq t_0, \tag{31}
$$

with probability one, so $\{\hat{X}(t), t \geq t_0\}$ is a (vector) Markov process. Note also that by (29) the discontinuities of the $\hat{X}$-process are precisely those of the B_n-process. In particular, if B_n is a martingale, then $\hat{X}$ can have only moving discontinuities, and thus if the B_n-process is a Brownian motion then $\hat{X}(t)$ will have almost all continuous sample paths. These remarks are summarized for reference in the following:

4. Proposition. *Let $L_n X(t) = \dot{B}_n(t)$ be an nth order linear SDE which is given in equivalent vector form by (28) where $\{B_n(t), t \geq t_0\}$ is an $L^{2,2}$-bounded noise. Then the initial value problem has a unique solution. If, moreover, A is a constant matrix (i.e., the a_i are independent of t), the vector solution process $\{\hat{X}(t), t \geq t_0\}$ of (28) associated with the $X(t)$-process is given explicitly by (29). If further the $B_n(t)$-process has independent increments, then the solution $\{\hat{X}(t), t \geq t_0\}$ is a vector Markov process (although the X_t-process itself will not be Markovian in general) whose discontinuities are those of the $B_n(t)$, and hence has continuous sample paths whenever the $B_n(t)$ has such. In particular, if the $B_n(t)$ is a Brownian motion, then the $\hat{X}(t)$ is a (vector) Markov process with continuous sample paths.*

Remarks. 1. The special form of A for (28) was used in the above proposition only in deducing that the components $X^{(i)}, i = 1, \dots, n-1$ of $\hat{X}$ are (mean-square) derivatives of $X^{(0)}$. The result also holds if (28) is given as a vector of (or system of n-)equations and $A(t)$ is a bounded measurable deterministic function. Then, with $X(t_0) = C$, one obtains

$$\hat{X}(t) = C + M(t) \int_{t_0}^{t} M(s)^{-1} \, dB(s), \tag{32}$$

where $M(t)$ is the fundamental (= invertible) $n \times n$-matrix solution of the homogeneous equation $d\hat{X}(t) = A(t)\hat{X}(t)dt$, which follows from Proposition 2 or a vector analog of Proposition 3. The argument thus implies that the $\hat{X}(t)$-process is a (vector) Markov process when $B(s)$ has independent increments, in addition, and moreover the continuity property holds if $A(\cdot)$ and $B(\cdot)$ are continuous.

2. Under specialized assumptions of the last part of the above proposition, Dym (1966) made a penetrating analysis of the solution of the SDE (28). A key part of the latter will be included here for comparison with a general class of nonlinear equations to be considered in the following sections.

When $B(t)$ is a Brownian motion, it is clear that the solution of (28) defined by (29) is a vector Gaussian (Markov) process. The conditional probability density p_{t,t_0} of $\hat{X}(t)$ given that $\hat{X}(t_0) = C$ can be expressed as follows. Let $\hat{B}(h) = B(t + h), \hat{Z}(h) = \hat{X}(t + h)$ for $h > 0$. Then (30) becomes

$$\hat{Z}(h) = e^{-Ah}[\hat{Z}(0) + \int_{t}^{t+h} e^{-A(u-t_0)} \, d\hat{B}(u)], \tag{33}$$

which is of the same form as (29), and since $\{\hat{B}(u), u \geq 0\}$ is again a Brownian motion starting at $u = 0$, the earlier reasoning implies that the

$\hat{Z}(h)$-process is again vector Markov, starting with the value $\hat{Z}(0) = \xi$, and depending only on 0 and h. Hence the conditional distribution, which is the transition probability function of the Markovian family, is stationary, and one has for any Borel set $B \subset \mathbb{R}^n$ if

$$F(\xi, t; B, t + h) = P(\hat{X}(t + h) \in B | \hat{X}(t) = \xi),$$

then $F(\xi, t; B, t + h) = F(\xi, 0; B, h) = \tilde{F}(\xi, h; B)$, (say). Thus taking $B = (-\infty, \lambda_1) \times \cdots \times (-\infty, \lambda_n)$ in the above Gaussian distribution, and differentiating it relative to the λ_i, the density p of $\tilde{F}$ is expressible as a function of $t - t_0$ as:

$$p_{t-t_0}(c, \lambda_1, \dots, \lambda_n) = (2\pi)^{-\frac{n}{2}} \det(R_{ij}(t - t_0))^{-\frac{1}{2}} \times$$
$$\exp\{-\langle \lambda - \mu(t), R^{-1}(t - t_0)(\lambda - \mu(t)) \rangle / 2\},$$

where $R = (R_{ij})$ is the covariance matrix and $\mu(t)$ is the mean vector of $\hat{X}(t)$, so that

$$\mu(t) = E(\hat{X}(t)) = e^{-At}C; \quad R_{ij}(t - t_0) = \int_{t_0}^t (e^{-Au})_{in}(e^{-Au})_{jn}\, du,$$

where $(e^{Au})_{ik}$ denotes the $(ik)^{th}$ element of the n-by-n matrix (e^{Au}), and the vector $\lambda = (\lambda_1, \dots, \lambda_n)$. The expression for p is then used to calculate the properties of the associated semi-group of contractive operators of the solution process $\{\hat{X}(t), t \geq t_0\}$. In fact let

$$(T_{t-t_0}f)(x) = \int_{\mathbb{R}^n} f(y)p_{t-t_0}(x, y)\, dy, \quad f \in B(\mathbb{R}^n), \tag{34}$$

where $B(\mathbb{R}^n)$ denotes the Banach space of bounded Borel functions on $\mathbb{R}^n$. Then by the Chapman-Kolmogorov equation, (34) gives $\{T_t, t \geq t_0\}$ as the desired semigroup. The adjoint semigroup $\{T_t^*, t \geq t_0\}$ is defined by

$$(T_{t-t_0}^* f)(y) = \int_{\mathbb{R}^n} f(x)p_{t-t_0}(x, y)\, dx, \quad f \in B(\mathbb{R}^n), \tag{35}$$

which is also a contraction. Let $C_b^{(i)}(\mathbb{R}^n), i \geq 0$, be the subspace of $B(\mathbb{R}^n)$ consisting of i-times continuously differentiable functions. Then the semi-groups associated with the solution process have the following important properties established by Dym (1966).

5. Theorem. *For each $f \in B(\mathbb{R}^n), T_t f$ and $T_t^* f$ are real analytic functions, and hence $T_t(B(\mathbb{R}^n)) \subset C_b^{(0)}(\mathbb{R}^n), T_t^*(B(\mathbb{R}^n)) \subset C_b^{(0)}(\mathbb{R}^n)$. Consequently, the process $\{\hat{X}(t), t \geq t_0\}$ of which $\{T_t, t \geq 0\}$ is the associated semigroup is* *a Feller process so that it is strongly Markovian. Moreover, the infinitesimal generators G and G^* of these semigroups are the following degenerate elliptic differential operators acting on $C_b^{(2)}(\mathbb{R}^n)$:*

$$G = \frac{1}{2}\frac{\partial^2}{\partial x_n^2} + \sum_{i=1}^{n-1} x_{i+1}\frac{\partial}{\partial x_i} - (x_1 a_n + \cdots + x_n a_1)\frac{\partial}{\partial x_n}, \qquad (36)$$

and

$$G^* = \frac{1}{2}\frac{\partial^2}{\partial x_n^2} - \sum_{i=1}^{n-1} x_{i+1}\frac{\partial}{\partial x_i} + (x_1 a_n + \cdots + x_n a_1)\frac{\partial}{\partial x_n} + a_1. \qquad (37)$$

Indeed, using the smoothness of the density $p_{t-t_0}(\cdot,\cdot)$, in all the variables, expanding it in power series around $(t_0, 0, 0), t_0 > 0$, and evaluating the integrals in (34) and (35), one obtains the real analytic assertions via the classical Hartogs theorem of several complex variables. Then one can evaluate the following strong limits (in the uniform norm) for each $f \in C_b^{(2)}(\mathbb{R}^n)$:

$$\lim_{t \to 0} \frac{T_t f - f}{t}(x) = (Gf)(x), \qquad (38)$$

and

$$\lim_{t \to 0} \frac{T_t^* f - f}{t}(y) = (G^* f)(y). \qquad (39)$$

The sample path behavior of the solution process is then determined by the transition probability family $\{p_t, t \geq t_0\}$ or equivalently by the semigroups $\{T_t, t \geq t_0\}$ and $\{T_t^*, t \geq t_0\}$, utilizing the finer structural analysis of the latter classes. The detailed theory was presented by Dym (1966) and the reader is referred to that paper.

The nonlinear analog of (28) will be considered in the next section, and a comparison of these cases will illuminate the work. It is important, however, to note a special problem even in the linear case, already observed by Dym (*ibid*, p.138), that the operators G and G^* expressed as

$$G = \frac{1}{2}\sum_{j,k=1}^{n} h_{jk}\frac{\partial^2}{\partial x_j \partial x_k} + \sum_{i=1}^{n} v_i\frac{\partial}{\partial x_i}, \qquad (40)$$

and similarly for G^* have the degeneracy property, i.e., the (constant) coefficient matrix (h_{jk}) of the second order differentials, is nonnegative definite, and has a vanishing determinant. As a result, almost none of the classical PDE theory can be directly invoked in the current analysis. The corresponding difficulty in the (higher order) nonlinear case is more pronounced as shown below. The preceding theorem is included here precisely for such a comparison.

4.4. HIGHER ORDER NONLINEAR EQUATIONS.

As noted in the Introduction, nonlinear extensions of the first order SDEs have been studied extensively in the literature, but the corresponding problems represented by Equations (7) and (8) of Section 1 for higher order systems have not received a similar treatment. This will be the subject of this and the following sections. Thus let $\{X(t), t \geq t_0\}$ be a process which is a solution of the equation:

$$dX(t) = q(t, X(t), \dot{X}(t)) \, dt + \sigma(t, X(t), \dot{X}(t)) \, dB(t). \tag{1}$$

In the integrated form this becomes:

$$
\begin{aligned}
X(t) &= A_1 + A_2(t - t_0) + \int_{t_0}^{t} [\int_{t_0}^{s} q(u, X(u), \dot{X}(u)) \, du \\
&\quad + \int_{t_0}^{s} \sigma(u, X(u), \dot{X}(u)) \, dB(u)] \, ds \\
&= A_1 + \int_{t_0}^{t} [A_2 + \int_{t_0}^{s} q(u, X(u), \dot{X}(u)) \, du \\
&\quad + \int_{t_0}^{s} \sigma(u, X(u), \dot{X}(u)) \, dB(u)] \, ds \\
&= A_1 + \int_{t_0}^{t} [A_2 + H(X(s)) + \int_{t_0}^{s} \sigma(u, X(u), \dot{X}(u)) \, dB(u)] \, ds,
\end{aligned}
\tag{2}
$$

where $X(t_0) = A_1, \dot{X}(t_0) = A_2$ and HX is the path integral of $q(\cdot, \cdot, \cdot)$ depending on the process X and t. The properties of the operator H, acting on processes, are abstracted and generalized to obtain an analog of Proposition 3.2 above, as follows.

For a given process $\{Y(t), \mathcal{F}_t, t \geq t_0\}$ and $t_0 \leq s < t$, let $\mathcal{F}_s^t(Y)$ denote the σ-algebra generated by $\{Y(u), s \leq u \leq t\}$, and completed for P, so that in the above notation the operator H satisfies the following conditions:

(Ai) $\Delta_s^t(HY) = (HY)(t) - (HY)(s)$, is $\mathcal{F}_s^t(Y)$-adapted. (Thus $\Delta_s^t(HY)$ depends only on the values of Y observed in the interval $[s, t]$ but not on the past values before s (or s-past), nor on the future values after t (or t-future).)

(Aii) For each measurable process Y, HY is an absolutely continuous process such that $(HY)(t_0) = 0$ a.e.

(Aiii) For each pair of processes $Y_i, i = 1, 2$, with a pathwise derivative Y_i' the following bounds hold with probability one,

(a) $|(HY_1) - (HY_2)|^2(t) \leq K\{|Y_1 - Y_2|^2(t_0) + \int_{t_0}^{t} |Y_1' - Y_2'|^2(s) \, d\alpha(s)\}$,

(b) $\qquad |HY_i|^2(t) \leq K\{1 + |Y_i|^2(t_0) + \int_{t_0}^{t} |Y_i'(s)|^2 \, d\alpha(s)\}$, for some constant $K > 0$ depending only on t_0 and t, and some nondecreasing function $\alpha : [t_0, t) \to \mathbb{R}^+$ (depending only on H). [In (b) $\alpha(\cdot)$ is the length function determing the Lebesgue measure and hereafter H is a general operator that satisfies the above three conditions and then it is specialized to the one describing the path integral in (2).]

Note that if X is an n-vector process (hence $\dot{X}$ is also an n-vector) in (1), $q(\cdot, \cdot, \cdot)$ will be a vector and $\sigma(\cdot, \cdot, \cdot)$ will be an $n \times n$ matrix, so that H of (2) takes vectors into vectors. In this case the absolute values in (Aiii) above denote Euclidean norms. An existence and uniqueness of the solutions of (2) with given initial condition can be established with minimal restrictions, by reducing it to a first order vector equation. However, the actual form of (2) or (1) and the conditions (Ai)-(Aiii) will be used in showing that the solution $X(\cdot)$ of (2) is an absolutely continuous process with pathwise derivative $\dot{X}(\cdot)$. The general result does not directly imply this. To illuminate the distinction, both statements will be separated and the second one analyzed later on.

Consider again (2) as a system of differential equations:

$$X(t) = A_1 + \int_{t_0}^t \dot{X}(s)\, ds,$$
$$\dot{X}(t) = A_2 + (HX)(t) + \int_{t_0}^t \sigma(u, X(u), \dot{X}(u))\, dB(u). \tag{3}$$

Since $\{B(t), t \geq t_0\}$ is an $L^{2,2}$-bounded process, the last integral is defined for all measurable and locally (i.e., on compact intervals) bounded σ. Denote the integrated process as $\{\bar{B}(t), t \geq t_0\}$ so that $\bar{B}(t_0) = 0$, and is also $L^{2,2}$-bounded. If the operator H operating on processes takes them into absolutely continuous processes (in t), then $\dot{X} - \bar{B}$ is also absolutely continuous. Thus (3) or (1) may be expressed as a first order ($2n$-vector) equation when X (hence $\dot{X}$) takes values in an n-dimensional space $\mathbb{R}^n$ (say). In fact let

$$\mathbf{X}(t) = (X(t), \dot{X}(t))^*, \quad \mathbf{Q}(t, x, y) = (y, q(t, x, y))^*, \quad \mathbf{A} = (A_1, A_2)^*,$$

$$\mathbf{S}(t, x, y) = \begin{pmatrix} 0 & 0 \\ 0 & \sigma(t, x, y) \end{pmatrix}, \quad \mathbf{B}(t) = (0, B(t))^*.$$

Again '$*$' denotes transposition of vectors or matrices. Thus (1) or (3) becomes

$$\mathbf{X}(t) = \mathbf{A} + \int_{t_0}^t \mathbf{Q}(s, \mathbf{X}(s))ds + \int_{t_0}^t \mathbf{S}(s, \mathbf{X}(s))\, d\mathbf{B}(s), \tag{4}$$

and certain standard Lipschitz conditions on q, σ, or equivalently on $\mathbf{Q}, \mathbf{S}$, give the following existence and uniqueness result corresponding to Proposition 3.2, obtained earlier. Here $\mathbf{X} : [t_0, b) \to \mathcal{X} \otimes \mathcal{X}$ [where $\mathcal{X} = L^2_{\mathbb{R}^n}(P)$], the Hilbert space of $\mathbb{R}^{2n}$-valued random variables, and $\|(x,y)\| = (\|x\|^2 + \|y\|^2)^{\frac{1}{2}}$, for $(x,y) \in \mathcal{X} \otimes \mathcal{X}$, is the norm in the tensor product space. First consider $\mathbf{X}(t)$ to be *any* vector process satisfying (4). [The latter may be canonically identified with $L^2_{\mathbb{R}^{2n}}(P)$.]

1.Theorem. *Suppose $\mathbf{Q}$ and $\mathbf{S}$ of (4) are measurable and locally bounded vector and matrix functions so that the integrals there are well defined, the last one forming an $L^{2,2}$-bounded process. Let $\mathbf{Q} : [t_0, b) \times \mathbb{R}^{2n} \to \mathbb{R}^n$ and $\mathbf{S} : [t_0, b) \times \mathbb{R}^{2n} \to B(\mathbb{R}^n, \mathbb{R}^n) = B(\mathbb{R}^n)$, the space of $n \times n$-matrices, be left continuous in t and satisfy a local Lipschitz condition: For each compact interval $[t_0, t] \subset [t_0, b)$ there is a constant K_t such that*

$$\|\mathbf{Q}(s, \mathbf{x}) - \mathbf{Q}(s, \mathbf{y})\| + \|\mathbf{S}(s, \mathbf{x}) - \mathbf{S}(s, \mathbf{y})\| \le K_t \|\mathbf{x} - \mathbf{y}\|,$$

for all $\mathbf{x}, \mathbf{y}$ in a ball of $\mathbb{R}^{2n}, t_0 \le s \le t, \| \cdot \|$ being the appropriate norms of $\mathbb{R}^n$ and $B(\mathbb{R}^n)$. Then there exists a unique process $\{\mathbf{X}(t), \mathcal{F}_t, t \in [t_0, b)\}$ that satisfies (4) with $\mathbf{X}(t_0) = \mathbf{A}$, so that it is the solution which is also an $L^{2,2}$-bounded process locally.

Proof. The argument is a modification of that given for Proposition 3.2, and hence a shortened version will suffice. To simplify, define the augmented (matrix and vector) functions $\tilde{\mathbf{S}}(t, \mathbf{x}) = (\mathbf{Q}, \mathbf{S})(t, \mathbf{x}), \tilde{\mathbf{B}}(t) = (t, \mathbf{B}(t))^*$, so that (4) becomes

$$\mathbf{X}(t) = \mathbf{X}(t_0) + \int_{t_0}^t \tilde{\mathbf{S}}(s, \mathbf{X}(s)) d\tilde{\mathbf{B}}(s), (\mathbf{X}(t_0) = \mathbf{A}). \tag{5}$$

For a given $\delta > 0$, consider the norm $\sup_{0 < t - t_0 < \delta} \|\mathbf{X}(t)\|_2 = \|\mathbf{X}(t)\|^\sim$, and $(T\mathbf{X})(t)$ for the right side of (5). Then the $L^{2,2}$-boundedness of $\tilde{\mathbf{B}}$ relative to a σ-finite μ on $\mathcal{B}([t_0, t]) \otimes \Sigma$, which on the "compact" set $[t_0, t] \times \Omega$ can and will be taken finite, implies

$$\|(T\mathbf{X})(t) - \mathbf{X}(t_0)\|_2^2 \le \int_{t_0}^t \int_\Omega \|\tilde{\mathbf{S}}(s, \mathbf{X}(s))\|^2 \, d\mu(s, \omega)$$
$$\le K_t^2 \mu([t_0, t] \times \Omega) < \infty,$$

for some constant $K_t > 0$. Then for a pair $\mathbf{X}_1, \mathbf{X}_2$ of processes the $L^{2,2}$-boundedness and the Lipschitz condition together yield the following inequality:

$$\|(T\mathbf{X}_1 - T\mathbf{X}_2)(t)\|_2 \le [\int_{t_0}^t \int_\Omega \|\tilde{\mathbf{S}}(s, \mathbf{X}_1(s)) - \tilde{\mathbf{S}}(s, \mathbf{X}_2(s))\|^2 \, d\mu(s, \omega)]^{\frac{1}{2}}$$
$$\le K_t [\int_{t_0}^t \int_\Omega \|\mathbf{X}_1(s) - \mathbf{X}_2(s)\|^2 \, d\mu(s, \omega)]^{\frac{1}{2}}$$
$$\le K_t \|\mathbf{X}_1 - \mathbf{X}_2\|^\sim (\mu(t, \Omega) - \mu(t_0, \Omega)). \tag{6}$$

Note that the given Lipschitz condition on $\mathbf{Q}, \mathbf{S}$, is equivalent to the one used on $\tilde{\mathbf{S}}$ since the topologies on the finite tensor products are equivalent. Now replacing T by T^2 and using the estimate (6), together with the fact that $\int_a^b f(s)\, dF(s) = \int_{F(a)}^{F(b)} f(F^{-1}(s))\, ds$ for a nondecreasing function $F(t) = \mu(t, \Omega)(= \mu([0, t], \Omega))$ (cf. Riesz-Nagy (1955), p.124), one gets

$$\|(T^2\mathbf{X}_1 - T^2\mathbf{X}_2)(t)\|_2 \leq \frac{1}{2!}(K_t[\mu(t, \Omega) - \mu(t_0, \Omega)])^2\|\mathbf{X}_1 - \mathbf{X}_2\|^{\sim}.$$

Hence by iteration,

$$\|(T^m\mathbf{X}_1 - T^m\mathbf{X}_2)(t)\|_2 \leq \frac{1}{m!}(K_t[\mu(t, \Omega) - \mu(t_0, \Omega)])^m\|\mathbf{X}_1 - \mathbf{X}_2\|^{\sim}.$$

Thus for large enough m, T^m is a contraction, and by the classical Banach Contraction Mapping Theorem there exists a unique $\mathbf{X}(t)$ in a ball of radius $\delta > 0$ centered at $\mathbf{X}(t_0)$ in $L^2_{\mathbb{R}^{2n}}(P)$ such that $(T\mathbf{X})(t) = \mathbf{X}(t)$. $\square$

The above argument uses only the fact that $\mathbf{X}$ is a $2n$-vector process that satisfies (5) or (4). However, the desired vector $\mathbf{X}$ for us must have the second half to be the derived process of the first half, and this is not obtained by the general method which is valid if $\mathbf{Q}(t, \mathbf{X}(t))$, and $\mathbf{S}(t, \mathbf{X}(t))$ depend on $\{X(s), s \leq t\}$ for all values without regard to conditions A(i)-A(iii). An alternative demonstration, restricting some generality of the above result, will now be presented establishing the stated additional property of the solution, and then it will be used with specializations for a refined analysis of the problem.

The following is a slightly generalized version of that given in (Rao (1995), Theorem VI.4.6 and Proposition VI.4.16). The necessary changes and relevant details (especially those that were omitted there) will be included. That result itself is an extension of the original one due to Borchers (1964) with the B-process as a square-integrable martingale, later specialized to a Brownian motion. Hereafter the $L^{2,2}$-bounded noise process is *always* assumed to be without fixed discontinuity points, and this will not be repeated.

2. Theorem. *Let H be an operator acting on absolutely continuous n-vector processes on (Ω, Σ, P) satisfying conditions $(Ai) - (Aiii)$ above, and $\{B(t), \mathcal{F}_t, t_0 \leq t < b\}$ be an $L^{2,2}$-bounded process. If $\sigma = (\sigma_{ij}) : [t_0, b) \times \mathbb{R}^n \times \mathbb{R}^n \to \mathbb{R}^{2n}$ is an $n \times n$ matrix satisfying (a Lipschitz condition):*

$$|\sigma_{ij}(t, \mathbf{x}_1, \mathbf{y}_1) - \sigma_{ij}(t, \mathbf{x}_2, \mathbf{y}_2)|^2 \leq K(\|\mathbf{x}_1 - \mathbf{x}_2\|^2 + \|\mathbf{y}_1 - \mathbf{y}_2\|^2),$$
$$|\sigma_{ij}(t, \mathbf{x}, \mathbf{y})|^2 \leq K(1 + \|\mathbf{x}\|^2 + \|\mathbf{y}\|^2), \tag{7}$$

where $\|\cdot\|$ is the Euclidean norm of $\mathbb{R}^n$ and $K > 0$ is a constant, then there exists a unique absolutely continuous process $\{X(t), \mathcal{F}_t, t_0 \leq t < b\}$ which is

250 CHAPTER 4. HIGHER ORDER SDEs

indeed a solution of (2), verifying the initial condition $X(t_0) = A$, $\dot{X}(t_0) = B$. Letting $V(t, \mathbf{x}) = (X(t), \dot{X}(t))^*$ to denote the solution of (2) with the given initial value $\mathbf{x}$, i.e., $V(t_0, \mathbf{x}) = (A, B)^*$, one has the estimate on the dependence of that value as:

$$E(\sup_{s \le t \le u} \|V(t, \mathbf{x}) - V(t, \mathbf{y})\|^2) \le C\|\mathbf{x} - \mathbf{y}\|^2, \tag{8}$$

where $C(= C_{s,u} > 0)$ is a constant.

Moreover, if the process $\{B(t), \mathcal{F}_t, t_0 \le t < b\}$ has independent increments, then the vector process $V(t)$ above is Markovian, and if the B-process has continuous paths, the $\{V(t), t \ge t_0\}$ has also the same property.

Proof. The existence proof is based on the classical Picard method of successive approximations. Thus if $X(t_0)$ is the initial value and $X^{(n)}(t)$ is the nth iterate, then one has to show that (i) $X^{(n)}(t) \to \bar{X}(t)$ a.e., (ii) $\dot{X}^{(n)}(t) \to \tilde{X}(t)$ a.e., (iii) $\dot{\bar{X}}(t) = \tilde{X}(t)$ a.e., and (iv) the pair $(\bar{X}(t), \dot{\bar{X}}(t))$ satisfies (2), so that $\{\bar{X}(t), t \ge t_0\}$ is a solution of (2). The uniqueness is established with a standard argument via Gronwall's inequality.

Now the hypotheses on HX and σ imply that the integrands in (2) are measurable relative to the predictable σ-algebra $\mathcal{O}$ determined by $\{\mathcal{F}_t, t \ge t_0\}$ in $\mathcal{B}([t_0, b)) \otimes \Sigma$ and then the $L^{2,2}$-bounded $B(t)$-process coincides with an $L^2(P)$-bounded semimartingale (cf., Rao (1995), Thm. VI.2.14). The σ-finite measure μ on $\mathcal{O}$ will be taken in the first instance to be finite (the general case being an easy extension). Then for any $\mathcal{O}$-measurable rectangle $C \times D$, one has:

$$\mu(C \times D) = \int_C \int_D d\mu(t, \omega) \le \int_C d\mu(t, \Omega) = \mu_1(C) \text{ (say)}.$$

Similarly

$$\mu(C \times D) \le \int_D d\mu([t_0, b), \omega) = \mu_2(D) \text{ (say)}.$$

Hence μ is dominated by both μ_1, μ_2 and then $\mu \ll \mu_1 \otimes \mu_2$ with a bounded density on $\mathcal{O}$. This condition shows that the proof of (Rao (1995), Thm. VI.4.6) applies verbatim here with β there replaced by μ_1 and P by μ_2. Since the rest of the hypothesis is the same as in the above reference, the existence and uniqueness follow. The uniqueness is also a simple consequence of (8), and the latter will be established.

Let us consider the dependence of the (vector) $V(t, \mathbf{x})$-process on the initial point. A proof of this was omitted in the above reference for space reasons. [It is one of the points which is not entirely trivial, and so the necessary detail will be added here.] Let $\mathbf{x} = (x_1, x_2), \mathbf{y} = (y_1, y_2)$ be the starting points of the vector processes $V(s, \mathbf{x}), V(s, \mathbf{y})$ on $[s, u]$. Then for $t \in [s, u]$ one has

$$
\begin{aligned}
X(t) - Y(t) &= (x_1 - y_1) + \int_s^t (\dot{X} - \dot{Y})(v) dv, \\
\dot{X}(t) - \dot{Y}(t) &= (x_2 - y_2) + (HX)(t) - (HY)(t) \\
&\quad + \int_s^t (\sigma(v, X(v), \dot{X}(v)) - \sigma(v, Y(v), \dot{Y}(v))) \, dB(v) \\
&= (x_2 - y_2) + R_1(t) + R_2(t), \text{(say)}.
\end{aligned}
\tag{9}
$$

Use of a simple Jensen-type inequality for these equations yields:

$$
\begin{aligned}
E(\sup_{t \le u} \|\dot{X}(t) - \dot{Y}(t)\|^2) \le 3[\|x_2 - y_2\|^2 + E(\sup_{t \le u} \|R_1(t)\|^2) \\
+ E(\sup_{t \le u} \|R_2(t)\|^2)].
\end{aligned}
\tag{10}
$$

By the conditions (Ai)-(Aiii) on the operator H of the theorem, it follows that

$$
E(\sup_{t \le u} \|R_1(t)\|^2) \le K\{\|x_1 - y_1\|^2 + \int_s^u E(\|\dot{X} - \dot{Y}\|^2(t)) \, d\alpha(t)\}, \tag{11}
$$

where $K(= K_{su} > 0)$ is a constant. Put $\Delta \sigma_{ij}(t) = \sigma_{ij}(t, X(t), \dot{X}(t)) - \sigma_{ij}(t, Y(t), \dot{Y}(t))$ and similarly $\Delta X(t), \Delta \dot{X}(t)$, with the matrix norm on σ, i.e., $\|\sigma\|^2 = \sum_{i,j=1}^n |\sigma_{ij}|^2$, to get a bound on $R_2(t)$:

$$
E(\sup_{t \le u} \|R_2(t)\|^2) \le n \sum_{i,j=1}^n E(\sup_{t \le u} (\int_s^t \Delta \sigma_{ij}(v) \, dB(v))^2). \tag{12}
$$

Since as noted already $B(t)$ can be identified with a semimartingale, so that $B(t) = M(t) + A(t)$ for a square integrable martingale and a process of bounded variation, one has $\|B(t)\| \le \|M(t)\| + |A|(t) = \tilde{M}(t)$ so that it is dominated by a positive submartingale. Hence (12) is bounded above (using the $L^2(P)$-submartingale maximal inequality), by the following:

$$
\begin{aligned}
LHS(12) &\le 4n \sum_{i,j=1}^n \int_s^u E(|\Delta \sigma_{ij}(t)|^2) \, d\tilde{\alpha}_j(t) \\
&\le 4Kn^2 \int_s^u E(\|\Delta X(t)\|^2 + \|\Delta \dot{X}(t)\|^2) \, d\bar{\alpha}(t), \text{ by (2)},
\end{aligned}
\tag{13}
$$

where $\bar{\alpha}(\cdot) = \sum_{j=1}^{n} \tilde{\alpha}_j(\cdot)$, and K is the Lipschitz constant. Also using Jensen's inequality again,

$$\|\Delta X(t)\|^2 \le 2(\|x_1 - y_1\|^2 + (t - s)\int_s^t \|\dot{X}(v) - \dot{Y}(v)\|^2 \, dv).$$

Substituting this in (12) one finds

$$E(\sup_{t \le u} \|R_2(t)\|^2) \le C_1\|x_1 - y_1\|^2 + C_2\int_s^u E(\|\Delta X(t)\|^2) \, d\alpha_0(t), \qquad (14)$$

where $d\alpha_0(t) = dt + d\bar{\alpha}(t), C_1 = 8K^2n^2(\alpha_0(u) - \alpha_0(s))$ and $C_2 = C_1 + 4K^2n^2$. With (14), (11) becomes

$$LHS(11) \le C_3\|\mathbf{x} - \mathbf{y}\|^2 + C_4\int_s^u E(\|\Delta\dot{X}(t)\|^2) \, d\alpha_0(t), \qquad (15)$$

where $C_3 = 3\max(1, C_1, K)$ and $C_4 = 3\max(K, C_2)$. From (15) it follows that

$$E(\|\Delta\dot{X}(t)\|^2) \le C_3\|\mathbf{x} - \mathbf{y}\|^2 + C_4\int_s^u E(\|\Delta\dot{X}(t)\|^2) \, d\alpha_0(t). \qquad (16)$$

By Gronwall's inequality, (16) implies for $s \le t \le u$,

$$E(\|\Delta\dot{X}(t)\|^2) \le C_5\|\mathbf{x} - \mathbf{y}\|^2,$$

where $C_5 = \max(C_3, C_4C_3 \exp(C_4(\alpha_0(u) - \alpha_0(s))))$. It therefore follows from (14)-(16) and (9) that

$$E[\sup_{t \le u} \|V(s, X) - V(s, Y)\|^2]$$
$$\le E[\sup_{t \le u} \|\Delta X(t)\|^2] + E[\sup_{t \le u} \|\Delta\dot{X}(t)\|^2]$$
$$\le C\|\mathbf{x} - \mathbf{y}\|^2,$$

with $C = 2 + C_5[1 + 2(u - s)^2]$ which depends only on u, s. This establishes the inequality about the dependence of the solution on the initial data.

Regarding the last part, the conditions on $\Delta_s^t(HX)$ being independent of the s-past and t-future ($s < t$), and on $B(\cdot)$ having independent increments imply that the solution process is Markovian. The simple proof of this fact is already included in (Rao (1995), Prop. VI.4.16) and need not be repeated. If $B(\cdot)$ has a.a. continuous paths, then it can be seen that the solution process also has the same property. When $B(\cdot)$ is a Brownian motion and (2) is a linear equation with constant coefficients, the solution process is a Feller (hence strongly Markovian) process, as discussed in the above reference. □

It is appropriate and useful to make an observation on the last part of the preceding result here which is stated in the following form:

3. Note. The solution process $U(t) = (X(t), \dot{X}(t))^*$ of (1), as a vector, is a Markov process which has stationary transition probabilities when the coefficients q and σ are independent of the time variable. Hence one can associate a semi-group of positive operators on $B(\mathbb{R}^{2n})$ of bounded Borel functions on $\mathbb{R}^{2n} \to \mathbb{R}$ with uniform norm. Moreover, as the work of the next sections shows, when $B(t)$ is a Brownian motion then (under mild conditions) this semi-group of operators maps continuous functions into themselves. This fact implies that the vector $U(t)$-process will be a Feller process (on the enlarged state space $\mathbb{R}^+ \times \mathbb{R}^{2n}$) and hence strongly Markovian there. Thus these specializations admit a more refined analysis of the solutions of equations of the form given by (1). It should be remarked that the component process $X(t)$ itself will *not* be Markovian even when $B(\cdot)$ is a Brownian motion.

To continue with a detailed analysis of the solution process of (1), several estimates on the second (mixed) moments of the process $\{X(t), t \in [t_0, b)\}$ and its derived process $\{\dot{X}(t), t \in [t_0, b)\}$ are needed. These will now be established in the next section and used for a finer path analysis of the solution process in subsequent work.

4.5. INFINITESIMAL CHARACTERISTICS OF SOLUTIONS.

From now on, solutions of equations of the form (1) of the preceding section will be considered. Consequently the resulting statement of Theorem 4.2 is given, for a convenient reference, as follows.

1. Theorem. *Suppose that $\{B(t), \mathcal{F}_t, t \in J = [t_0, b]\}$ is an $L^{2,2}$-bounded (n-vector) process on (Ω, Σ, P) relative to a σ-finite measure μ on $\mathcal{B}(J) \otimes \Sigma$, and $q = (q_i) : J \times \mathbb{R}^n \times \mathbb{R}^n \to \mathbb{R}^n, \sigma = (\sigma_{ij}) : J \times \mathbb{R}^n \times \mathbb{R}^n \to \mathbb{R}^{2n}$ are Borel functions satisfying the following Lipschitz-type conditions:*

$$|q_i(t, \mathbf{x}_1, \mathbf{y}_1) - q_i(t, \mathbf{x}_2, \mathbf{y}_2)|^2 + |\sigma_{ij}(t, \mathbf{x}_1, \mathbf{y}_1) - \sigma_{ij}(t, \mathbf{x}_2, \mathbf{y}_2)|^2$$
$$\leq K(\|\mathbf{x}_1 - \mathbf{x}_2\|^2 + \|\mathbf{y}_1 - \mathbf{y}_2\|^2), \qquad (1)$$
$$|q_i(t, \mathbf{x}, \mathbf{y})^2 + |\sigma_{ij}(t, \mathbf{x}, \mathbf{y})|^2 \leq K(1 + \|\mathbf{x}\|^2 + \|\mathbf{y}\|^2), \ \mathbf{x}, \mathbf{y} \in \mathbb{R}^n,$$

where $\| \cdot \|$ is the Euclidean norm of $\mathbb{R}^n$ and $K > 0$ is an absolute constant. Then there exists a unique absolutely continuous process $\{X(t), \mathcal{F}_t, t \in J\}$ with its derived process $\{\dot{X}(t), \mathcal{F}_t, t \in J\}$ satisfying the initial value problem:

$$d\dot{X}(t) = q(t, X(t), \dot{X}(t))dt + \sigma(t, X(t), \dot{X}(t))dB(t)$$
$$X(t_0) = A, \ \dot{X}(t_0) = B, \qquad (2)$$

for given $\mathcal{F}_{t_0}$-measurable random variables $A, B \in L^2(P)$. The dependence on the initial values is continuous in that if $A = \mathbf{x}, B = \mathbf{y}$ are constants, and $X^{\mathbf{x}}, X^{\mathbf{y}}$ denote the solutions of (2) starting at $\mathbf{x}, \mathbf{y}$ then

$$E(\sup_{t \in J} \|X^{\mathbf{x}}(t) - X^{\mathbf{y}}(t)\|^2) \leq C\|\mathbf{x} - \mathbf{y}\|^2, \qquad (3)$$

for some constant $C(= C_J > 0)$. If further $B(\cdot)$ has independent increments, then the solution $\{X(t), t \in J\}$ of (2) is Markovian with a.a. continuous paths when B has the last property.

On the other hand, if $q_i(t, \cdot, \cdot)$ and $\sigma_{ij}(t, \cdot, \cdot)$ are continuously differentiable, then $X^{\mathbf{x}}(t) = X(t, \mathbf{x})$ is also continuously (partially) differentiable relative to $x_i, i = 1, \dots, n$ where $\mathbf{x} = (x_i) \in \mathbb{R}^n$ besides being left continuous with right limits on J.

Outline of proof. This is a specialization and restatement of the earlier result, and only the last assertion needs a verification. But by the first part, (2) can be expressed as a first order $2n$-vector differential equation and, in fact, as that given by Theorem 4.1. Consequently one only has to deal with the statement for $\mathbf{X}(t) = (X(t), \dot{X}(t))$-process. But the present hypothesis implies that used for the solution of (the $2n$-vector) Equation (4) of Section 4. Since the $B(t)$-process does not depend on $\mathbf{x}$, the classical differentiation process can be applied to (2), because q_i, σ_{ij} have continuous partial derivatives, the $\mathbf{X}(t)$-process has a weak (or distributional) derivative relative to each x_i, and that the latter are continuous with probability one implying the strong derivative. The details are tedious but straightforward (cf., e.g., the discussion preceding Prop. 3.4, and also Protter (1990), p.250). This terminates our sketch. $\square$

Suppose that $\{B(t), \mathcal{F}_t, t \in J = [t_0, b)\}$ is an $L^{2,2}$-bounded process with independent increments, and consider the differential Equation (2) with conditions on q, σ such that the initial value problem has a unique solution. By the above result the process $\{\mathbf{X}(t), \mathcal{F}_t, t \in J\}$, where $\mathbf{X}(t) = (X(t), \dot{X}(t))^*$, $X(t)$ being the solution of (2), is a Markov process with range, also called state space, $\mathbb{R}^{2n}$ and $E(\|\mathbf{X}(t)\|_2^2) < \infty$. Let the conditional probability functions of this system, denoted $p(\cdot, \cdot; \cdot, \cdot)$, be given by (for $s \leq t$ and $\mathbf{x} \in \mathbb{R}^{2n}, A \subset \mathbb{R}^{2n}$ a Borel set):

$$p(s, \mathbf{x}; t, A) = P[\mathbf{X}(t) \in A | \mathbf{X}(s) = \mathbf{x}]. \qquad (4)$$

Assume that one can select a version of p such that it is a regular conditional probability, i.e., the following conditions hold: (i) $p(s, \mathbf{x}; t, \cdot)$ is a measure on $\mathcal{B}(\mathbb{R}^{2n})$, (ii) $p(s, \cdot; t, A)$ is $\mathcal{B}(\mathbb{R}^{2n})$-measurable, and (iii) $p(t, \mathbf{x}; t, A) = \chi_A$. As a consequence of the Markovian property of $\mathbf{X}(t)$, one has the Chapman-Kolmogorov equation:

$$p(s, \mathbf{x}; t, A) = \int_{\mathbb{R}^{2n}} p(u, \mathbf{y}; t, A) p(s, \mathbf{x}; u, d\mathbf{y}), \qquad (5)$$

for each $s \leq u \leq t$. This translates into the following operator equation on $B(\mathbb{R}^{2n})$, the space of bounded Borel (real) functions on $\mathbb{R}^{2n}, s \leq t$.

$$
\begin{aligned}
(U_{st}f)(\mathbf{x}) &= \int_{\mathbb{R}^{2n}} f(\mathbf{y})p(s, \mathbf{x}; t, dy) \\
&= E(f(\mathbf{X}(t))|\mathbf{X}(s) = \mathbf{x}), \ f \in B(\mathbb{R}^{2n}),
\end{aligned} \tag{6}
$$

and then for $s \leq v \leq t$,

$$
(U_{sv}U_{vt}f)(\mathbf{x}) = (U_{st}f)(\mathbf{x}), \ \mathbf{x} \in \mathbb{R}^{2n}. \tag{7}
$$

Thus $\{U_{st}, t_0 \leq s \leq t < b, U_{ss} = id\}$ is a family of positive (i.e., $U_{st}f \geq 0$ if $f \geq 0$) evolution operators, which is a contraction if uniform norm is used on $B(\mathbb{R}^{2n})$. The regularity of the conditional probabilities of (4) is automatic if the process is given a canonical representation, i.e., $\Omega = (\mathbb{R}^{2n})^J, \Sigma$ is the cylinder σ-algebra, and P is the induced probability measure from the original space (cf., e.g., Rao (1993), Ch.5). This can be taken for convenience. However, it must be observed that there is in general no algorithm or other constructive methods to find such regular conditional probabilities and it remains a nontrivial unresolved problem in the subject. As discussed in (Rao (1993), Sec. 5.4) "proper" regular conditioning is essentially equivalent to the existence of a strong lifting, and lifting operation depends on the choice axiom so that a general method of construction does not seem simple. So particular cases of interest should be considered for this purpose. In fact this is essentially the same as constructing Radon-Nikodým derivatives of differentiation theory. Nevertheless, it is a common practice in the literature to term such *regular* conditional measures of a Markov process a family of *transition probability functions*, and are assumed as part of the "given" data. [In the preceding reference, this point on construction, which was not emphasized in the existing literature, has been discussed in considerable detail.] It will also be seen from the following work that when $B(\cdot)$ has stationary independent increments and q, σ do not involve the time variable, then the solution process $X(t)$ of (2) is a Markov process with stationary transition probabilities.

Now the procedure to be used for a sharper analysis of the solution process is to calculate the infinitesimal operator of the evolution Equation (7) which (in the most interesting cases) turns out to be a second order differential operator on a suitable function space and through it one studies the path behavior of the $X(t)$-process using the ideas and methods of Feller (1952). The next result introduces the desired property.

2. Proposition. *Let $\{U_{st}, t_0 \leq s \leq t < \infty\}$ be a family of (bounded) evolution operators on a Banach function space $\mathbf{B}$ of Borel functions on $\mathbb{R}^n$. If $\{X(t), \mathcal{F}_t, t \geq 0\}$ is a Markov process with $\{p(s, \xi; t, d\zeta), t_0 \leq s \leq t < \infty\}$,*

as the transition probability family and U_{st} as the associated evolutions, then
$\{(U_{st}f)(X(t)), \mathcal{F}_t, t_0 \leq s \leq t < \infty\}$ *is a martingale for all f in the domain*
of the U_{st}-family for which the indicated random variables are integrable.

Proof. Consider for each $t_0 \leq s \leq t$, and $f \in Dom(U_{st})$,

$$
\begin{aligned}
Y_{st}^f &= (U_{st}f)(X(s)) = E^{\sigma(X(s))}(f(X(t))) \\
&= \int_{\mathbb{R}^n} f(y)p(s, X(s); t, dy) \\
&= E^{\mathcal{F}_s}(f(X(t))), \text{ since the process is Markov.} \tag{8}
\end{aligned}
$$

Applying the operator $E^{\mathcal{F}_s}, s \leq s_1 \leq t$, to both sides of (8), one obtains

$$
\begin{aligned}
E^{\mathcal{F}_s}(Y_{s_1 t}^f) &= E^{\mathcal{F}_s}(U_{s_1 t}f)(X(s_1)) \\
&= E^{\mathcal{F}_s}(E^{\mathcal{F}_{s_1}}(f(X(t))), \text{ by (8)}, \\
&= E^{\mathcal{F}_s}(f(X(t))), \text{ since } \mathcal{F}_s \subset \mathcal{F}_{s_1}, \\
&= Y_{st}^f, \text{ by (8)}.
\end{aligned}
$$

Thus $\{Y_{st}^f, \mathcal{F}_s, t_0 \leq s \leq t < \infty\}$ is a martingale whenever f is such that
$(U_{t_0 s}f)(X(s))$ is integrable for $t_0 \leq s \leq t$. $\square$

This property of $\{(U_{st}f)(X(t)), t \geq s\}$ for fixed $s \geq t_0$ is of interest in
this work. Moreover, using the evolution operator Equation (7), and the
fact that $U_{ss} = id.$, one has

$$
\lim_{t \downarrow s} \frac{U_{st}f - U_{ss}f}{t - s} = G_s f, \quad f \in Dom(U_s) \tag{9}
$$

the limit taken in the strong operator topology, the G_s being the infinites-
imal operator of the family at s. Hence $\frac{\partial U_{st}}{\partial t} f = U_{st}G_t f$. One can also
express (9), again with the evolution identity, as:

$$
U_{st}f = f + \int_s^t U_{sv}G_v f dv, \ f \in \cap_{s \leq v \leq t} Dom(G_v), \tag{10}
$$

from the general theory, where the integral here is of multiplicative Riemann
type (cf., e.g., Dalecky and Krein (1974), p.102). (Our U_{st} is denoted as
U_{ts} in this reference.) In the case of stationary transitions of the Markov
process $X(t)$, one has $U_{st} = U_{t-s}$ and G_s is independent of s; it commutes
with U_{t-s}. The latter fact is not valid for the general case. In the stationary
context (10) becomes, letting $V_s = U_{t_0 s} = U_{s-t_0}$ for $s \geq t_0$,

$$
V_s f = f + \int_{t_0}^s V_u G f du, \ f \in Dom(G), \tag{11}
$$

the family $\{V_s, s \geq t_0\}$ being a (strongly continuous) semi-group of bounded operators on $B(\mathbb{R}^n)$. Hence one has the following refinement.

3.   Corollary.   *If $\{X(t), \mathcal{F}_t, t \geq t_0\}$ is a Markov process with stationary transition probabilities and $\{V_s, s \geq t_0\}$ as the associated semi-group of bounded linear perators on $B(\mathbb{R}^n)$ with G as its infinitessimal operator, then for each bounded $f \in Dom(G)$, the process $\{f(X(t)) + \int_{t_0}^t (V_s G) f(X(t)) ds, \mathcal{F}_t, t \geq t_0\}$ is a martingale.*

The preceding result, along with (10) (or (11)) shows the necessity of calculating the generator G (or G_s) and deciding precisely the corresponding domains of definitions of these operators in order to use their properties here. This will take the bulk of our ensuing work in which estimates of integrals are first obtained for the general case; then it is refined when the driving force is a Brownian motion and the coefficients q, σ are independent of the time variable.

4. Theorem. *Suppose the hypothesis of Theorem 1 holds for Equation (2), and the $B(t)$-process has no discontinuities, so that there is a unique separable solution $\{X(t), \mathcal{F}_t, t \geq t_0\}$ for the given initial data. Then the following (conditional) moment estimates hold (a.e.) for the increments of X and $\dot{X}$ on $I_h^t = [t, t+h]$ as $h \downarrow 0$: ($|\cdot|$ denoting the norm of $\mathbb{R}^n$)*

1.(a) $E[\sup_{u \in I_h^t} |X(u) - \dot{X}(t)|^2] = O(h^2)[1 + E(|X(t)|^2) + E(|\dot{X}(t)|^2)],$

(b) $E[\sup_{u \in I_h^t} |\dot{X}(u) - \dot{X}(t)|^2] = O(h)[1 + E(|X(t)|^2) + E(|\dot{X}(t)|^2)];$

(a') $E^{\sigma(X(t))}[\sup_{u \in I_h^t} |X(u) - X(t)|^2] = O(h^2)[1 + |(X(t), \dot{X}(t))|^2],$

(b') $E^{\sigma(X(t))}[\sup_{u \in I_h^t} |\dot{X}(u) - \dot{X}(t)|^2] = O(h)[1 + |(X(t), \dot{X}(t))|^2].$

2.(a) $E^{\sigma(X(t))}[X(t+h) - X(t)] = \dot{X}(t)h$

$\qquad + O(h^{3/2})[1 + |(X(t), \dot{X}(t))|^2]^{1/2},$

(b) $E^{\sigma(X(t))}[\dot{X}(t+h) - \dot{X}(t)] = -\int_t^{t+h} q(u, X(t), \dot{X}(t)) du$

$\qquad + O(h^{3/2})[1 + |(X(t), \dot{X}(t))|^2]^{1/2},$

(c) $E^{\sigma(X(t))}[|X(t+h) - X(t)|^2] = O(h^2)[1 + |(X(t), \dot{X}(t))|^2],$

(d) $E^{\sigma(X(t))}[(X(t+h) - X(t))(\dot{X}(t+h) - \dot{X}(t))] = O(h^{3/2})$

$\qquad [1 + |(X(t), \dot{X}(t))|^2],$

(e) $E^{\sigma(X(t))}[|\dot{X}(t+h) - \dot{X}(t)|^2] = \int_t^{t+h} \sigma(u, X(t), \dot{X}(t))^2 \mu(du, \cdot)$

$$+ O(h^{3/2})[1 + |(X(t), \dot{X}(t))|^2],$$

where μ is the dominating measure of $B(\cdot)$ in the $L^{2,2}$-bounded hypothesis. Here the $O(h)$ terms are uniform in t for compact intervals, and assertions 2(a)–2(e) hold with probability 1.

Proof. For notational simplicity the (scalar) case $n = 1$ will be discussed. First consider 1(a). Since $X(\cdot)$ is absolutely continuous with derivative $\dot{X}(\cdot)$ one has on using a separable version of the derived process, with $u > s$,

$$(\Delta X)(u) = X(u) - X(s) = (u - s)\dot{X}(s) + \int_s^u [\dot{X}(r) - \dot{X}(s)]dr, \qquad (12)$$

and since $(a + b)^2 \le 2(a^2 + b^2)$,

$$|(\Delta X)(u)|^2 \le 2h^2|\dot{X}(s)|^2 + 2h^2 \int_s^{s+h} |\dot{X}(r) - \dot{X}(s)|^2 dr, \qquad (12')$$

$$\le 2h^2|\dot{X}(s)|^2 + 2h^2 \sup_{r \in I_h^s} |\dot{X}(r) - \dot{X}(s)|^2 (\text{a.e.}). \qquad (13)$$

Hence taking expectations and substituting the bound from 1(b) for the second term proves 1(a), so that it is necessary to establish 1(b) now.

Let $\Delta \dot{X}(u)$ be the increment of the derived process, and similarly define

$$\Delta q(u) = q(u, X(u), \dot{X}(u)) - q(u, X(t), \dot{X}(t)),$$
$$\Delta \sigma(u) = \sigma(u, X(u), \dot{X}(u)) - \sigma(u, X(t), \dot{X}(t))$$

for $u > t$. Since $B(\cdot)$ is $L^{2,2}$-bounded, there is a σ-finite measure μ associated with it which is finite on "compact" sets $I_h^t \times \Omega$, (cf., e.g., the proof of Theorems 4.1 and 4.2 above) and in this case it can be taken finite. Let $F(t) = \mu(I_h^t \times \Omega)$ so that $\mu \ll F \otimes P$ and $F(\cdot)$ is even continuous since $B(t)$ has no discontinuities. Then one can use the argument of (Rao (1995), p.518), but the detail will be included here for completeness. With a modified μ if necessary, one can assume that F is continuous.

Now from (2), after integration and subtraction, one gets

$$\Delta \dot{X}(t) = -\int_s^t q(u, X(u), \dot{X}(u))du + \int_s^t \sigma(u, X(u), \dot{X}(u))dB(u)$$

$$= -\int_s^t \Delta q(u)du + \int_s^t \Delta \sigma(u)dB(u)$$

$$- \int_s^t q(u, X(s), \dot{X}(s))du + \int_s^t \sigma(u, X(s), \dot{X}(s))dB(u).$$

$$(14)$$

Hence using the Jensen inequality, one finds for $u \in I_h^t$ in lieu of $[s, u]$, and fixed $t, h > 0$:

$$|\Delta \dot{X}(t)|^2 \leq 4h\Big(\int_{I_h^t} [|\Delta q(u)|^2 + |q(u, X(t), \dot{X}(t))|^2]du\Big)$$

$$+ 4[(\int_t^u \Delta \sigma(v)dB(v))^2 + (\int_t^u \sigma(v, X(t), \dot{X}(t))dB(v))^2],$$

$$= 4h(\alpha_1 + \alpha_2) + 4(\alpha_3(u) + \alpha_4(u)), \text{ (say)}. \tag{15}$$

Let us now estimate the α_i separately. Using the Lipschitz conditions, the boundedness of q (and σ), and (13), one has:

$$E(\alpha_1 + \alpha_2) \leq K_1^2\Big\{\int_{I_h^t} E[|\Delta X(u)|^2 + |\Delta \dot{X}(t)|^2$$

$$+ 1 + |X(t)|^2 + |\dot{X}(t)|^2]du\Big\}$$

$$\leq K_1^2\{2h^3 E(|\dot{X}(t)|^2) + hE[1 + |X(t)|^2 + |\dot{X}(t)|^2]$$

$$+ (1 + 2h^2)\int_{I_h^t} E(|\Delta \dot{X}(u)|^2)du\}. \tag{16}$$

Next a bound should be found for α_3 and α_4 which involve integrals relative to $B(\cdot)$. The conditions on q and σ imply that they are predictable and the $L^{2,2}$-bounded $B(\cdot)$ agrees with a semimartingale in that $|B| = |B_1 + A_1| \leq |B_1| + |A_1|$, where A_1 is a process of bounded variation and B_1 a (local) martingale (and both are in $L^2(P)$ by our assumptions on B). Hence using the domination measure $F \otimes P$ on the latter submartingale, its maximal inequality, and the given Lipschitz conditions one obtains the following bounds on α_3 and α_4:

$$E(\sup_{u \in I_h^t} \alpha_3(u)) \leq 4\int_{I_h^t} E(|\Delta \sigma(u)|^2)dF(u)$$

$$\leq 4K_2^2 \int_{I_h^t} E[|\Delta X(u)|^2 + |\Delta \dot{X}(u)|^2]dF(u)$$

$$\leq 8K_2^2 h^2 \Delta F\{E(|\dot{X}(t)|^2) + \int_{I_h^t} E(|\Delta \dot{X}(u)|^2)du\}$$

$$+ 4K_2^2 \int_{I_h^t} E(|\Delta \dot{X}(u)|^2 dF(u), \text{ (by (12'))},$$

$$= K_2^2 h^2 \Delta F\{E(|\dot{X}(t)|^2) + \int_{I_h^t} E(|\Delta \dot{X}(u)|^2)du\}$$

$$+ 4K_2^2 \int_{I_h^t} E(|\Delta \dot{X}(u)|^2)dF(u). \tag{17}$$

Similarly,

$$E(\sup_{u\in I_h^t} \alpha_4(u)) \le 4\int_{I_h^t} E(\sigma(u, X(t), \dot{X}(t))^2)dF(u)$$

$$\le 4K_2^2 \int_{I_h^t} E(1 + |X(t)|^2 + |\dot{X}(t)|^2)dF(u)$$

$$= 4K_2^2 \Delta F(1 + E(|X(t)|^2 + |\dot{X}(t)|^2)). \tag{18}$$

Substituting (16)–(18) in (15), one gets

$$E(\sup_{u\in I_h^t} |\Delta\dot{X}(u)|^2) \le (4K_1^2 h^2 + 16K_2^2\Delta F) + (4K_1^2 h^2 + 16K_2^2\Delta F)$$

$$\times E(|X(t)|^2) + (8K_1^2 h^4 + 32K_2^2 h^2\Delta F + 4K_1^2 h^2$$

$$+ 16K_2^2\Delta F)E(|\dot{X}(t)|^2) + (8K_1^2 h^2 + 4K_1^2 h\Delta F$$

$$+ 16K_2^2)\int_{I_h^t} E(|\Delta\dot{X}(u)|^2)(du + dF(u))$$

$$\le C_1\Delta G(1 + E(|X(t)|^2 + |\dot{X}(t)|^2)$$

$$+ C_2\int_{I_h^t} E(|\Delta\dot{X}(u)|^2)dG(u), \tag{19}$$

where C_1 is the largest of the constants in the first three terms, C_2 is that multiplying the integral, and $G(t) = t + F(t)$. Note that C_1, C_2 depend only on the compact interval I_h^t, and all quantities in (19) are finite. It now follows by the Gronwall inequality applied to (19) that

$$E(|\Delta\dot{X}(u)|^2) \le C_1\Delta G(1 + E(|X(t)|^2 + |\dot{X}(t)|^2))\exp(C_2\Delta G),$$

for all $u \in I_h^t$. Substituting this in the right side of (19) and applying again Gronwall's inequality for the resulting expression it follows that

$$E(\sup_{u\in I_h^t} |\Delta\dot{X}(u)|^2) \le C\Delta G(1 + E(|X(t)|^2 + |\dot{X}(t)|^2)), \tag{20}$$

with $C = C_1 \sup_{u\in I_h^t}(1 + [G(u+h) - G(u)]\exp(C_2[G(u+h) - G(u)])) < \infty$. Since $G(t+h) - G(t) = o(h)$, uniformly for t in compact intervals, this establishes 1(b), and hence also 1(a), as noted earlier.

The estimates of 1(a') and 1(b') can be obtained from the preceding analysis as follows. Replacing the expectation operator $E(\cdot)$ with the conditional expectation $E^{\sigma(X(t))}(\cdot)$, and noting that $E^{\sigma(X(t))}(Z) = Z$, a.e., for any $\sigma(X(t))$-measurable random variable $Z \in L^1(P)$, one gets the desired bounds immediately, since $E^{\sigma(X(t))}(|X(t)|^2 + |\dot{X}(t)|^2) = |(X(t), \dot{X}(t))|^2$ in the Euclidean norm. The same constants obtain in this case also. So all parts of 1. are established, and we now proceed to prove those of 2.

For simplicity, let us set $X(t) = x_1$ and $\dot{X}(t) = x_2$. So (2) can be expressed as a system on the compact interval $I_h^t = [t, t+h]$ as:

$$X(t+h) - x_1 = x_2 + \int_{I_h^t} [\dot{X}(t) - x_2] du, \tag{21}$$

and

$$\begin{aligned}
\dot{X}(t+h) - x_2 = &- \int_{I_h^t} [q(u, X(u), \dot{X}(u)) - q(u, x_1, x_2)] du \\
&+ \int_{I_h^t} [\sigma(u, X(u), \dot{X}(u)) - \sigma(u, x_1, x_2)] dB(u) \\
&- \int_{I_h^t} q(u, x_1, x_2) du + \int_{I_h^t} \sigma(u, x_1, x_2) dB(u).
\end{aligned} \tag{22}$$

In this form, 2. of the statement will be established, using the results of part 1. Thus from (21) and the conditional CBS inequality one has

$$\begin{aligned}
|E^{\sigma(X(t))}(X(t+h) - x_1) - x_2 h| &\leq \int_{I_h^t} E^{\sigma(X(t))}(|\dot{X}(u) - x_2|) du \\
&\leq h[E^{\sigma(X(t))}(\sup_{u \in I_h^t} |\dot{X}(u) - x_2|^2)]^{\frac{1}{2}} \\
&\leq h(C\Delta G)^{\frac{1}{2}}(1 + |(x_1, x_2)|^2)^{\frac{1}{2}}.
\end{aligned}$$

Since $\Delta G = O(h)$, and $C > 0$ is a constant, this implies 2(a) with probability 1.

Next consider (22) together with the Lipschitz condition, the local boundedness of q, σ, (cf. (1)), the $L^{2,2}$-boundedness of $B(\cdot)$, and the estimates of 1(b). One can simplify it as follows.

$$\begin{aligned}
|E^{\sigma(X(t))}(\dot{X}(t+h) - x_2) &+ \int_{I_h^t} q(u, x_1, x_2) du| \\
&\leq |\int_{I_h^t} E^{\sigma(X(t))}(\Delta q(u)) du| + |\int_{I_h^t} E^{\sigma(X(t))}(\Delta \sigma(u)) dB(u)| \\
&+ |\int_{I_h^t} \sigma(u, x_1, x_2) dB(u)|, \\
&\leq [h \int_{I_h^t} E^{\sigma(X(t))}(\Delta q(u))^2]^{\frac{1}{2}} \\
&+ [\Delta F \int_{I_h^t} E^{\sigma(X(t))}(\Delta \sigma(u))^2 dF(u)]^{\frac{1}{2}}
\end{aligned}$$

$$+ (\Delta F \int_{I_h^t} \sigma(u, x_1, x_2)^2 dF(u))^{\frac{1}{2}},$$

$$\leq [hK_1 \int_{I_h^t} E^{\sigma(X(t))} \{|X(u) - x_1|^2 + |\dot{X}(u) - x_2|^2\} du]^{\frac{1}{2}}$$

$$+ [\Delta F K_1 \int_{I_h^t} E^{\sigma(X(t))} (|X(u) - x_1|^2 + |\dot{X}(u) - x_2|^2) dF(u)]^{\frac{1}{2}}$$

$$+ [(\Delta F) K_2 (\Delta F (1 + |(x_1, x_2)|^2))]^{\frac{1}{2}}$$

$$\leq [hK_1 hO(h)(1 + |(x_1, x_2)|^2)]^{\frac{1}{2}}$$

$$+ [(K_1 \Delta F)(\Delta F) O(h)(1 + |(x_1, x_2)|^2)]^{\frac{1}{2}}$$

$$+ (\Delta F K_2)^{\frac{1}{2}} (\Delta F (1 + |(x_1, x_2)|^2))^{\frac{1}{2}},$$

$$= O(h^{3/2})(1 + |(x_1, x_2)|^2)^{\frac{1}{2}},$$

since $\Delta F = O(h)$ and $K_1, K_2 > 0$ are constants of the Lipschitz conditions on q, σ. This gives 2(b) with probability 1.

The estimate 2(c) follows from the corresponding case in part 1(a') and 2(d) via the conditional CBS inequality along with 1(a') and 1(b'). As for 2(e), it may be expressed as:

$$E^{\sigma(X(t))}(|\dot{X}(t) - x_2|^2) = \sum_{i=1}^{4} E^{\sigma(X(t))}(\alpha_i^2)$$

$$+ 2 \sum_{1 \leq i < j \leq 4} E^{\sigma(X(t))}(\alpha_i \alpha_j), \qquad (23)$$

where the α_j stand for

$$\alpha_1 = -\int_{I_h^t} \Delta q(u) du, \qquad \alpha_2 = \int_{I_h^t} \Delta \sigma(u) dB(u),$$

$$\alpha_3 = -\int_{I_h^t} q(u, x_1, x_2) du, \qquad \alpha_4 = \int_{I_h^t} \sigma(u, x_1, x_2) dB(u). \qquad (24)$$

Note that $E^{\sigma(X(t))}(\alpha_4^2) \leq \int_{I_h^t} \sigma(u, x_1, x_2)^2 d\mu$, by the $L^{2,2}$-boundedness and $\mu = F \otimes P$ may be taken, where F is continuous. Then one has

$$|E^{\sigma(X(t))}(|\dot{X}(t) - x_2|^2) - \int_{I_h^t} \sigma(u, x_1, x_2)^2 d\mu|$$

$$\leq \sum_{j=1}^{3} E^{\sigma(X(t))}(\alpha_j^2) + 2 \sum_{1 \leq i < j \leq 4} E^{\sigma(X(t))}(|\alpha_i \alpha_j|),$$

and each term on the right can be bounded by the preceding work in 2(a) and 2(b), together with the conditional CBS inequality for the second term on the right. Here one also uses the (local) boundedness of σ. Substituting these estimates in (23) one gets 2(e) as stated.

The preceding computations show that on compact I_h^t all the bounds are uniform, and hence the last statement is a consequence. Thus all assertions of the theorem are established. $\square$

This result and the estimates enable one to study certain "measure-valued" differential equations and their infinitesimal characteristics in the subject. However the details are not completed at this time.

Although the hypotheses of the preceding result are all satisfied if $B(\cdot)$ is a Brownian motion, one can refine the work in the latter case utilizing the additional information that the process is also Gaussian with independent and stationary increments. Then as noted before, the solution process is Markovian and one can present the estimates on the conditional distribution itself and not just only the two moments. This is the subject of the next section.

4.6. REFINED ANALYSIS WITH BROWNIAN NOISE.

Throughout this section $B(\cdot)$ will be a standard Brownian motion, i.e., a Gaussian process with independent and stationary increments, $B(0) = 0$ a.e., $E(B(s)) = 0$, and $E(|B(t) - B(s)|^2) = |t - s|$. Then the estimates of the conditional moments of Theorem 5.4 can be improved. The following computations, with slight modifications, are essentially adapted from the unpublished work of Borchers (1964). For simplicity only scalar-valued processes will be considered in what follows.

1. Proposition. *Let the hypothesis of Theorem 5.4 hold so that there is a unique solution of equation (2) of Sec. 5, where $B(\cdot)$ is now the Brownian motion. For the solution process $\{X(t), \mathcal{F}_t, t \geq t_0\}$, given $\varepsilon > 0$ and $I_h^t = [t, t + h], t \geq t_0$, one has (with prob. 1):*

$$1(a) \ P^{\sigma(X(t))}[\sup_{u \in I_h^t} |X(u) - x_1| > \varepsilon] = O(h^2)(1 + |\mathbf{x}|^2),$$

$$(b) \ P^{\sigma(X(t))}[\sup_{u \in I_h^t} |\dot{X}(u) - x_2| > \varepsilon] = O(h^{\frac{3}{2}})(1 + |\mathbf{x}|^2)^{\frac{3}{2}},$$

and hence for $V(t) = (X(t), \dot{X}(t))^*$, the vector Markov process, one has:

$$1(c) \ P^{\sigma(X(t))}[\sup_{u \in I_h^t} |V(u) - \mathbf{x}| > \varepsilon] = O(h^{\frac{3}{2}})(1 + |\mathbf{x}|^2)^{\frac{3}{2}},$$

where $\mathbf{x} = (x_1, x_2)^*$ is the value of V at the "initial point" t. The $O(h)$-terms are uniform on compact sets I_h^t.

2. If further q, σ are bounded, then each factor $(1 + |\mathbf{x}|^2)^\alpha$ may be replaced by $(1 + |x_2|^2)^\alpha$.

Proof. The argument is a refinement of that given for Theorem 5.4 using the additional hypothesis here. Thus for 1(a), consider, with the form of Eq. (12) of the preceding section (all statements holding with prob. 1):

$$\tilde{X}_{t,h} = \sup_{u \in I_h^t} |X(u) - x_1| \leq |x_2|h + \int_{I_h^t} |\dot{X}(u) - x_2| du. \tag{1}$$

Then one has for any $\varepsilon > 0$, using (1) and (conditional) Markov's inequality:

$$
\begin{aligned}
P^{\sigma(X(t))}[\tilde{X}_{t,h} > \varepsilon] &\leq \frac{1}{\varepsilon^2} E^{\sigma(X(t))}[(|x_2|h + \int_{I_h^t} |\dot{X}(u) - x_2| du)^2] \\
&\leq \frac{2}{\varepsilon^2} \{x_2^2 h^2 + h^2 E^{\sigma(X(t))}[\sup_{u \in I_h^t} |\dot{X}(u) - x_2|^2]\} \\
&= \frac{2h^2}{\varepsilon^2} \{(1 + |\mathbf{x}|^2) + E^{\sigma(X(u))}[\sup_{u \in I_u^t} |\dot{X}(u) - x_2|^2]\}.
\end{aligned}
\tag{2}
$$

Hence 1(a) will follow if the bound of 1(b) is established and substituted in the second term on the right above. So we turn to 1(b).

Again using the decomposition in (14) of the preceding section, with $X(t) = x_1, \dot{X}(t) = x_2$, one obtains

$$\sup_{u \in I_h^t} |\dot{X}(t) - x_2| \leq \sum_{i=1}^{4} \sup_{u \in I_h^t} |\alpha_i(u)|, \tag{3}$$

where the $\alpha_i(t)$ are defined by (24) there. Now, given $\varepsilon > 0$, choose $h, 0 < h < h_0$, such that the deterministic $\alpha_3(u)$, satisfies for the conditional measure:

$$\sup_{u \in I_h^t} |\alpha_3(u)| \leq K_1 h (1 + |\mathbf{x}|^2)^{\frac{1}{2}} < \frac{\varepsilon}{4}, \tag{4}$$

where K_1 is a bound coming from the constants of (1) in Sec. 5, and $h(= h(t, \mathbf{x}, \varepsilon))$ is chosen to satisfy the right side inequality of (4). The other α_i are random variables and are estimated as follows. Let $A_i = [\sup_{u \in I_h^t} |\alpha_i| > \frac{\varepsilon}{4}]$. Then from (4) it follows that $P^{\sigma(X(t))}(A_3) = 0$, and hence (3) becomes a.e.,

$$P^{\sigma(X(t))}[\sup_{u \in I_h^t} |\dot{X}(t) - x_2| > \varepsilon] \leq \sum_{i=1}^{4} P^{\sigma(X(t))}(A_i). \tag{5}$$

Each of the (nontrivial) terms on the right in (5) is estimated as:

$$
\begin{aligned}
P^{\sigma(X(t))}(A_1) &\leq P^{\sigma(X(t))}\Big(\int_{I_h^t} |\Delta q(u)|du > \frac{\varepsilon}{4}\Big) \\
&\leq \frac{16h}{\varepsilon^2} E^{\sigma(X(t))}\Big(\int_{I_h^t} |\Delta q(u)|^2 du\Big) \\
&\leq \frac{16h^2 K_1^2}{\varepsilon^2}\{E^{\sigma(X(t))}(\sup_{u\in I_h^t}[|\Delta X(u)|^2 + |\Delta \dot{X}(t)|^2])\} \\
&= \frac{16h^2 K_1^2}{\varepsilon^2} O(h)(1+|\mathbf{x}|^2) \\
&= O(h^3)(1+|\mathbf{x}|^2).
\end{aligned}
\tag{6}
$$

As for A_2 one gets similarly,

$$
\begin{aligned}
P^{\sigma(X(t))}(A_2) &\leq \frac{16}{\varepsilon^2}\int_{I_h^t} E^{\sigma(X(t))}[|\Delta\sigma(u)|^2]du,\ (B(\cdot)\ \text{Brownian motion!}) \\
&\leq \frac{16hK_2^2}{\varepsilon^2}\{E^{\sigma(X(t))}(\sup_{u\in I_h^t}[|\Delta X(u)|^2 + |\Delta \dot{X}(t)|^2])\} \\
&= \frac{16hK_2^2}{\varepsilon^2} O(h)(1+|\mathbf{x}|^2) \\
&= O(h^2)(1+|\mathbf{x}|^2).
\end{aligned}
\tag{7}
$$

Here the Lipschitz conditions on q, σ and the fact that $\{|\alpha_2|^2, \mathcal{F}_u, u \in I_h^t\}$ is a separable submartingale are used. A slightly more detailed analysis is needed to bound A_4.

Since $\sigma(u, x_1, x_2)$ is nonstochastic and $B(\cdot)$ is a Brownian motion, it follows that $\{\alpha_4(u), \mathcal{F}_u, u \in I_h^t\}$ is a Gaussian process with mean zero and covariance $r(s, t) = p(s\wedge t)q(s\wedge t)$ where $p(v) = \int_t^v \sigma(u, x_1, x_2)^2 du$, $q(v) = 1$, and $p(\cdot)$ is nondecreasing. Hence $\{(\alpha_4 \circ p^{-1})(v), v \in p^{-1}(I_h^t)\}$ is a Brownian motion which is obtained from $\alpha_4(\cdot)$ with a change of time (cf., e.g., Rao (1995), Lemma V.3.22). From this and the symmetry of the distribution of the increments together with the reflection principle one obtains:

$$
\begin{aligned}
P^{\sigma(X(t))}[A_4] &= P^{\sigma(X(t))}[\sup_{v\in p^{-1}(I_h^t)} |(\alpha_4 \circ p^{-1})(v)| > \frac{\varepsilon}{4}] \\
&= \frac{1}{2} P^{\sigma(X(t))}[\sup_{v\in p^{-1}(I_h^t)} (\alpha_4 \circ p^{-1})(v) > \frac{\varepsilon}{4}] \\
&= P^{\sigma(X(t))}(\alpha_4(t+h) > \frac{\varepsilon}{4}) \\
&\leq \sqrt{\frac{2}{\pi}}\frac{2}{a^3} = O(h^{\frac{3}{2}})(1+|\mathbf{x}|^2)^{\frac{3}{2}},
\end{aligned}
\tag{8}
$$

where $a = \frac{\varepsilon}{4}(p(t+h))^{-\frac{1}{2}} \geq \frac{\varepsilon}{4}(K_2 h(1+|\mathbf{x}|^2))^{-\frac{1}{2}}$, and $K_2 > 0$ coming from the constant of the Lipschitz hypothesis on $\sigma(\cdot)$. The last bound in (8) also uses a property of the Gaussian integral (cf., Doob (1953), p.285). Thus (4)–(8) establish 1(b), and hence 1(a) also.

Next 1(c) is deduced from 1(a) and 1(b) as follows:

$$P^{\sigma(X(t))}[\sup_{u \in I_h^t} |V(u) - \mathbf{x}| > \varepsilon] \leq P^{\sigma(X(t))}[\sup_{u \in I_h^t} |X(u) - x_1| > \frac{\varepsilon}{\sqrt{2}}]$$

$$+ P^{\sigma(X(t))}[\sup_{u \in I_h^t} |\dot{X}(u) - x_2| > \frac{\varepsilon}{\sqrt{2}}]$$

$$= O(h^{\frac{3}{2}})(1 + |\mathbf{x}|^2)^{\frac{3}{2}}.$$

Finally for 2., if q, σ are bounded, then (1) and the ensuing work show that all the estimates depend on the $\dot{X}(t)$-process which in turn are controlled by $(1 + |x_2|^2)$. Hence the above computations imply that the terms $(1 + |\mathbf{x}|^2)^\beta$ may be replaced by $(1 + |x_2|^2)^\beta$ and the $O(h)$ terms are uniform on compact t-intervals. $\square$

For further analysis it will be useful to have estimates on higher (e.g., fourth) order moments of the solution process when q, σ are suitably bounded. For this purpose the following modified version of a lemma of Maruyama (1955) is employed and hence is detailed. For simplicity the initial condition at $t_0(t_0 = 0)$ of the solution is taken to be a constant, since otherwise these will be conditional values given the solution at t_0.

2. Lemma. *Let $\{X(t), t \geq 0\}$ be the solution of the second order SDE (2) of Sec. 5 with Brownian noise and constant initial conditions so that $\{V(t) = (X(t), \dot{X}(t)), t \geq 0\}$ is a Markov process and $V(0) = \mathbf{x}$ a.e. Then for any bounded measurable $f : \mathbb{R}^+ \times \mathbb{R}^2 \to \mathbb{R}$, one has $(v \in \mathbb{R})$:*

(a) $E(\exp[v \int_0^t f(u, V(u))dB(u) - \frac{1}{2}v^2 \int_0^t f(u, V(u))^2 du]) = 1$

(b) $E(\exp[v \int_0^t \int_0^s f(u, V(u))dB(u)ds$

$$- \frac{1}{2}v^2 \int_0^t (t-u)^2 f(u, V(u))^2 du]) = 1,$$

for any $t \geq 0$.

Proof. Since the dominated convergence theorem holds for $L^{2,2}$-bounded integrators, and $B(\cdot)$ is one such, it suffices to establish (a) for simple f. So let $f = \sum_{j=0}^{n-1} \varphi_j \chi_{[u_j, u_{j+1})}$, where $0 = u_1 < u_2 < \cdots < u_n = t$ and φ_j is an $\mathcal{F}_{u_j}$-adapted bounded random variable for each j. Then the integrals in (a) become sums. Consider

$$J_n = \sum_{j=0}^{n-1} \varphi_j \Delta_{j+1} B(u), \quad L_n = \sum_{j=0}^{n-1} \varphi_j^2 \Delta_{j+1} u, \qquad (9)$$

where $\Delta_{j+1}B(u) = B(u_{j+1}) - B(u_j)$ and $\Delta_{j+1}u = u_{j+1} - u_j$. Now by the independent increments property, $\mathcal{F}_{u_j}$ and $\Delta_{j+1}B(u)$ are independent. Setting $Q_n = vJ_n - \frac{1}{2}v^2 L_n$, and $r_n = v\varphi_{n-1}\Delta_n B(u) - \frac{1}{2}v^2\varphi_{n-1}^2\Delta_n u$, one finds that $Q_n = Q_{n-1} + r_n$ and Q_n is $\mathcal{F}_{u_n}$-adapted; whence for any $c \in \mathbb{R}$,

$$E[\exp(cQ_n)] = E[E^{\mathcal{F}_{u_n-1}}(\exp(cQ_n))]$$
$$= E[\exp(cQ_{n-1})E^{\mathcal{F}_{u_n-1}}(\exp(cr_n))]. \qquad (10)$$

Since $\mathcal{F}_{u_n}$ (hence $\mathcal{F}_{u_{n-1}}$) and $\Delta_n B(u)$ are independent, the inside conditional expectation can be evaluated using the well-known moment-generating function of the Gaussian variable $\Delta_n B(u)$. This gives

$$E^{\mathcal{F}_{u_n-1}}[\exp(cr_n)] = \exp[\frac{1}{2}c(c-1)v^2\varphi_{n-1}^2\Delta_n u],$$

with probability one. If $c \geq 1$, this quantity is ≥ 1. Hence (10) becomes (since $|\varphi_j| \leq K < \infty$, for all j, for some constant K)

$$1 \leq E(\exp(cQ_n)) \leq \exp[\frac{1}{2}c(c-1)v^2 K^2\Delta_n u]E(\exp(cQ_{n-1})). \qquad (11)$$

Iterating the procedure and using $\sum_{j=0}^{n-1}\Delta_{j+1}u = t$, (10)-(11) yield

$$1 \leq E(\exp(cQ_n)) \leq \exp[\frac{1}{2}c(c-1)v^2 K^2 t]. \qquad (12)$$

Taking $c = 1$ this proves the first assertion of the lemma for simple, and then, by the initial reduction, for general f.

Next (b) will be deduced from (a). Indeed let $\tilde{f}(u,x) = (t - u)f(u,x)\chi_{[0,t]}(u)$. Then the double integral of (6) becomes

$$\int_0^t \int_0^s f(u,V(u))dB(u)\,ds = \int_0^t \tilde{f}(s,V(s))dB(s), \qquad (13)$$

which is true for simple functions f and the general case follows by approximation as in the preceding case. [In fact (13) is even true for an $L^{2,2}$-bounded process $B(\cdot)$.] This gives the lemma. $\square$

As noted before, the result holds on any interval $[t_0,\infty)$ when $E(\cdot)$ is replaced by the conditional expectation $E^{\mathcal{F}(X(t_0))}$, and $V(t_0)$ denotes the initial value of the solution process expressed as a vector function, with $B(\cdot)$ still as Brownian motion.

If the coefficients q,σ are bounded then the above lemma allows us to obtain the following inequalities for $X(t)$ and $\dot{X}(t)$:

3. Lemma. *Under the hypothesis of Lemma 2, the following bounds hold for compact t-intervals:*

$$(a)\; E(\exp[vX(t)]) \leq \frac{1}{2}K\exp[v(x_1 + x_2 t)],\; v \in \mathbb{R},$$

$$(b)\; E(\exp[v|X(t)|]) \leq K\exp[v|x_1 + x_2 t|],\; v \geq 0,$$

$$(c)\; E(\exp[v\dot{X}(t)]) \leq \frac{1}{2}\tilde{K}\exp[vx_2],\; v \in \mathbb{R},$$

$$(d)\; E(\exp[v|\dot{X}(t)|]) \leq \tilde{K}\exp[v|x_2|],\; v \geq 0,$$

where $K = 2\exp[\frac{1}{2}|v|K_1 t^2 + \frac{1}{6}v^2 K_1^2 t^3]$, *and* $\tilde{K} = 2\exp[\frac{1}{2}(|v|K_1 t + v^2 K_1^2 t], K_1$ *being a constant depending on the compact t-sets and the Lipschitz bounds on* q, σ.

Proof. Define the functions $Y(\cdot), Z(\cdot)$ as:

$$Y(t) = \int_0^t \int_0^s q(u, V(u))du\, ds;\; Z(t) = \int_0^t \int_0^s \sigma(u, V(u))dB(u)ds.$$

Then using the Lipschitz conditions on q, σ, one gets

$$E[\exp(vZ(t))] \leq \exp[\frac{1}{2}v^2 K_1^2 \int_0^t (t-s)^2 ds],\, \text{by Lemma 2},$$

$$= \exp[\frac{1}{6}v^2 K_1^2 t^3]. \tag{14}$$

Similarly since $|vY(t)| \leq K_1|v|t^2$, one gets from Eq. (2) of Sec. 5, in its integrated form with constant initial values $X(0) = x_1, \dot{X}(0) = x_2$ inserted:

$$E[\exp(vX(t))] = E[\exp(v(Y(t) + Z(t)))]\exp[v(x_1 + x_2 t)]$$

$$\leq \frac{1}{2}K\exp[v(x_1 + x_2 t)], \tag{15}$$

using (14) and the bound on $Y(t)$, with K as given. If $v \geq 0$, since $e^u \leq (e^u + e^{-u})$ for any $u \in \mathbb{R}$, one gets from (15):

$$E[\exp(v|X(t)|] \leq E[\exp(vX(t))] + E[\exp(-vX(t))].$$

Thus (a) and (b) hold.

Inequalities (c) and (d) are similarly obtained, since $E[\exp(vZ(t))] \leq \exp[\frac{1}{2}K_1^2 v^2 t]$, and $|vY(t)| \leq |v|K_1 t$ so that $\tilde{K}$ is as in the statement. $\square$

Using the limiting values (i.e., integrals) of J_n, L_n of (9), and Lemma 2, higher order moments of the corresponding random variables can often be obtained. The following evaluations will be needed in the later calculations. Thus for a bounded measurable f, set

$$J = \int_0^t f(u, V(u))dB(u); \quad L = \int_0^t f(u, V(u))^2 du,$$

$$\tilde{Q}(v) = vJ - \frac{1}{2}v^2 L = v(J - \frac{1}{2}vL) = vQ(v), \text{ say.} \tag{16}$$

Then Lemma 2 implies $1 = E(\exp[\tilde{Q}(v)])$, $v \in \mathbb{R}$, (cf. (12)) and expanding the exponential and identifying the coefficients of v, one gets

$$\sum_{n=1}^{\infty} a_n v^n = \sum_{n=1}^{\infty} \frac{1}{n!} v^n E(Q(v)^n) = 0, \ v \in \mathbb{R}. \tag{17}$$

Hence $a_n = 0$ for all n, and if $b_k = \frac{1}{k!}E(Q(v)^k)$, then $b_4 = \frac{1}{24}v^4(J - \frac{1}{2}vL)^4$. Substituting this in (17) and simplifying for $a_4 = 0$, it is seen that

$$\frac{1}{8}E(L^2) - \frac{1}{2}E(J^2 L) + \frac{1}{24}E(J^4) = 0. \tag{18}$$

With the expressions for J, L from (16), this gives

$$E[(\int_0^t f(u, V(u))dB(u))^4] = 6E[(\int_0^t f(u, V(u))dB(u))^2 \times$$
$$\int_0^t f(u, V(u))^2 du] - 3E[(\int_0^t f(u, V(u))^2 du)^2]. \tag{19}$$

Again the result is valid for compact intervals of $[t_0, \infty)$ if (conditional expectation) $E^{\sigma(X(t))}$ replaces E above. Thus our previous estimates of Theorem 5.4 can be utilized to obtain the corresponding bounds. Indeed, assuming the Lipschitz condition (1) of Sec. 5, Equation (19) takes the form with $f = \sigma, I_h^t = [t, t + h]$:

$$E^{\sigma(X(t))}[(\int_{I_h^t} \sigma(u, V(u))dB(u))^4]$$

$$\leq 6K^2 h E^{\sigma(X(t))}[(\int_{I_h^t} \sigma(u, V(u))dB(u))^2] + 3K^4 h^2$$

$$\leq 6K^4 h^2 + 3K^4 h^2 = 9K^4 h^2. \tag{20}$$

Similarly, using Jensen's inequality for the convex function $\varphi : u \mapsto u^4$, one gets with the same notation as before,

$$E^{\sigma(X(t))}[|\dot{X}(t + h) - x_2|^4] \leq 8E^{\sigma(X(t))}[(\int_{I_h^t} q(u, V(u))du)^4$$

$$+ (\int_{I_h^t} \sigma(u, V(u))dB(u))^4]$$

$$\leq 8K^4(h^4 + 9h^2) = O(h^2). \tag{21}$$

Regarding the $X(t)$-process one has,

$$|X(t+h) - x_1|^4 \leq 16h^4 x_2^4 + 4h^3 \int_{I_h^t} |\dot{X}(u) - x_2|^4 du,$$

so that using (21),

$$E^{\sigma(X(t))}[|X(t+h) - x_1|^4] \leq 16h^4 x_2^4 + 32h^4(K^4 h^2(9 + h^2))$$
$$\leq 32K^4 h^4(1 + x_2^4)$$
$$= O(h^4)(1 + x_2^2)^2. \tag{22}$$

The $O(h)$ terms are uniform for compact t-intervals.

These estimates are used in bounding the second (mixed) truncated moments of the vector Markov process $\{V(T), t \geq t_0\}$. Thus for any given $\varepsilon_i > 0, i = 1, 2$, let

$$R = R(\varepsilon_1, \varepsilon_2, \mathbf{x}) = \{\mathbf{y} = (y_1, y_2) \in \mathbb{R}^2 : |y_i - x_i| < \varepsilon_i, i = 1, 2\}$$

and define the truncated moments $\hat{\mu}_{ij}$, using the transition probability function p of the $V(t)$-process, as:

$$\hat{\mu}_{ij} = \iint_R (y_1 - x_1)^i (y_2 - x_2)^j p(t, \mathbf{x}; t+h, d\mathbf{y}), \tag{23}$$

and set $\mu_{ij} = \hat{\mu}_{ij}$ if R is replaced by $\mathbb{R}^2$.

The analog of Theorem 5.4 for the truncated moments can be given when the noise process is a Brownian motion and q, σ satisfy a Lipschitz condition:

4. Theorem. *Suppose that the hypothesis of Lemma 1 holds and that q, σ are bounded in compact t-intervals. Then the following estimates of the truncated moments $\hat{\mu}_{ij}(= \hat{\mu}_{ij}(t, h; \varepsilon_1, \varepsilon_2, \mathbf{x}))$ obtain:*

$$(a)\ \hat{\mu}_{10} = x_2 h + O(h^{\frac{3}{2}})(1 + x_2^2);$$

$$(b)\ \hat{\mu}_{01} = -\int_{I_h^t} q(u, \mathbf{x}) du + O(h^{\frac{5}{2}})(1 + x_2^2);$$

$$(c)\ \hat{\mu}_{20} = O(h^2)(1 + x_2^2)^{\frac{3}{2}};$$

$$(d)\ \hat{\mu}_{02} = \int_{I_h^t} \sigma(u, \mathbf{x})^2 du + O(h^{3/2})(1 + x_2^2);$$

$$(e)\ \hat{\mu}_{11} = O(h^{\frac{3}{2}})(1 + x_2^2),$$

and each O-term is uniform on compact t-intervals.

Proof. As observed in Proposition 1, the factor $(1 + |\mathbf{x}|^2)^\beta$ can be replaced by $(1 + x_2)^{\beta'}, \beta' \leq \beta$ in the present case. Consequently, the bounds of Theorem 5.4 can be restated simply, on writing R^c for $\mathbb{R}^2 - R$, as:

$$\iint_{R^c} (y_1 - x_1)^2 p(t, \mathbf{x}; t + h, d\mathbf{y}) \leq E^{\sigma(X(t))}(|X(t + h) - x_1|^2)$$
$$= O(h^2)(1 + x_2^2).$$

Similarly,

$$\iint_{R^c} (y_2 - x_2)^2 p(t, \mathbf{x}; t + h, d\mathbf{y}) \leq K^2 h + O(h^{\frac{3}{2}})(1 + x_2^2)$$
$$= O(h)(1 + x_2^2),$$

$$\iint_{R^c} (y_1 - x_1)^4 p(t, \mathbf{x}; t + h, d\mathbf{y}) \leq O(h^4)(1 + x_2^2), \text{ by } (22),$$

$$\iint_{R^c} (y_2 - x_2)^4 p(t, \mathbf{x}; t + h, d\mathbf{y}) \leq O(h^2), \text{ by } (21).$$

Let $\varepsilon = \min(\varepsilon_1, \varepsilon_2)$. Then using the CBS inequality, Proposition 1, and the above bounds, the following integrals are estimated to obtain bounds for $\hat{\mu}_{ij}$:

$$|I_1|^2 = |\iint_{R^c} (y_1 - x_1) p(t, \mathbf{x}; t + h, d\mathbf{y})|^2$$
$$\leq \iint_{R^c} (y_1 - x_2)^2 p(t, \mathbf{x}; t + h, d\mathbf{y}) P^{\sigma(X(t))}[|V(t + h) - \mathbf{x}| > \varepsilon]$$
$$\leq [O(h^2)(1 + x_2^2)][O(h^{\frac{3}{2}})(1 + x_2^2)]$$
$$= O(h^{\frac{7}{2}})(1 + x_2^2)^2.$$

Similarly

$$|I_2|^2 = (\iint_{R^c} (y_2 - x_2)^2 p(t, \mathbf{x}; t + h, d\mathbf{y}))^2$$
$$\leq O(h^{\frac{5}{2}})(1 + x_2^2)^2.$$

$$|I_3|^2 = (\iint_{R^c} (y_1 - x_1)^2 p(t, \mathbf{x}; t + h, d\mathbf{y}))^2$$
$$\leq \iint_{R^c} (y_1 - x_1)^4 p(t, \mathbf{x}; t + h, d\mathbf{y}) P^{\sigma(X(t))}[|V(t + h) - \mathbf{x}| > \varepsilon]$$
$$= [O(h^4)(1 + x_2^2)^2][O(h^{\frac{3}{2}})(1 + x_2^2)]$$
$$= O(h^{\frac{11}{2}})(1 + x_2^2)^3.$$

$$|I_4|^2 = (\iint_{R^c} (y_2 - x_2)^2 p(t, \mathbf{x}; t + h, d\mathbf{y}))^2 = O(h^{\frac{7}{2}})(1 + x_2^2)$$

$$|I_5|^2 = (\iint_{R^c} (y_1 - x_1)(y_2 - x_2) p(t, \mathbf{x}; t + h, d\mathbf{y}))^2$$

$$\leq I_3 I_4 \leq O(h^{\frac{9}{2}})(1 + x_2^2)^2.$$

These bounds are uniform on compact t-intervals, under the given hypothesis. With these expressions and Theorem 5.4, one gets the estimates:

$$\hat{\mu}_{10} = \mu_{10} - I_1 = x_2 h + O(h^{\frac{3}{2}})(1 + x_2^2),$$

$$\hat{\mu}_{01} = \mu_{01} - I_2 = -\int_{I_h^t} q(u, \mathbf{x}) du + O(h^{\frac{5}{4}})(1 + x_2^2),$$

$$\hat{\mu}_{11} = \mu_{11} - I_5 = O(h^{\frac{3}{2}})(1 + x_2^2),$$

$$\hat{\mu}_{20} = \mu_{20} - I_3 = O(h^2)(1 + x_2^2)^{\frac{3}{2}},$$

$$\hat{\mu}_{02} = \mu_{02} - I_4 = \int_{I_h^t} \sigma(u, \mathbf{x})^2 du + O(h^{\frac{3}{2}})(1 + x_2^2),$$

as asserted. $\square$

The significance of the preceding work is appreciated when one finds conditions on an absolutely continuous process to be the solution of a second order (nonlinear) stochastic differential equation driven by a Brownian noise with prescribed initial values; hence the given and its derived process form a vector Markov process by Theorem 5.1. The following result presents such a characterization which generalizes a result, in the first order SDE case, given in Doob (1953, p.287).

5. Theorem. *Let $\{V(t) = (X(t), \bar{X}(t)), \mathcal{F}_t, t \in [0, a]\}$ be an adapted measurable vector process, with $E(|V(t)|^2) < \infty, t \in [0, a]$ for some $a > 0$. Suppose that: (i) $X(\cdot)$ has absolutely continuous, and $\bar{X}(\cdot)$ has continuous, sample paths; (ii) the process $\{|V(t)|^2, t \geq 0\}$ is conditionally uniformly integrable relative to the family $\{\mathcal{F}_t, t \geq 0\}$, in the sense that $\lim_{k \to \infty} E^{\mathcal{F}_s}(|V(t)|^2 \chi_{[|V(t)| \geq k]}) = 0$ uniformly in t for each $s \leq t$; (iii) there exist Borel functions q, σ on $[0, a] \times \mathbb{R}^2$, continuous in $\mathbf{x}$, where $\sigma \geq 0$, such that for some $K > 0$, and $V(0) = \mathbf{x}$,*

$$|q(t, \mathbf{x})| + \sigma(t, \mathbf{x}) \leq K(1 + |\mathbf{x}|^2)^{\frac{1}{2}};$$

and (iv) there is a nondecreasing function f such that $f(h) \searrow 0$ as $h \searrow 0$ for which the following relations hold a.e.

(a) $E^{\mathcal{F}_s}(|X(s + h) - X(s) - h\bar{X}(s)|) \leq hf(h)(1 + |V(s)|^2)$

(b) $|E^{\mathcal{F}_s}(\bar{X}(s + h)) - \bar{X}(s) + \int_{I_h^s} q(u, V(u)) du| \leq hf(h)(1 + |V(s)|^2)$

(c) $|E^{\mathcal{F}_s}(\bar{X}(s + h) - \bar{X}(s))^2 - \int_{I_h^s} \sigma(u, V(u))^2 du| \leq hf(h)(1 + |V(s)|^2).$

Then the derived process $\{\dot{X}(t), t \geq 0\}$ of $\{X(t), t \geq 0\}$ exists, $\bar{X}(t) = \dot{X}(t)$, a.e., and there is a Brownian motion process (perhaps on an enlarged probability space) such that the $X(t)$-process is a unique solution of a second order SDE with the q, σ as the pair of its coefficients and initial value $V(0) = \mathbf{x} = (x_1, x_2)$, i.e.,

$$d\dot{X}(t) + q(t, X(t), \dot{X}(t))dt = \sigma(t, X(t), \dot{X}(t))dB(t), \tag{24}$$

with $V(t) = (X(t), \dot{X}(t))$, in the present notation.

Proof. It may be noted that the conditional uniform integrability is always satisfied if there is an $\mathcal{F}_s$-adapted integrable g such that $E^{\mathcal{F}_s}(|V(t)|^2) \leq g$, a.e., but the present condition is weaker (cf., e.g., Rao (1984), pp.108–109). The proof is an extension of the first order case discussed in Doob (1953). It will be first shown that $\{Y(t) = \bar{X}(t) + \int_0^t q(u, V(u))du, \mathcal{F}_t, t \geq 0\}$ is a continuous square integrable martingale. Then this process is verified to be a solution of (24) for a Brownian motion. It is an interesting application of a general theorem stating that a continuous square integrable martingale is (essentially) obtained from a Brownian motion by a continuous strict time change transformation. The latter result is due to Dambis-Dubins-Schwartz (see, e.g. Rao (1995), Theorem V.3.23, and Corollary V.3.26). Now here are the details.

Regarding the martingale property of the $Y(t)$-process, let $t_1 < t_2$ be fixed and consider a partition $t_1 = s_0 < s_1 < \cdots < s_n = t_2$ with $\delta = \max_j(s_{j+1} - s_j)$. Then

$$|E^{\mathcal{F}_{t_1}}(Y(t_2) - Y(t_1)| = |E^{\mathcal{F}_{t_1}}(\bar{X}(t_2) - \bar{X}(t_1) + \int_{t_1}^{t_2} q(u, V(u))du)|$$

$$= |E^{\mathcal{F}_{t_1}}\{\sum_{j=0}^{n-1} E^{\mathcal{F}_{s_j}}[\bar{X}(s_{j+1}) - \bar{X}(s_j)$$

$$+ \int_{s_j}^{s_{j+1}} q(u, V(u))du]\}|$$

$$\leq \sum_{j=0}^{n-1}(s_{j+1} - s_j)f(s_{j+1} - s_j)[1 + E^{\mathcal{F}_{t_1}}(|V(s_j)|^2)]$$

$$+ |E^{\mathcal{F}_{t_1}}(\sum_{j=0}^{n-1} \int_{s_j}^{s_{j+1}} [q(u, V(u)) - q(u, V(s_j))]du)|, \tag{25}$$

by condition (iv)(b). If $\delta > 0$ is chosen so that $f(\delta)(1 + E^{\mathcal{F}_{t_1}}(|V(s_1)|^2))$ is small (since $f(\delta) \downarrow 0$ as $\delta \downarrow 0$) then the first term on the right can be made small. Since $q(s, \cdot)$ is continuous, the quantity in brackets of the second term tends to zero a.e. by the path continuity of $V(\cdot)$ and the continuity of $q(s, \cdot)$ so that a conditional Vitali argument applies. Thus as $\delta \downarrow 0$, the right side of (25) tends to zero with probability one, which implies that the $Y(t)$-process is a continuous martingale.

Next observe that the $Y(t)$-process is square integrable since from the Lipschitz conditions one has

$$|Y(t)|^2 \le 2|\bar{X}(t)|^2 + 2 \int_0^a q(u, V(u))^2 du$$

$$\le 2|\bar{X}(t)|^2 + K^2 \int_0^a (1 + |V(u)|^2) du,$$

and the terms on the right of this expression are integrable where $K > 0$ is a constant. With this the quadratic variation of the $Y(t)$-process is obtained as follows. For a partition of $[t_1, t_2]$ as above consider

$$|E^{\mathcal{F}_{t_1}}[(Y(t_2) - Y(t_1))^2 - \int_{t_1}^{t_2} \sigma(u, V(u))^2 du]|$$

$$= |E^{\mathcal{F}_{t_1}} (\sum_{j=0}^{n-1} [(Y(s_{j+1}) - Y(s_j))^2 - \int_{s_j}^{s_{j+1}} \sigma(u, V(u))^2 du])|$$

$$\le |E^{\mathcal{F}_{t_1}} [\sum_{j=0}^{n-1} (\bar{X}(s_{j+1}) - \bar{X}(s_j))^2 - \int_{s_j}^{s_{j+1}} \sigma(u, V(u))^2 du)]|$$

$$+ E^{\mathcal{F}_{t_1}} [\sum_{j=0}^{n-1} 2|((\bar{X}(s_{j+1}) - \bar{X}(s_j)) \int_{s_j}^{s_{j+1}} q(u, V(u)) du|$$

$$+ (\int_{s_j}^{s_{j+1}} q(u, V(u)) du)^2]$$

$$\le |E^{\mathcal{F}_{t_1}} (Z_1)| + 2 E^{\mathcal{F}_{t_1}} (|Z_2|) + E^{\mathcal{F}_{t_1}} (Z_3) \text{ (say).} \qquad (26)$$

The right side is now shown to go to zero as $\delta \downarrow 0$ using (iv)(c) again. Here the first and the last terms tend to zero by the argument already used for (25) and the middle term follows the same method after the conditional CBS inequality. Let us sketch the detail for completeness.

By (iv)(a), which has not yet been used, and the conditional uniform integrability as well as the fact that $E(E^{\mathcal{F}_t}(\cdot)) = E(\cdot)$, one has

$$\lim_{h \downarrow 0} E[|\frac{X(s + h) - X(s)}{h} - \bar{X}(s)|] = 0, \ 0 \le s \le a,$$

and since $\bar{X}(\cdot)$ is continuous, the derivative exists in mean and is a.e. equal to a continuous process. Hence $\bar{X}(\cdot) = \dot{X}(\cdot)$ a.e., and $\{\dot{X}(t), t \ge 0\}$ is a derived process of $X(\cdot)$ by Proposition 2.5. Thus $V(t) = (X(t), \bar{X}(t)) =$

$(X(t), \dot{X}(t))$, and this will be used from now on. Also $X(t) = X(0) + \int_0^t \dot{X}(u) du$, a.e.

Next consider each term of (26). By conditional Jensen's inequality and the hypothesis on q,

$$E^{\mathcal{F}_{t_1}}(Z_3) \le \sum_{j=0}^{n-1} (s_{j+1} - s_j) E^{\mathcal{F}_{t_1}} \left(\int_{s_j}^{s_{j+1}} q(u, V(u))^2 du \right)$$

$$\le K^2 \sum_{j=0}^{n-1} (s_{j+1} - s_j) \int_{s_j}^{s_{j+1}} (1 + E^{\mathcal{F}_{t_1}}(|V(u)|^2)) du$$

$$\le K^2 \delta^2 (t_2 - t_1)(1 + E^{\mathcal{F}_{t_1}}(|V(t_1)|^2)), \text{a.e.} \tag{27}$$

This tends to zero a.e. as $\delta \downarrow 0$.

Similarly

$$|E^{\mathcal{F}_{t_1}}(Z_1)| \le \left| E^{\mathcal{F}_{t_1}} \left(\sum_{j=0}^{n-1} \int_{s_j}^{s_{j+1}} (\sigma(u, V(u))^2 - \sigma(u, V(s_j))^2) du \right) \right|$$

$$+ \sum_{j=0}^{n-1} (s_{j+1} - s_j) f(s_{j+1} - s_j)(1 + E^{\mathcal{F}_{t_1}}(|V(s_j)|^2)).$$

The right side tends to zero as $\delta \downarrow 0$ since $f(\delta) \downarrow 0$, by the same argument as before. Finally the middle term uses the conditional CBS inequality to obtain:

$$(E^{\mathcal{F}_{t_1}}(Z_2))^2 \le \sum_{j=0}^{n-1} E^{\mathcal{F}_{t_1}} (\dot{X}(s_{j+1}) - \dot{X}(s_j))^2$$

$$\times E^{\mathcal{F}_{t_1}} \left(\int_{s_j}^{s_{j+1}} q(u, V(u))^2 du \right)$$

$$\le E^{\mathcal{F}_{t_1}} \left(\sum_{j=0}^{n-1} E^{\mathcal{F}_{s_j}} (\dot{X}(s_{j+1}) - \dot{X}(s_j))^2 \right) \cdot E^{\mathcal{F}_{t_1}}(Z_3). \tag{28}$$

Since the second factor on the right was shown to tend to zero as $\delta \downarrow 0$, it suffices to show that the first factor there is finite a.e. For this, using (iv)(c) and the Lipschitz bounds, one notes that it is dominated by

$$E^{\mathcal{F}_{t_1}} \left(\sum_{j=0}^{n-1} \int_{s_j}^{s_{j+1}} \sigma(u, V(u))^2 du \right.$$

$$+ (s_{j+1} - s_j) f(s_{j+1} - s_j)(1 + |V(s_j)|^2))$$

$$\le \sum_{j=0}^{n-1} (s_{j+1} - s_j)(1 + E^{\mathcal{F}_{t_1}}(|V(s_j)|^2)(K^2 + f(s_{j+1} - s_j))$$

$$\le (K^2 + f(\delta))(t_2 - t_1)(1 + E^{\mathcal{F}_{t_1}}(|V(s_1)|^2)),$$

which is finite a.e. This shows that the right side of (28) (hence of (26)) tends to zero a.e., because of (27),(and also in mean) as desired.

Now the quadratic variation of the continuous square integrable martingale $\{Y(t), \mathcal{F}_t, t \geq 0\}$ can be calculated as:

$$[Y]_t = \lim_{n \to \infty} \sum_{k=0}^{2^n-1} (Y(\frac{k+1}{2^n} \wedge t) - Y(\frac{k}{2^n} \wedge t))^2$$

$$= \int_0^t \sigma(u, V(u))^2 du, \text{ by (26)},$$

(cf., e.g., Rao (1995), p.385). Taking $\varphi(u) = \sigma(u, V(u))^2$ in Corollary V.3.26 of the same reference, we conclude that, when $\varphi > 0$ a.e.,

$$B(t) - B(0) = \int_0^t [\varphi(u)]^{-\frac{1}{2}} dY(u),$$

defines a Brownian motion, and hence one gets the representation (using the Itô calculus of these functions) as

$$Y(t) - Y(0) = \int_0^t \sigma(u, V(u)) dB(u), \text{ a.e.} \tag{29}$$

If $\varphi(u) = 0$ on a set of positive measure $(Leb. \otimes P)$, then the representation still holds after a standard enlargement of the space and an adjunction procedure which need not be repeated here (cf., e.g., Doob (1953), p.450).

Since

$$Y(t) = \bar{X}(t) + \int_0^t \sigma(u, V(u)) du$$

$$= \dot{X}(t) + \int_0^t \sigma(u, X(u), \dot{X}(u)) du,$$

one gets from (29) that $X(\cdot)$ is a solution of the SDE

$$d\dot{X}(t) = dY(t) - q(t, V(t)) dt$$

$$= -q(t, X(t)\dot{X}(t)) dt + \sigma(t, X(t), \dot{X}(t)) dB(t), \tag{30}$$

with initial values $X(0) = x_1, \dot{X}(0) = x_2$ a.e. The uniqueness of the solution follows from Theorem 5.1, and the result holds as stated. □

After the work on existence, uniqueness, and (for Brownian disturbance) the Markovian property of the vector process $\{(X(t), \dot{X}(t)), t \geq 0\}$ one should consider the behavior of the solution process for large time. A finer analysis of this aspect can be obtained on using an operator representation of the associated Markov process and utilizing the sharper tools of the corresponding results with estimates of various quantities developed here. The next section will be devoted to this investigation.

4.7. OPERATOR REPRESENTATION OF THE V-PROCESS.

Let $X(t)$ be the (unique) solution of the second order SDE, given by Theorem 5.1, with the noise process $B(t)$ as Brownian motion, and hence the vector process $\{V(t) = (X(t), \dot{X}(t)), t \geq t_0\}$ is Markovian whose transition probability function is written as:

$$p(s, \mathbf{x}; t, \mathbf{y}) = P(X(t) < y_1, \dot{X}(t) < y_2 | V(s) = \mathbf{x}), \tag{1}$$

where $V(s) = \mathbf{x} = (x_1, x_2)$ and $s \leq t$. This $p(\cdot, \cdot; \cdot, \cdot)$ will not be stationary. However, as noted by Dynkin (1961, p.102), one can associate a stationary transition function on a different (somewhat complicated) state space, sometimes termed a *space-time process representation*. In the present context, this is stated as follows.

If $X(t)$ is the solution, and $\dot{X}(t)$ its derived process, of the SDE considered above, then the vector $V(t) = (X(t), \dot{X}(t))$ has $\mathbb{R}^2$ as its state space. Although $\{V(t), t \geq 0\}$ is Markovian, its transition probabilities need not be stationary. However, one can associate a new process $\{Z(t) = (V(t), t), t \geq 0\}$ with state space $\tilde{S} = [0, \infty) \times \mathbb{R}^2$ and the family of transition functions $\bar{p} : \mathbb{R}^+ \times \tilde{S} \times \mathcal{B}(\tilde{S}) \to [0, 1]$ for $\tilde{s} = (s, v) \in \tilde{S}$ and $C = I \times A \in \mathcal{B}(\tilde{S}) = \mathcal{B}(\mathbb{R}^+) \otimes \mathcal{B}(\mathbb{R}^2)$, as:

$$\bar{p}(t, \tilde{s}; C) = p(s, v; s + t, A)\chi_I(s + t), \tag{1'}$$

where $\mathcal{B}(\mathbb{R}^+)$ and $\mathcal{B}(\tilde{S})$ are the Borel σ-algebras of the respective spaces, $p(s, v; t, \cdot)$ being the transition probability function of the $V(t)$-process. Then $\bar{p}$ can be extended uniquely, denoted by the same symbol, to $\mathcal{B}(\tilde{S})$ and it inherits all the measurability properties from p including the satisfaction of the Chapman-Kolmogorov equation. This fact and the definition $\bar{p}$ itself are noted in Dynkin (1961, p.102). The $Z(t)$ is the space-time process alluded to above and it has the stationary transition probabilities. Thus with the product topology of $\tilde{S}$, if $C(\tilde{S})$ denotes the Banach space of bounded continuous functions with the uniform norm, then the family $\{\bar{T}_t, t \geq 0\}$ of operators defined for $\tilde{s}_0 = (s_0, v_0)$ and $f \in C(\tilde{S})$ by

$$(\bar{T}_t f)(\tilde{s}_0) = \int_{\tilde{S}} f(s, v)\bar{p}(t, (s_0, v_0); d\delta_{s_0} dv),$$

$$= \int_{\mathbb{R}^2} f(s_0, v)p(s, v_0; s_0 + t, dv), \tag{2}$$

forms a semi-group of positive contractions because of the Chapman-Kolmogorov identity for the p and hence $\bar{p}$ functions. (Here $\delta_{(\cdot)}$ is the usual point mass function.) The precise domain of definition of the infinitesimal generator of the semi-group should be given for an application in analyzing the path properties of the V-process. Now comparing this family with the evolution collection $\{U_{st}, 0 < s \leq t\}$ considered in Proposition 5.2 earlier, it is seen that for $h > 0, U_{s(s+h)} = \bar{T}_h$ on the (suitably) matching sets in their domain spaces.

So far the definitions of $\bar{p}$ and the associated family $\{\bar{T}_t, t \geq 0\}$ hold for any (general) Markov process. If $\mathcal{A}$ is the infinitesimal generator of such a $\bar{T}_t$-family, then a detailed analysis for the space-time process was given by Lai (1973) for a one-dimensional Markov (diffusion) process. For us, the V-process is (at least) two-dimensional and *what is more*, it is obtained from a solution of a second order SDE in which the coefficients q, σ play a decisive role. Because of this the detailed analysis given by Lai is not directly applicable. In fact his work assumes that the generator is a strictly elliptic partial differential operator, whereas in the present case it will be seen that it is a degenerate operator, as already noted in Theorem 3.5 for the (even) linear equation. Consequently, it is necessary to proceed differently, and in fact we need to obtain the infinitesimal generator on a larger space than $C(\tilde{S})$, by admitting unbounded functions.

Let us start with the following observation:

1. Proposition. *Suppose that the coefficients q, σ do not involve the time variable t, so that $q(\mathbf{x}), \sigma(\mathbf{x})$ are defined only on the state space $\mathbb{R}^2$. Then for the unique solution process $X(t)$ of the SDE above with the Brownian noise, the associated vector process $\{V(t) = (X(t), \dot{X}(t)), t \geq 0\}$ is Markovian with stationary transition probability functions.*

Proof. The argument is classical and in fact is analogous to that of Proposition 3.4. It is sketched here for completeness. The simplest way to proceed is to represent the (second order) SDE in the vector form as in Theorem 4.1. Thus the desired equation becomes

$$V(t)^* = \mathbf{A}^* + \int_0^t \mathbf{Q}(s, V(s))^* ds + \int_0^t \mathbf{S}(s, V(s)) d\mathbf{B}(s), \qquad (3)$$

where $\mathbf{A} = V(0), \mathbf{Q}(s, \mathbf{x}) = (x_2, q(s, \mathbf{x}))$ with $\mathbf{x} = (x_1, x_2)$, and

$$\mathbf{S}(t, \mathbf{x}) = \begin{pmatrix} 0 & 0 \\ 0 & \sigma(t, \mathbf{x}) \end{pmatrix}, \qquad \mathbf{B}(t) = (0, B(t))^*.$$

Once again $\{\mathbf{B}(t), t \geq 0\}$ is a (two-dimensional) Brownian motion, and $\mathbf{Q}, \mathbf{S}$ satisfy the corresponding Lipschitz conditions so that (3) has a unique vector solution which is $V(t) = (X(t), \dot{X}(t))$. Let $h > 0$, and (3) can be expressed as:

$$V(t + h)^* = V(t)^* + \int_t^{t+h} \mathbf{Q}(u, V(u))^* du + \int_t^{t+h} \mathbf{S}(u, V(u)) d\mathbf{B}(u). \quad (4)$$

By the preceding work this equation has a unique solution, for given $v(t)$ as the initial value. Let this be denoted by $V(t+h)^{v(t)}$, and let $V(t)^{\mathbf{x}}$ be the solution of (3). Since (4) can be expressed as

$$V(t+h)^* = V(t)^* + \int_0^h \mathbf{Q}(u, V(u+t))^* du + \int_0^h \mathbf{S}(u, V(u+t)) d\tilde{\mathbf{B}}(u), \quad (5)$$

where $\tilde{\mathbf{B}}(u) = \mathbf{B}(u+t), u \geq 0$ is again a Brownian motion starting afresh at t whose increments are independent of $\mathbf{B}(s), s \leq t$, so that $\tilde{\mathbf{B}}(u) - \tilde{\mathbf{B}}(0)$ is independent of $\mathcal{F}_t$, the conditional distributions of $V(h)$ given $V(0) = \mathbf{x}$ and $V(t+h)$ given $V(t) = \mathbf{x}$ must agree a.e. by uniqueness. Consequently for any Borel set $C \subset \mathbb{R}^2$, one has a.e.

$$P[V(t+h) \in C | \mathcal{F}_t](\mathbf{x}) = P[V(t+h) \in C | V(t) = \mathbf{x}]$$
$$= p[V(h) \in C | V(0) = \mathbf{x}]. \quad (6)$$

But the left side is $p(t, \mathbf{x}; t+h, C)$ and the right side of (6) is $p(h, \mathbf{x}; C)$. The equality between these two quantities (for all $h > 0, t \geq 0, C$, and initial values $\mathbf{x}$) implies the stationarity of the transition functions, as desired. $\square$

2. Remarks. 1. It was shown by Kuo (1972) that even for infinite-dimensional (first order) SDEs the same argument holds, and in fact replacing h here by a stopping time one can deduce the same conclusion so that the solution in this case is a strong Markov process.

2. In the above argument, the only point essential to the proof is that, if **B** is an $L^{2,2}$-bounded process, its increments should be independent [for the Markov property (see Theorem 5.1)] and have stationary increments for (6) to hold. But it need not be necessarily a Brownian motion. For instance, it can be a Poisson process with q, σ suitably chosen.

3. In analogy with the first order case, q, σ (or more properly the matrices $\mathbf{Q}, \mathbf{S}$ of (3)) will be called *drift* and *diffusion* coefficients for convenience, even though the $X(t)$-process is not a diffusion in the exact sense. [Perhaps the $V(t)$-process may be regarded a "degenerate" diffusion. The $X(t)$-process itself will be termed a *second order Itô process* as in Borchers (1964).]

In view of the preceding proposition and (2), we consider the stationary case (i.e., q, σ are independent of t), in detail and calculate the infinitesimal generator of the semi-group in this section. This gives an essential structure of the problem under consideration for this important case. The general aspect will be discussed later, in so far as an extension can be made using (2), and the stationary results.

It is known from earlier studies that if the (contraction) semi-group is assumed to operate on the space $B(\mathbb{R}^2)$ of bounded continuous functions on $\mathbb{R}^2$, then the infinitesimal generator is a complicated integro-differential operator (even for the first order equations). To get the latter as a differential operator, it will be necessary to admit unbounded functions in the domain of the semi-group, introducing a weighted norm. In this situation, the semi-group itself will not be contractive and it should be modified to get the desired results. We follow the ideas of Doob (1955) and a further study due to Rosenkrantz (1974) in this respect. Here one uses the estimates of the preceding section and some simplified estimates based on the work due to Borchers (1964). Our aim is to get the corresponding generated martingales, as in Corollary 5.3, for utilizing it in the asymptotic analysis of the $X(t)$ and $\dot{X}(t)$-processes.

Let $\varphi(\cdot) : \mathbb{R}^2 \to \mathbb{R}^+$ be a positive continuous function such that $\inf_{\mathbf{x}} \varphi(\mathbf{x}) \geq \delta > 0$, and $\lim_{|\mathbf{x}| \to \infty} \varphi(\mathbf{x}) = \infty$, a weight function. If $B(\mathbb{R}^2)$ is the space of real continuous functions as before, let $C(\mathbb{R}^2) = \{f : f/\varphi \in B(\mathbb{R}^2)\}$, with norm $\|f\|_\varphi = \sup\{|f/\varphi|(\mathbf{x}) : \mathbf{x} \in \mathbb{R}^2\}$. Then the conventional proof shows that $\{C(\mathbb{R}^2), \|\cdot\|_\varphi\}$ is a Banach space. For the following work, however, we can choose $\varphi(\mathbf{x}) = \exp|x_2|$ where $\mathbf{x} = (x_1, x_2) \in \mathbb{R}^2$.

As before, let us define a family of operators T_t by using the (stationary) transition probability functions of the V-process similar to (1) but without a 'bar' because of Proposition 1, and with $\varphi(\cdot)$ as just chosen:

$$(T_t f)(\mathbf{x}) = \int_{\mathbb{R}^2} f(\mathbf{y}) p(t, \mathbf{x}; d\mathbf{y}), \quad f \in C(\mathbb{R}^2). \tag{7}$$

Now by the Chapman-Kolmogorov equation, this satisfies $T_s(T_t f) = T_{s+t} f$, for all such f and $s, t \geq 0$. To see that (7) is well-defined, it suffices to verify that $T_t(C(\mathbb{R}^2)) \subset C(\mathbb{R}^2)$ which is seen as follows:

$$
\begin{aligned}
|(T_t f)(\mathbf{x})| &= |E^{\sigma(V(0))}(f \circ V(t))(\mathbf{x})|, \text{ by (2)}, \\
&= |\int_{\mathbb{R}^2} \varphi(\mathbf{y})(f/\varphi)(\mathbf{y}) p(t, \mathbf{x}; d\mathbf{y})| \\
&\leq \|f\|_\varphi \int_{\mathbb{R}^2} e^{|y_2|} p(t, \mathbf{x}; d\mathbf{y}) \\
&= \|f\|_\varphi E(e^{|\dot{X}(t)|} | V(0) = \mathbf{x}) \\
&\leq \|f\|_\varphi \tilde{K} \exp(|x_2|), \text{ by Lemma 6.3(d)},
\end{aligned}
$$

where $\tilde{K} = 2 \exp[\frac{1}{2}(K_1 + K_1^2)t]$ is a finite constant for t in compact intervals. Thus $\{T_t, t \geq 0\}$ is a bounded semi-group. But since the dependence of $V(t)$ on the initial value $\mathbf{x}$ is continuous (cf. Theorem 4.2) and by Lemma 6.3(d), $\{f(V^{\mathbf{x}}(t)), \|\mathbf{x} - \mathbf{x}_0\| \leq \delta\}$ is bounded in a ball of $L^2(P)$ so that it is uniformly integrable as $\mathbf{x}$ varies in the neighborhoods of $\mathbf{x}_0 \in \mathbb{R}^2$, the integrals in $\mathbf{x}$ result in continuous functions. It therefore follows from (7) that $(T_t f)(\cdot)$ is continuous for each continuous f. Thus $T_t f \in C(\mathbb{R}^2)$ for each t, as asserted. Thus the semi-group is positivity preserving, continuous, but not necessarily a contractive family of operators on $C(\mathbb{R}^2)$. Let us record this result for reference as follows.

3.   Proposition.   *Let $\{V(t), t \geq 0\}$ be the associated (vector) Markov process of the second order Itô process $X(t)$, i.e., the solution process of the second order SDE. If the "drift" and "diffusion" coefficients are independent of the time variable, and if $\{T_t, t \geq 0\}$ is the family of operators determined by the transition probabilities of the V-process by (7), then it is a bounded positivity preserving (but not contractive) semi-group on $(C(\mathbb{R}^2), \|\cdot\|_\varphi)$, with weight $\varphi(\mathbf{x}) = e^{|x_2|}, (\mathbf{x} = (x_1, x_2))$, into itself. [However it may not also be a strongly continuous semi-group.]*

Because of the lack of some standard properties for the semi-group defined by (7), one can not utilize the well-known theorems of the subject to deduce some desired consequences. In the one-dimensional case a simplification and extension of the original argument in Doob (1955) has been accomplished by Rosenkrantz (1976). But that method does not seem to extend easily to the present case. These classes are sometimes referred to as "semi-groups of type Γ", following Doob (1955).

Using the definition of generators of semi-groups, we now calculate the same for the present family utilizing the estimates of the last section. Now for each $f \in C(\mathbb{R}^2)$, let $f_{ij} = \frac{\partial^{i+j} f}{\partial x_1^i \partial x_2^j}$ be the $(i+j)^{th}$ partial derivative when it exists where $0 \leq i, j \leq 3$ with $f_{0j} = \frac{\partial^j f}{\partial x_2^j}$ and similarly f_{i0} (and $f_{00} = f$). Let $C_2 \subset C(\mathbb{R}^2)$ be the set such that $f \in C_2$ iff there exist a compact set $S(= S_f) \subset \mathbb{R}^2$ and a constant $0 \leq c(= c_f) < 1$ satisfying the following two conditions for the $\|\cdot\|_\varphi$-norm:

$$(i) \ \|f_{ij}\|_{\varphi_c} < \infty, \ 0 \leq i + j \leq 2, \ \varphi_c = (\varphi)^c,$$

$$(ii) \ \|f_{ij}\chi_{S^c}\|_{\varphi_c} < \infty, \ i + j = 3, \ (S^c = \mathbb{R}^2 - S).$$

Clearly C_2 is a linear manifold which need not be complete. We now calculate the generator of the semi-group $\{T_t, t \geq 0\}$ on $C(\mathbb{R}^2)$ as follows. This extends Doob's one-dimensional result, but the present analysis is considerably more involved than the former work.

4.   Theorem.   *Let $\{V(t), t \geq 0\}$ be the (vector) Markov process associated with the second order Itô-process with bounded "drift" and "diffusion" coefficients (q, σ) on $\mathbb{R}^2$ as in Proposition 2. If $\{T_t, t \geq 0\}$ is the corresponding semi-group, defined by (7), then its generator D is a degenerate elliptic operator given by:*

$$D = \frac{1}{2}\sigma(\mathbf{x})^2 \frac{\partial^2}{\partial x_2^2} - q(\mathbf{x})\frac{\partial}{\partial x_2} + x_2 \frac{\partial}{\partial x_1}, \tag{8}$$

whose domain contains C_2 introduced above so that

$$\lim_{h \searrow 0} \left\| \frac{(T_h - I)f}{h} - Df \right\|_\varphi = 0, \ f \in C_2, \tag{9}$$

where φ is the particular weight used in our study, namely $\varphi(\mathbf{x}) = \exp|x_2|$ for $\mathbf{x} = (x_1, x_2) \in \mathbb{R}^2$.

Proof. The argument is, as usual, based on the Taylor series expansion of f and then verifying that $\|\frac{1}{h}(T_h - I)f - Df\|_\varphi \to 0$ in the particular norm topology using the estimates of Section 6. Here are the detailed computations.

Using the notations as for Theorem 6.4 (and its conclusions later on) let $R = R(\varepsilon_1, \varepsilon_2; \mathbf{x}) = \{\mathbf{y} = (y_1, y_2) : |y_i - x_i| < \varepsilon_i, i = 1, 2\}$ be an open rectangle in $\mathbb{R}^2$ about $\mathbf{x}$ and side lengths $2\varepsilon_i > 0$. If $f \in C_2$, let $S(= S_f)$ be the compact set in the definition and take $\mathbf{x} \in S$. Then by the compactness of S, we can take $\varepsilon_i > 0$ small enough so that for each $\mathbf{y} \in R$ Taylor's expansion around $\mathbf{x}$ can be obtained as:

$$f(\mathbf{y}) = f(\mathbf{x}) + (\mathbf{y} - \mathbf{x}) \cdot (\nabla f)(\mathbf{x})$$
$$+ \frac{1}{2} \sum_{\substack{0 \le i, j \\ i+j=2}} f_{ij}(\mathbf{x})(y_i - x_i)(y_j - x_j) + r(\mathbf{x}, \mathbf{y})\|\mathbf{y} - \mathbf{x}\|^2, \qquad (10)$$

where $\lim_{\|\mathbf{y}-\mathbf{x}\|\to 0} r(\mathbf{x}, \mathbf{y}) = 0$, uniformly for $\mathbf{x} \in S$, $\nabla f = (f_{10}, f_{01})$ being the gradient of f. Using (10) and the definition of truncated moments as in Theorem 6.4, one obtains:

$$\left(\frac{(T_h - I)}{h}f\right)(\mathbf{x}) = h^{-1}[f_{10}(\mathbf{x})\hat{\mu}_{10} + f_{01}(\mathbf{x})\hat{\mu}_{01}]$$
$$+ \frac{1}{2}h^{-1}[f_{20}(\mathbf{x})\hat{\mu}_{20} + 2f_{11}(\mathbf{x})\hat{\mu}_{11} + f_{02}(\mathbf{x})\hat{\mu}_{02}]$$
$$+ h^{-1}\left[\iint_R r(\mathbf{x}, \mathbf{y})|\mathbf{x} - \mathbf{y}|^2 p(h, \mathbf{x}; d\mathbf{y})\right.$$
$$+ \left.\iint_{R^c}(f(\mathbf{y}) - f(\mathbf{x}))p(h, \mathbf{x}; d\mathbf{y})\right]. \qquad (11)$$

Denoting the last two integrals by $\alpha_1(h, \mathbf{x}), \alpha_2(h, \mathbf{x})$ for short, and using the asymptotic estimates of $\hat{\mu}_{ij}$ from Theorem 6.4 (in the stationary case) one immediately gets the following:

$$\frac{(T_h - I)}{h}f(\mathbf{x}) - (Df)(\mathbf{x}) = \sum_{i=0}^{2} \alpha_i(h, \mathbf{x}), \qquad (12)$$

where $\alpha_0(h, \mathbf{x})$ denotes:

$$\alpha_0(h, \mathbf{x}) = O(h^{\frac{1}{4}})(f_{10} + f_{01} + \frac{1}{2}[f_{20} + 2f_{11} + f_{02}])(\mathbf{x})(1 + x_2^2)^{\frac{3}{2}}. \qquad (13)$$

It is to be shown that the $\alpha_i(h, \mathbf{x}), (i = 0, 1, 2)$ tend to zero in $\|\cdot\|_\varphi$-norm as $h \to 0$ to complete the demonstration. The rest of the proof is devoted to establish these three assertions.

First consider α_2. Observe that with $\bar{R} = \mathbb{R}^2 - R(= R^c)$

$$\int_{\bar{R}} p(h, \mathbf{x}; d\mathbf{y}) \leq P[|V^*(h) - \mathbf{x}| \geq \varepsilon_0], \ (\varepsilon_0 = \varepsilon_1 \wedge \varepsilon_2),$$

$$= O(h^{\frac{3}{2}})(1 + x_2^2)^{\frac{3}{2}}, \tag{14}$$

the O-term being uniform on $[0, k] \times \mathbb{R}^2$, for a given k, by Proposition 6.1. Also

$$\int_{\bar{R}} |f(\mathbf{y})|^4 p(h, \mathbf{x}; d\mathbf{y}) \leq E(|f(V^*(h))|^4)$$

$$\leq \|f\|_{\varphi_c}^4 E(e^{4c|\dot{X}(h)|})$$

$$\leq \|f\|_{\varphi_c}^4 L(4c, h)e^{4c|x_2|} < \infty, \tag{15}$$

by Lemma 6.3(d). As before, here again $\varphi_c = (\varphi)^c$, with $0 \leq c(= c_f) < 1$. These two estimates and the Hölder inequality imply:

$$|\alpha_2(h, \mathbf{x})| \leq h^{-1}\Big[\int_{\bar{R}} |f(\mathbf{y})|p(h, \mathbf{x}; d\mathbf{y}) + \int_{\bar{R}} p(h, \mathbf{x}; d\mathbf{y})\Big]$$

$$\leq h^{-1}\Big(\int_{\bar{R}} |f(\mathbf{y})|^4 p(h, \mathbf{x}; d\mathbf{y})\Big)^{\frac{1}{4}}\Big(\int_{\bar{R}} p(h, \mathbf{x}; d\mathbf{y})\Big)^{\frac{3}{4}}$$

$$+ h^{-1}|f(\mathbf{x})|\int_{\bar{R}} p(h, \mathbf{x}; d\mathbf{y})$$

$$\leq \|f\|_{\varphi_c} L(4c, h)^{\frac{1}{4}} e^{c|x_2|} O(h^{\frac{1}{8}})(1 + x_2^2)^{\frac{9}{8}}$$

$$+ \|f\|_{\varphi_c} O(h^{\frac{1}{2}})(1 + x_2^2)^{\frac{3}{2}} e^{c|x_2|}$$

$$\leq O(h^{\frac{1}{8}})e^{c|x_2|}(1 + x_2^2)^{\frac{3}{2}},$$

and the O-terms are uniformly bounded on $[0, k] \times \mathbb{R}^2$. From this it follows that

$$\|\alpha_2\|_\varphi = O(h^{\frac{1}{8}}), \tag{16}$$

since $\sup_{\mathbf{x}}(1 + x_2^2)^{\frac{3}{2}} e^{-(1-c)|x_2|} < \infty$. Thus one gets $\lim_{h\searrow 0} \|\alpha_2\|_\varphi = 0$.

We next consider α_1 which needs a more elaborate computation as it should be evaluated on S_f and $\mathbb{R}^2 - S_f$ for suitable ε_is as $h \searrow 0$ for each f. By the property of $r(\mathbf{x}, \mathbf{y})$ and the compactness of $S(= S_f)$, for $\varepsilon_0 > \varepsilon > 0$, one can choose $\varepsilon_i > 0$, small enough, such that for any $\mathbf{y} \in R(= R(\varepsilon_1, \varepsilon_2, \mathbf{x}))$, one has $|r(\mathbf{x}, \mathbf{y})| < \varepsilon$ for all $\mathbf{x} \in S$. Thus for $\mathbf{x} \in S$:

$$|\alpha_1(h, \mathbf{x})| \leq \varepsilon h^{-1} \int_R |\mathbf{x} - \mathbf{y}|^2 p(h, \mathbf{x}; d\mathbf{y})$$

$$= \varepsilon h^{-1}[\hat{\mu}_{20} + \hat{\mu}_{02}], \text{ by definition of } \hat{\mu}_{ij},$$

$$= \varepsilon[O(h)(1 + x_2^2)^{\frac{3}{2}} + (\sigma(\mathbf{x})^2 + O(h^{\frac{1}{2}})(1 + x_2^2))]$$

$$\leq \varepsilon(1 + x_2^2)^{\frac{3}{2}}[O(1) + O(h^{\frac{1}{2}})],$$

since σ is uniformly bounded on $\mathbb{R}^2$, and the estimates from Theorem 6.4 are used. Consequently for all $\mathbf{x} \in S$,

$$\|\alpha_1\|_\varphi = \varepsilon[O(1) + O(h^{\frac{1}{2}})] \to 0, \text{ as } h \searrow 0, \tag{17}$$

since $\varepsilon > 0$ is arbitrary.

Next for $\mathbf{x} \in \mathbb{R}^2 - S_f$ and $\mathbf{y} \in R$, $f \in C_2$, consider the remainder term $r(\cdot, \cdot)$ in α_1 which can be simplified with $\Delta y_i = (y_i - x_i)$, $\mathbf{x}' = \mathbf{x} + \theta(\mathbf{y} - \mathbf{x})$, $\theta \in [0, 1)$ as follows:

$$
\begin{aligned}
|r(\mathbf{x}, \mathbf{y})| &= \frac{1}{3!}|[(\Delta y_1 \frac{\partial}{\partial y_1} + \Delta y_2 \frac{\partial}{\partial y_2})^3 f](\mathbf{x}')| \\
&= \frac{1}{6}|[(\Delta y_1)^3 f_{30} + 3(\Delta y_1)^2(\Delta y_2)f_{21} + 3(\Delta y_1)(\Delta y_2)^2 f_{12} \\
&\quad + (\Delta y_2)^3 f_{03}](\mathbf{x}')| \\
&\leq \frac{1}{6}Me^{c|x_2'|}[|\Delta y_1| + |\Delta y_2|]^3 \\
&\leq \frac{1}{6}Me^{c|x_2'|}[\varepsilon_1|\Delta y_1|^2 + 3\varepsilon_2|\Delta y_1|^2 + 3|\Delta y_2|^2 + \varepsilon_2|\Delta y_2|^2] \\
&\leq \frac{1}{2}(\varepsilon_1 + \varepsilon_2)Me^{c|x_2'|}|\mathbf{x} - \mathbf{y}|^2,
\end{aligned}
$$

where $|f_{ij}|(\mathbf{x}) \leq e^{c|x_2|}\|f_{ij}\|_{\varphi_c} \leq Me^{c|x_2|}$, is used from the definition of C_2 and $c(= c_f) \in [0, 1)$. But $x_2' = \theta(y_2 - x_2)$ implies $|x_2'| \leq |x_2| + \varepsilon_2$ so that (17) becomes for $\mathbf{x} \in \mathbb{R}^2 - S_f$

$$|r(\mathbf{x}, \mathbf{y})| \leq \frac{1}{2}(\varepsilon_1 + \varepsilon_2)Me^{\varepsilon_2 + c|x_2|}|\mathbf{x} - \mathbf{y}|^2.$$

It follows from the definition of α_1 that for $\mathbf{x} \in \mathbb{R}^2 - S$,

$$
\begin{aligned}
\|\alpha_1\|_\varphi &\leq \frac{1}{2}(\varepsilon_1 + \varepsilon_2)Me^{\varepsilon_2} \sup_{\mathbf{x}} e^{-(1-c)|x_2|} \\
&\quad + h^{-1} \int_R |\mathbf{x} - \mathbf{y}|^2 p(h, \mathbf{x}; dy) \\
&\leq \frac{1}{2}(\varepsilon_1 + \varepsilon_2)[O(1) + O(h^{\frac{1}{2}})] \to 0, \tag{18}
\end{aligned}
$$

as $h \searrow 0$, since $\sup_{\mathbf{x}}(1 + x_2^2)^{\frac{3}{2}} e^{-(1-c)|x_2|} < \infty$, the O-terms being uniformly bounded, and $\varepsilon_i > 0$ arbitrary. Hence (17) and (18) imply that $\lim_{h \searrow 0}\|\alpha_1\|_\varphi = 0$, as desired.

Finally α_0 can be simplified in the same manner, namely

$$
\begin{aligned}
|\alpha_0|e^{-|x_2|} &= O(h^{\frac{1}{4}})(1 + x_2^2)^{\frac{3}{2}} e^{-(1-c)|x_2|} \\
&= O(h^{\frac{1}{4}}),
\end{aligned}
$$

and hence $\|\alpha_0\|_\varphi \to 0$ as $h \searrow 0$ for all $\mathbf{x} \in \mathbb{R}^2$. Thus the right side of (12) tends to zero as $h \searrow 0$, as asserted. $\square$

It is now of interest to discuss an extension of the preceding result to the family $\{\bar{T}_t, t \geq 0\}$ defined in (2), and as indicated in Remarks 2(3). The state space for this family (of a space-time process) is $\tilde{S} = \mathbb{R}^+ \times \mathbb{R}^2$. However all the necessary estimates of Section 6 were carried out for $q(t, \mathbf{x}), \sigma(t, \mathbf{x})$ for $t \geq t_0$. Taking $t_0 = s$, arbitrarily fixed, one can verify that $\bar{T}_t : C(\tilde{S}) \to C(\tilde{S})$ where $C(\tilde{S}) = \{f : f/\varphi \in B(\tilde{S})\}$ and $\|f\|_\varphi = \sup\{|f/\varphi|(\mathbf{x}) : f \in B(\tilde{S})\} < \infty, B(\tilde{S})$ being the space of bounded continuous functions on $\tilde{S}$. If $\tilde{C}_2 = \{f \in C(\tilde{S}) : \|f_{ij}\|_{\varphi_c} < \infty, 0 \leq i + j \leq 2,$ and $\|f_{ij}\chi_{\tilde{S}-S_f}\| < \infty, i+j = 3\}$ for compact S_f and $\varphi_c = (\varphi)^c, c \in [0, 1)$ as in the definition of C_2 before, then the preceding work holds for the new (non closed) subspace with the same weight function φ. The evaluations hold with q, σ depending on time and satisfying the other conditions, and so the corresponding generator $\tilde{D}_s$ of (8) will be valid for these coefficients starting at s. Consequently Theorem 4 can be stated in this context as follows.

5.   Theorem. *Let $\{V(t), t \geq 0\}$ be the (vector) Markov process of the second order Itô process with bounded 'drift' and 'diffusion' coefficients $q(t, \mathbf{x}), \sigma(t, \mathbf{x})$. If $\{\bar{T}_t, t \geq 0\}$ is the associated semi-group, then its generator $\tilde{D}$ has a domain containing $\tilde{C}_2$ and on the latter it coincides with the differential operator:*

$$\tilde{D}_t = \frac{1}{2}\sigma(t, \mathbf{x})\frac{\partial^2}{\partial x_2^2} - q(t, \mathbf{x})\frac{\partial}{\partial x_2} + x_2\frac{\partial}{\partial x_1}, \tag{19}$$

so that

$$\lim_{h \searrow 0} \|\frac{(\bar{T}_h - I)f}{h} - \tilde{D}_t f\|_\varphi = 0, \ f \in \tilde{C}_2, \tag{20}$$

where $\varphi(\tilde{\mathbf{x}}) = \exp|x_2|$ for $\tilde{\mathbf{x}} = (s, x_1, x_2) \in \tilde{S}$, and $\bar{T}_h = U_{t,t+h}$ in the notation of Proposition 5.2.

One can derive (19) directly using the general estimates of Section 6, except that, notationally, it is cumbersome and the calculations have to be repeated in the new context. This will be omitted. It should also be noted that if T_t and $\bar{T}_t$ are uniformly bounded (e.g., contractions) then $\bar{C}_2$, the closure, may be taken as the new domain of the semi-group T_t (similarly for $\bar{T}_t$), as in Feller (1952), p.476, which will be strongly continuous. But this is not the case here. Using the generators (8) or (19) appropriately, the sample path analysis of the $X(t)$-process can and will be discussed in the next section.

4.8 SAMPLE PATH BEHAVIOR FOR LARGE TIME.

Consider a second order SDE given in a vector form, as in Eq. (4) of Section 4, i.e.,

$$X(t) = A_1 + \int_{t_0}^t \dot{X}(u)du,$$

$$\dot{X}(t) = A_2 + \int_{t_0}^t q(u, X(u), \dot{X}(u))du \qquad (1)$$

$$+ \int_{t_0}^t \sigma(u, X(u), \dot{X}(u))dB(u),$$

expressed with

$$V(t) = (X(t), \dot{X}(t)), \quad \mathbf{Q}(t) = (t, V(t)), \quad \mathbf{A} = (A_1, A_2)^*,$$
$$\mathbf{B}(t) = (0, B(t))^*, \quad \mathbf{S}(t) = ((0,0)^*, (0, \sigma(t, V(t))^*)),$$

as follows:

$$V(t)^* = \mathbf{A} + \int_{t_0}^t \mathbf{Q}(u)^* du + \int_{t_0}^t \mathbf{S}(u) d\mathbf{B}(u). \qquad (2)$$

Here $V(0) = \mathbf{A}$. If $\{B(u), u \geq t_0\}$ is an $L^{2,2}$-bounded process and q, σ satisfy a Lipschitz condition, then the integrals in (1) as well as (2) are well defined. Moreover, they represent (locally) $L^{2,2}$-bounded processes so that the (vector) V and the $X(t)$-processes are (locally) $L^{2,2}$-bounded. In (2), the first integral involving just the 'drift' vector has (locally) finite variation since the path-wise integral is in Lebesgue's sense. But the second integral, denoted temporarily as $Y(t)$, gives a (local) $L^{2,2}$-bounded process for which the quadratic variation exists, and the latter is expressible as the matrix:

$$[Y]_t = \int_{t_0}^t \mathbf{S}(u)\mathbf{S}^*(u)d[\mathbf{B}](u), \qquad (3)$$

where $[\mathbf{B}]_u$ is the matrix "increasing" function of the $\mathbf{B}$-process which always exists (cf., Rao (1995), Proposition VI.2.10). Here it is a 2×2-matrix process. Expanding (3) one finds that it is given by $\int_{t_0}^t \sigma^2(u, V(u))d[B](u)$ which does not vanish if $\sigma \neq 0$, a.e., and the latter implies that this component of the $Y(t)$-process has unbounded variation on each non-degenerate interval which is charged by the measure determined by $[B](u)$. It follows that the $V(t)$-process of (2) cannot have finite variation in any open nondegenerate interval that is charged by $[\mathbf{B}]$. This may be stated for reference as follows.

1. Proposition. *Let* $\{V(t), t \geq t_0\}$ *be the vector process defined by (2) relative to an* $L^{2,2}$-*bounded process* $\{\mathbf{B}(t), t \geq 0\}$ *and* $(\mathbf{Q}, \mathbf{S})$ *satisfying a local Lipschitz and boundedness conditions. Then the* $V(t)$-*process is locally* $L^{2,2}$-*bounded and its (matrix) quadratic variation exists on each compact t-interval. Hence its variation on each nondegenerate interval charged by the* *quadratic variation of the $L^{2,2}$-bounded process $\{B(t), t \geq t_0\}$ (in the matrix norm) is unbounded with probability 1. In particular if the $B(t)$-process is a Brownian motion, then the $V(t)$-process has the last property.*

It may be noted that the variation of a vector SDE satisfying (2), but which is not necessarily a second order equation, also has an unbounded variation on such intervals provided one assumes conditions for the existence of its solutions. This fact has been directly proved by Goldstein (1969) when (2) is a (first order) n-vector SDE with $B(t)$ as Brownian motion. The $L^{2,2}$-boundedness integrators and the corresponding result given by Theorem 4.1 allow us to obtain a painless extension to the most general statement on the variation properties of the paths of these solutions.

The asymptotic analysis of sample paths (for large t) is a more delicate problem. Not much has been done on this topic. We restrict $B(t)$ to Brownian motion and discuss the subject. The main work here is due to Goldstein (1969), and since then the progress in PDE has not yet helped to fully resolve the problems raised in our subject.

The ordering relationships of pairs of solutions $X_i(t)$ for different coefficients q_i, σ_i, $i = 1, 2$, are determined by the domination properties of the latter. The following result is a specialization of one due to Goldstein (1969) and it illustrates the possibilities.

2. Theorem. *Suppose that $\{X_i(t), t \geq 0\}$ is the solution of (1) for the coefficients $(q_i, \sigma_i), i = 1, 2$, with the same Brownian driving force, or noise, $B(t)$, where the Lipschitz and boundedness conditions are assumed for the existence of solutions. Suppose also that (q_i, σ_i) are continuous, $q_1(t, \mathbf{x}) < q_2(t, \mathbf{x}), \sigma_1 = \sigma_2(= \sigma$ say$)$ and that the common σ satisfies a Lipschitz condition of order $\alpha > \frac{1}{2}$ on each closed ball of $\mathbb{R}^2$ in the $\mathbf{x}$-variable (uniformly in t). If $\tau : \Omega \to \bar{\mathbb{R}}^+$ is a finite stopping time of $\{\sigma(B(s), s \leq t), t \geq 0\}$ and $X_1 \circ \tau = X_2 \circ \tau$, a.e., so that the paths agree until time τ, then there is a positive random variable $h : \Omega \to \mathbb{R}^+$ such that*

$$X_1(t, \omega) < X_2(t, \omega), \text{ and } \dot{X}(t, \omega) < \dot{X}(t, \omega), \ \tau(\omega) < t < h(\omega), \qquad (4)$$

for a.a. $\omega \in \{\omega : \tau(\omega) < \infty\}$, and if moreover $\sigma(t, x_1, x_2) = \sigma(t, x_2)$ (so σ is independent of x_1) and $\tau = 0$, then (4) holds for all $t > 0$ for the X_i and the weaker inequality relation holds for $\dot{X}_i$s for $t \geq h(\omega)$, with probability 1.

Further, if $q_1 = q_2(= q$, say$)$ so that $X_1 = X_2(= X$, say$)$ (by the uniqueness of solutions), and if $|q| \geq \delta > 0$ with $\sigma > 0$ such that $\lg \lg \int_0^t \sigma^2(u, \mathbf{x}) du \leq (\delta - \gamma)^2 t^2$ for large t and some $\gamma > 0$, then $\dot{X}(t)$ as well as $X(t)$ visit $\pm\infty$ infinitely often with probability 1.

These properties are established with standard but nontrivial computations using the fact that $\int_0^t \sigma(u, X(u), \dot{X}(u)) dB(u)$ is a Brownian motion after a time change. There are, however, more specialized properties obtainable after using the generator of the associated semi-group which we now consider.

The following result depends on solutions of the differential equations by using the infinitesimal operator of the semi-group, given by Theorem 7.4 (or 7.5). Its interest comes from the connections with martingale tools established in Corollary 5.3.

4. Theorem. *Let $\{X(t), \mathcal{F}_t, t \geq 0\}$ be the solution of our SDE system (1) where the q, σ satisfy the boundedness and Lipschitz conditions (assumed throughout for the existence and uniqueness of solutions), and q, σ are independent of the time variable t. Suppose that $p(x_2) = q(\mathbf{x})/\sigma(\mathbf{x})$ depends only on the second component of $\mathbf{x}$, and that p is bounded with a bounded continuous derivative on the deleted line $\{x_2 : |x_2| \geq \delta > 0\}$. Define a function g as:*

$$g : (x_1, x_2) \mapsto \int_0^{x_2} \exp[-2 \int_0^t p(u)du]dt. \tag{5}$$

Then $g \in C_2$ and $Dg = 0$ where D and C_2 are given by Theorem 7.4. Moreover, if for each stopping time τ of $\{\mathcal{F}_{t_i}, t_i \geq 0\}$, where $\{t_i \leq t_{i+1} \leq \cdots\}$ is any subsequence of $\mathbb{R}^+$, we have $E(g^+(\dot{X}(\tau))) < \infty$, then $\dot{X}(t) \to Z$ a.e., as $t \to \infty$, (Z can take infinite values on sets of positive probability, however) and for a separable version of the X-process:

$$P[\lim_{t \to \infty} (X(s+t) - X(t)) = sZ] = 1, \tag{6}$$

for each $s > 0$. Thus the increments of the solution (of the second order Itô) process become "asymptotically stationary".

From our earlier work it follows that $\{g(\dot{X}(t)), \mathcal{F}_t, t \geq 0\}$ is a martingale, but it may not be $L^1(P)$-bounded. However, the stopping time condition assures us, by an old theorem of Y. S. Chow's (cf., e.g., Rao (1995), p.282), the martingale convergence, and the same property holds for the $\dot{X}(t)$-process since $g(\cdot)$ is strictly monotone. The rest of the proof uses the (time changed) Brownian motion property of the stochastic integral term in the equation for $\dot{X}(t)$ in (1), which is the reason for the eventual "stationarity" of the increments. The details need care and additional computation, and can be obtained from the paper by Goldstein (1969).

This result indicates that other reasonable conditions on q, σ give valuable information on the sample path analysis of the second order Itô processes. The first order case has been studied in detail in Doob (1955), see also Mandl (1968). These depend on the solutions of the PDE $Dg = f$ where D is the (degenerate) elliptic (or elliptic-parabolic) differential operator, with various boundary conditions. For instance, one may verify that the solution of $Dg = \lambda g$ for g continuous and vanishing at infinity, is the trivial solution $g = 0$. Thus for a more interesting analysis allowing solutions to be unbounded as above is essential in our case of degenerate operators since then $\{e^{-\lambda t}g(V(t)), \mathcal{F}_t, t \geq t_0\}$ may be shown to be a martingale if this set is in $L^1(P)$ for each $g \in C_2, (Dg = \lambda g)$. A study of the latter is an intricate PDE subject in itself, as seen, for instance, from the monograph by Oleĭnik and Radkevič (1973). However, an analysis of our operator D of (8) (or $\tilde{D}_t$ of (19)) in Section 7 does not obtain directly from the theorems of the above volume and additional nontrivial study of these equations is necessary to use it profitably in the second order SDE work.

We now turn to some considerations of stochastic flows related to these higher order equations, and note again that some interesting new phenomena occurs compared to the first order case.

4.9. ASPECTS OF STOCHASTIC FLOWS.

In the classical theory of differential equations, if $\frac{dX(t)}{dt} = F(t, X(t))$ satisfies a continuous Lipschitz condition in the second variable where $F(\cdot, \cdot)$ is (jointly) continuous, then it has a unique solution in a neighborhood of the initial point $(t_0, X(t_0) = x)$ (by an ancient theorem due to Picard). Denoting the solution by $f_{t_0 t}(x), t \geq t_0$ one has: (i) $f_{t_0, t_0}(x) = x$, (ii) f is continuous in all variables, and (by the uniqueness property), (iii) for $t_0 \leq s \leq t$ $(f_{st}(f_{t_0 s}))(x) = f_{t_0 t}(x)$ for all x in the range or phase space of $X(\cdot)$. If M denotes the phase space of $X(t)$, then one also proves (a deeper property) that (iv) $f_{st} : M \to M$ to be a homeomorphism. An f with properties (i)–(iv) is said to determine a *phase flow* of the motion on the manifold M (cf. Arnold (1973), p. 250 ff).

Now let $X(t), t \geq t_0$, be a stochastic process which is a solution of an SDE of a similar nature with $F(t, X(t))dt(= dX(t))$ representing the generalized Itô form satisfying the Lipschitz-type conditions. Then there is a unique solution which is termed an Itô process in our study. By Theorem 5.1 the dependence of the solution on the initial conditions is continuous. With this motivation one can introduce the following.

1.   Definition. Let $\{f_{t_0 t}(\mathbf{x}), t \geq t_0\}$ be a family of $\mathbb{R}^n$-valued random vectors indexed by the triple $(s, t, \mathbf{x}) \in \mathbb{R}^+ \times \mathbb{R}^+ \times \mathbb{R}^n$. It is called a *stochastic flow of homeomorphisms* if for a.a.(ω) the $f_{st}(\mathbf{x}, \omega)$ satisfies

$$(i) \ f_{ss}(\mathbf{x}) = \mathbf{x}, \ s \in \mathbb{R}^+, \ \mathbf{x} \in \mathbb{R}^n,$$
$$(ii) \ f_{ut}(f_{su}(\mathbf{x}, \omega), \omega) = f_{st}(\mathbf{x}, \omega), \text{ for } s \leq u \leq t,$$
$$(iii) f_{st}(\cdot, \omega) : \mathbb{R}^n \to \mathbb{R}^n \text{ is a homeomorphism onto.}$$

Moreover, for a.a. (ω), if the vector $f_{st}(\cdot, \omega)$ is in $C^k(\mathbb{R}^n)$ so that it has continuous k^{th} order (partial) derivatives, then f is termed a *stochastic flow of C^k-diffeomorphisms*.

Consider the second order SDE given by Eq. (4) of Section 4. Under the local Lipschitz and boundedness conditions, on the "drift" vector and "diffusion" matrix coefficients of Theorems 4.1 and 4.2 there is a unique (vector) solution process having continuous dependence on the initial conditions. Moreover, since this is a first order vector SDE, its solution is a (vector) Markov process when the noise $\mathbf{B}$ is an $L^{2,2}$-bounded, independent increment process. Suppose now that $\mathbf{B}(t) = (0, B(t))$ is replaced by $\bar{\mathbf{B}}(t) = (B_1(t), B_2(t))$ where the $B_i(t)$ are independent Brownian motions. By using the adjunction procedure, if necessary, one can assume that they are all defined on the same underlying probability space, and that B_1 is also independent of the initial values $X(0)$ and $\dot{X}(0)$. Consequently the second order Itô process $X(t)$ is independent of B_1. Now $\bar{\mathbf{B}}$ charges every nondegenerate interval. Moreover, in this case the stochastic integral gives a (local) martingale, and the first integral defines a process of (local) bounded variation. Hence in the treminology of Kunita's book ((1990), p.79) the $V(t)$-process has $(\mathbf{Q}, \mathbf{SS}^*, dt)$ as the "local characteristics" which satisfy the boundedness hypothesis, Lipschitz, and a linear growth condition there. Then the SDE has a unique solution with the initial value $\mathbf{x}$ at time t_0, and has a modification which is a stochastic flow of homeomorphisms by Kunita's result.

This may be summarized as follows.

2. Theorem. *If $\{V_{t_0}(t), t \geq t_0\}$ is the unique (vector) solution of the SDE (4) of Section 4, under the (local) Lipschitz and linear growth conditions, the process starting at t_0 with the initial value $\mathbf{x}$, then it has a modification which is a ('forward') stochastic flow of homeomorphisms a.e.*

Using some stand time change techniques as discussed in Kunita (1990), the result can be extended to the case that $\mathbf{B}$ is a continuous $L^{2,2}$-bounded process, which under the current conditions of continuity on $\mathbf{Q}, \mathbf{S}$, or equivalently on q, σ, reduces to semi-martingales. This will not be detailed here. Note however that, in our case, the second component of $V(t)$ is a stochastic derivative of the first. So if one goes back to the second order Itô process $\{X(t), t \geq t_0\}$ itself, this additional information appears to indicate that it will be a C^1-diffeomorphism as an additional property of the solution, and in general the nth order SDE should thus have a C^{n-1} diffeomorphic solution. The exact verification needs further computations extending Kunita's work ((1990), Sec. 4.6), to establish this statement.

The preceding result brings up a converse implication analogous to that of Theorem 6.5. Namely, under what conditions a given stochastic flow of homeomorphisms (or diffeomorphisms) arises from a (second order) SDE relative to some $L^{2,2}$-bounded (or more particularly Brownian motion) noise process? For the first order case the problem has been solved (cf. Kunita (1990), p.184 ff). The estimates derived for Theorem 6.5 should be useful for this problem also; but since all the necessary details are not available, we have to leave the subject here. It is, however, interesting (and possible) to present the exact conditions for a stochastic C^{k-1} diffeomorphic flow to arise as a solution of a kth *order* SDE. Both the direct and the converse problems have meaningful analogs if the state space is a differentiable manifold. In the first order case, much work is available from the books of Kunita (1990) and Ikeda-Watanabe (1989), (see also Emry(1989))and its extension to higher order SDEs is a reasonable research project in the area.

4.10. MULTIPARAMETER ANALOGS AND SPDEs.

A natural problem now is to consider the case for random fields, meaning processes with a multidimensional parameter index. These can be studied from two different points of view. First consider the process $\{X(t), t \in T \subset \mathbb{R}^k\}$ where $k > 1$. For simplicity take $k = 2$, and $T = \mathbb{R}_+^2$. Then the corresponding (first order) SDE is written as:

$$dX(t) = a(t, X(t))dt + b(t, X(t))dB(t), \qquad (1)$$

which is understood again in the integrated form so that it becomes, on setting $t = (t_1, t_2) \geq 0$ (i.e., $t_1 \geq 0, t_2 \geq 0$):

$$
\begin{aligned}
X(t_1, t_2) = {} & X(t_1, 0) + X(0, t_2) - X(0, 0) \\
& + \int_0^{t_1} \int_0^{t_2} a(u_1, u_2, X(u_1, u_2)) du_1 du_2 \\
& + \int_0^{t_1} \int_0^{t_2} b(u_1, u_2, X(u_1, u_2)) dB(u_1, u_2), \qquad (2)
\end{aligned}
$$

where the last symbol is a suitable double stochastic integral to be defined. This will be discussed presently. This point of view was developed primarily by X. Guyon-B. Prum (1981) and J. Yeh (1986). The second approach is to consider the classical PDEs which are perturbed by a random noise. More explicitly, let, for instance, $u(x, t)$ denote the displacement of a vibrating string from equilibrium, depending on the position x and time t. If the height of vibration is relatively small, then by the classical theory u satisfies the one-dimensional wave equation subject to a random disturbence:

$$
\begin{aligned}
u_{tt} &= \alpha^2 u_{xx} + \dot{B}(x, t), \ t > 0, \ x \in \mathbb{R}, \\
u(x, 0) &= f(x); \ u_t(x, 0) = g(x),
\end{aligned} \qquad (3)
$$

the second line denoting the initial conditions, and $\dot{B}(x, t)$ is analogous to white noise. Here α^2 is a physical constant ($=$ ratio of the tension to the linear density of the string, to be taken as unity for convenience here). Since $\dot{B}$ typically does not exist, (3) is converted into an integral (or a weak or distributional) form by integrating the equation after multiplying by an infinitely differentiable compactly based function φ to get (formally integrating by parts):

$$\int_0^T \int_{\mathbb{R}} u(x, t)[\varphi_{tt} - \varphi_{xx}(x, t)] dx dt = \int_0^t \int_{\mathbb{R}} \varphi(x, t) dB(x, t), \qquad (4)$$

if $\varphi(x,T) = 0 = \varphi_t(x,T)$ for all $x \in \mathbb{R}$. Again, one has to define the right-side (two-dimensional) stochastic integral to proceed further. Thus (3) or its equivalent integrated form (4) is the stochastic PDE for study in this context. This view is developed by E. Wong and M. Zakai and especially by J. B. Walsh (cf. the latter's lectures (1984) for a good account with references to earlier work), and many authors thereafter follow this approach.

Noting that $\{B(t_1, t_2), t_i \geq 0, i = 1, 2\}$ of (2), or $\{B(x,t), t \geq 0, x \in \mathbb{R}\}$ of (4), is a multiparameter Brownian (or semi-martingales that include the former) process, one has to define the corresponding integral by extending either the Itô-procedure, or (more inclusively) the $L^{2,2}$-boundedness concept and the resulting method. The thus obtained integral can be used in both contexts. Classically, restricting to Brownian motion in the plane, Cairoli and Walsh (1975) have presented a detailed structure theory for these integrals and this work with a component-wise (= lexicographic type) ordering allows one to obtain sharp results compared to general methods based on abstract partial ordering. The latter view was developed by Hürzeller (1985), (see also Dozzi (1989) for the n-parameter extension). We include an outline of these integrals to present the higher order problems thereafter.

As in the case of one-dimensional time, consider a stochastic field $X : T \to \mathbb{R}$ where $T \subset \mathbb{R}^2_+$, the positive quadrant of the plane $\mathbb{R}^2$. If there exists a σ-finite measure $\alpha : \mathcal{O} \subset \mathcal{B}(\mathbb{R}^2_+) \otimes \Sigma \to [0, \infty]$ such that for each simple function $f = \sum_{i=1}^n a_i \chi_{A_i}$, $A_i \in \mathcal{O}$, a σ-subalgebra, one has ($f : \Omega' = T \times \Omega \to \mathbb{R}$)

$$E(|\tau(f)|^2) = E(|\sum_{i=1}^n a_i X(A_i)|^2)$$
$$\leq C \int_{\Omega'} |f|^2 d\alpha(t, \omega), \tag{5}$$

for some absolute constant $C > 0$, then the integrator X is termed $L^{2,2}$-*bounded (relative to $\mathcal{O}$ and α)*. Then τ is a continuous linear mapping on the dense subset of simple functions of $L^2(\alpha)(= L^2(\Omega', \mathcal{O}, \alpha))$ into $L^2(P)$ and hence has a unique continuous extension onto $L^2(\alpha)$. This extended τ is the stochastic integral written, using the same symbol, as:

$$\tau(f) = \int_T f dX, \quad f \in L^2(\alpha), \tag{6}$$

and it satisfies the dominated convergence criterion. This is an extension of Bochner's boundedness principle in the multiparameter context, (cf. Rao (1995), Section VI.3). It will now be shown that (5) is always satisfied for the two-parameter Wiener-Brownian motion with $T = \mathbb{R}^2_+$, if $\mathcal{O}$ is suitably chosen so that $\alpha = \mu \otimes P$, where μ is the planar Lebesgue measure (the $\mathcal{O}$ will be determined by certain (predictable) σ-subalgebras of Σ and $\mathcal{B}(\mathbb{R}_+^2)$).

Recall that a two-dimensional Brownian motion of the above type is a real Gaussian process $\{X_t, t \in \mathbb{R}_+^2\}$, starting at the origin, having independent increments, with mean zero, and covariance $E(X_s X_t) = \min(s_1, s_2)\min(t_1, t_2)$ where $s = (s_1, s_2)$ and $t = (t_1, t_2)$. [Again X_t and $X(t)$ are synonymous.] The existence of such a process follows immediately from a classical projective limit theorem after noting that the corresponding distributions (or their characteristic functions) satisfy the Kolmogorov consistency conditions (cf., e.g., Rao (1995), Theorem I.2.4). Alternatively this existence can be obtained from a direct calculation with the Wiener measure space, as was done by Yeh (1986). It then follows [see the one-dimensional time case] that such an X_t is $L^{2,2}$-bounded so that (5) holds with $\mathcal{O} = \sigma\{(s, t] \times A : A \in \mathcal{F}_s\}$, $\mathcal{F}_t = \sigma(X_s : s \le t)$. Here $s \le t$ means $s_i \le t_i, i = 1, 2$ when $s = (s_1, s_2), t = (t_1, t_2)$, (the lexicographic ordering) X_t being the $\mathcal{F}_t$-adapted Brownian motion, and $\alpha = \mu \otimes P$, noted above. The standard properties of this process, which is a ("strong") martingale and whose "marginals" have a conditional independence property, can be abstracted and employed in more general situations as follows (cf., Cairoli and Walsh (1975)).

For the probability space (Ω, Σ, P), a family $\{\mathcal{F}_t, t \in T \subset \mathbb{R}_+^2\}$ is a *standard filtration* if :(i) each $\mathcal{F}_t$ is P-complete, (ii) it is increasing, i.e., $s \le t \Rightarrow \mathcal{F}_s \subset \mathcal{F}_t$ for the coordinatewise ordering, and (iii) right (order) continuous, i.e., $\mathcal{F}_s = \cap_{t > s}\mathcal{F}_t$. It is *strongly filtered* if, moreover, for each $t = (t_1, t_2), \mathcal{F}_t^1$ and $\mathcal{F}_t^2$ are conditionally independent relative to $\mathcal{F}_t$, where $\mathcal{F}_t^1 = \sigma(\cup_r \mathcal{F}_{t_1, r})$ and $\mathcal{F}_t^2 = \sigma(\cup_r \mathcal{F}_{r, t_2})$ are the (marginal) σ-subalgebras of Σ thus defined. In other words, $E^{\mathcal{F}_t^1}$ and $E^{\mathcal{F}_t^2}$ commute and $E^{\mathcal{F}_t} = E^{\mathcal{F}_t^1} E^{\mathcal{F}_t^2}$. It is seen that the two-parameter Brownian motion (also termed a *Brownian sheet*), $\{X_t, \mathcal{F}_t, t \in T\}$ described above is automatically strongly filtered for $\{\mathcal{F}_t, t \in \mathbb{R}_+^2\}$.

In contrast to the one-dimensional time problem, the boundry $\partial \mathbb{R}_+^2$ of $\mathbb{R}_+^2$ is not a point, but consists of the x_1- and x_2-axes, and the boundary value of a process $X_t, t \in \mathbb{R}_+^2$ is denoted as $X(\partial \mathbb{R}_+^2)$ or simply ∂X. Also there are several types of martingales for multiparameters, namely, weak, strong, ordinary, and i^{th}-coordinate versions (cf., Cairoli and Walsh (1975)), and thus the analysis gets considerably more involved than the one-dimensional time. This general account will not be discussed further.

Here we first present a solution of the SDE of the type (1) in the form (2). The reader is referred to the paper by Yeh (1986) for the many details.

1. Theorem. *Let $\{B(t), \mathcal{F}_t, t \in \mathbb{R}_+^2\}$ be a (strongly filtered) Brownian sheet with $\partial B = 0$, a.e. on (Ω, Σ, P). Suppose the coefficients a, b of (1) satisfy the following (local) Lipschitz and boundedness conditions as t varies on* *compact squares $R_T \subset \mathbb{R}_+^2$ of side length T:*

(1) $|a(t, x) - a(t, x')|^2 + |b(t, x) - b(t, x')|^2 \le C_t |x - x'|^2$

(2) $|a(t, x)|^2 + |b(t, x)|^2 \le C_T(|x|^2 + 1), \quad (x, x') \in \mathbb{R}^2,$

where $C_T > 0$ is a constant depending only on R_T.

Then there exists a (pathwise) unique solution X of (2) satisfying the given boundary condition $\partial X = Z$ which is a continuous square integrable process $\{Z_t, \mathcal{F}_t, t \in \partial \mathbb{R}_+^2\}$.

The above Lipschitz and boundedness conditions (1) and (2) can be weakened slightly by replacing the x's on the right side with an average value (or an integral) relative to some Radon measure on $\mathbb{R}_+^2$. The present case is obtained if the measure concentrates on suitable points. The proof, with the weaker assumptions, is given by Yeh (*op.cit*), and it follows by a two-dimensional version of the Picard approximation, which uses several moment estimates and is quite long. Perhaps a suitable modification of the Banach Contraction Mapping principle invoked in Theorem 4.1 can be followed to simplify the work, but this has not been carried through. We discuss the second order analog of the problem instead, to point out some new questions.

In analogy with (1) or (2) for the second order case one may consider the partially derived processes $D_i X, i = 1, 2$ with $D_i = \frac{\partial}{\partial t_i}$, whose existence should be obtained as in Proposition 2.5; namely, $D_i X$ is to exist in the strong or $L^p(P)$-topology and that it has no (fixed) points of discontinuity. Thus (1) becomes a system (symbolically written again) as:

$$d(D_i X)(t) = a_i(t, X(t), (D_1 X)(t), (D_2 X)(t)) dt$$
$$+ b_i(t, X(t), (D_1 X)(t), (D_2 X)(t)) dB(t), \tag{7}$$

for $i = 1, 2$. These may be converted to a pair of integral equations:

$$X(t_1, t_2) - X(0, t_2) - X(t_1, 0) + X(0, 0)$$
$$- [(D_1 X)(0, t_2) + (D_1 X)(0, 0)] t_1$$
$$= \int_0^{t_1} [\int_0^r \int_0^{t_2} a_1(u, v, (D_1 X)(u, v), (D_2 X)(u, v)) du dv] dr$$
$$+ \int_0^{t_1} [\int_0^r \int_0^{t_2} b_1(u, v, (D_1 X)(u, v), (D_2 X)(u, v)) dB(u, v)] dr, \tag{8}$$

and a similar equation with t_1, t_2 interchanged with a_2, b_2 in place of a_1, b_1. The conditions on a_i, b_i and the moment estimates necessary for further analysis have not been studied. Moreover, one should restrict these four structural coefficients in order that, under appropriate initial and boundary conditions, both equations determine a single random field X so that the system (7) admits a unique solution. Thus the higher order equations lead to *systems* of SDEs, and essentially nothing is known about their solutions and this aspect is an area for future exploration. A related multiple stochastic integration problem when B is a multiindexed $L^{2,2}$-bounded process has yet to be explored. A recent contribution in this direction, due to Green (1996), is of interest.

Let us now consider the second point of view indicated by (3) or (4), since this is based on the classical PDE theory. Thus a general (nonlinear) wave equation is given in the symbolic form as:

$$u_{tt}(x,t) = u_{xx}(x,t) + g(u(x,t),t) + f(u(x,t),t)\dot{X}(x,t) \qquad (9)$$

with initial and bounding conditions:

$$u(x,0) = u_0(x), \; ; u_t(x,0) = v_0(x), \; \forall x \in \mathbb{R}. \qquad (10)$$

Here X is typically a Brownian sheet or more generally a "martingale measure", which is replaced by an $L^{2,2}$-bounded measure in what follows to include semi-martingale measures, to be recalled below. This will conform to our general viewpoint, but the argument of Walsh's (1984) will be adapted to this context.

Since $\dot{X}$ does not usually exist, (9) will be interpreted in the weak sense after formally integrating by parts. This gives for each compactly based real smooth function φ on $[0,T] \times \mathbb{R}$ for $T > 0$, and $\varphi(x,T) = 0 = \varphi_t(x,T)$, the following:

$$\int_0^T \int_{\mathbb{R}} \{u(x,t)[\varphi_{tt} - \varphi_{xx}](x,t) - g(u(x,t),t)\varphi(x,t)\}dxdt$$
$$= \int_0^T \int_{\mathbb{R}} f(u(x,t),t)X(dxdt), \qquad (11)$$

where subscripts (such as in φ_{tt}) denote partial differentiation relative to the indicated variable, as usual. Here the left side of (11) is the (double) Lebesgue integral, but that on the right, relative to the $L^{2,2}$-bounded measure X, has to be defined in an appropriate manner and then the solution of (9) subject to (10) is defined as the weak solution of the integral Equation (11), subject to (10), for each smooth φ of compact support satisfying the additional boundary condition prior to (11) and with suitable Lipschitz conditions on the functions f, g. The existence and uniqueness of solutions of (11) will be through the classical Picard method as in the PDE theory. For this idea to succeed, it is now necessary to define the (double) integral on the right of (11) for the $L^{2,2}$-bounded X. This is done along the following lines, based on the work of martingale measures and the trimeasure integrals (taking care of the new problems that are absent in the bimeasure theory, cf., Graham and Ylinen (1991)).

Recall that a vector measure $\nu : \mathcal{A} \to \mathcal{X}$ ($\mathcal{A}$ a σ-algebra of a set A and $\mathcal{X}$ a Banach space) is bounded and has a dominating or control measure $\lambda : \mathcal{A} \to \mathbb{R}^+$, using which a vector (or Dunford-Schwartz) integral of bounded $\mathcal{A}$-measurable scalar functions can be defined (cf., Dunford-Schwartz (1958), Sec. IV.10). Here 'bounded vector measure' ν means $|\nu|(A) = \sup\{\|\nu(B)\| : B \in \mathcal{A}\} < \infty$, $|\nu|(\cdot)$ being the semi-variation. The corresponding statement for multimeasures is not true. (A function $\nu : \mathcal{A} \times \cdots \times \mathcal{A} \to \mathcal{X}$ where $\nu(A_1, A_2, \cdots, A_n)$ is individually σ-additive in A_i for each $i = 1, \ldots, n$ when the rest are held fixed, is a *multimeasure*.) The case $n = 2$ gives bimeasures and $n > 2$ the multimeasures; but these cases have sharp distinctions and, for instance, control measures need not exist for the latter in contrast to the vector measure case. This problem appears early in defining the stochastic integral for (11), and it will be clarified.

2.   Definition. If $\{\mathcal{F}_t, t \geq 0\}$ is a standard filtration of a probability space (Ω, Σ, P) and $(A, \mathcal{A})$ is a measurable space, then the family $X_t : \mathcal{A} \to L^2(\Omega, \mathcal{F}_t, P), t \geq 0$, of vector measures is called an $L^{2,2}$-bounded measure relative to an indexed (by $\mathcal{A}$) family of σ-finite measures, starting at '0', if:

(i) $X_t(\cdot)$ is a vector measure for each $t \geq 0$,

(ii) $X_0(B) = 0$, a.e., $\forall B \in \mathcal{A}$,

(iii) $\{X_t(B), \mathcal{F}_t, t \geq 0\}$ is a locally $L^{2,2}$-bounded process for each $B \in \mathcal{A}$, relative to a σ-finite measure ν_B, i.e., for each $T > 0$, and simple function $f = \sum_{i=0}^n a_i \chi_{A_i}, A_i = (t_i, t_{i+1}], 0 = t_1 < t_2 < \cdots < t_n \leq T$, if $\tau(f)(B) = \sum_{i=1}^n a_i(X_{t_{i+1}} - X_{t_i})(B)$, then one has:

$$E(|\tau(f)(B)|^2) \leq C \int_0^T |f(t)|^2 d\nu_B(t), \ B \in \mathcal{A},$$

where $C(= C_T > 0)$ is an absolute constant.

If $\{X_t(B), \mathcal{F}_t, t \geq 0, B \in \mathcal{A}\}$ is a martingale measure, then the function $\nu_B(\cdot)$ is the Doléan-Dade measure (cf., e.g., Rao (1995), p.466). Now in the case at hand one needs to define integrals relative to X, regarded as a measure function of t and B, i.e., $X : \mathcal{B}(0,T) \otimes \mathcal{A} \to L^2(\mathcal{F}_T, P)$ is a vector bimeasure.

Since by (iii) of the above definition, $\{X_t(B), \mathcal{F}_t, t \geq 0\}$ is locally $L^{2,2}$-bounded, it has finite quadratic variation on $[0, T]$ for each $T > 0$, and hence (via polarization) its covariation measure K_t (say) is given by

$$[X_t(B), X_t(C)] = K_t(B, C), \ t \geq 0, \tag{12}$$

which exists and defines a bimeasure (cf., Rao, *ibid*, p.472) and which has (locally) finite variation for each fixed B and C. Consequently, the function Q defined by

$$Q(B, C, D) = K_t(B, C) - K_s(B, C), \ D = (s, t], \tag{13}$$

and extended to the class $\mathcal{A} \times \mathcal{A} \times \mathcal{B}_0$, where $\mathcal{B}_0$ is the ring of all rectangles $D \subset \mathbb{R}^+$, gives a *trimeasure associated with* X, (cf., discussion prior to Definition 2). Also as noted before, such a trimeasure may not have a dominating measure; and so those which possess the latter property will be termed controlled $L^{2,2}$-bounded measures for which the desired integral is definable. More explicitly one has the following.

3. Definition. Let $X = \{X_t(B), \mathcal{F}_t, t \geq 0, B \in \mathcal{A}\}$ be an $L^{2,2}$-bounded measure with the associated trimeasure Q as given by (13). Then X is called a *controlled* $L^{2,2}$-*bounded measure* if there existed a locally bounded trimeasure $L : \sigma(\mathcal{A} \times \mathcal{A} \times \mathcal{B}_0) \to (L^2(P))^+$ which extends to a vector measure, such that $L(\cdot, \cdot, B)$ is a covariance bimeasure for each $B \in \mathcal{B}_0$, that majorizes Q pointwise, i.e.,

(i) $L(A_1, A_2, B) = L(A_2, A_1, B)$, $\forall B \in \mathcal{B}_0$,

(ii) L is positive definite in the sense that for each bounded measurable function $f : A \times \mathbb{R}^+ \to \mathbb{R}$, one has (the Dunford-Schwartz integral) $\int_{A \times A \times \mathbb{R}^+} f(x, s) f(y, s) L(dx, dy, ds) \geq 0$, so that this integral defines a semi-inner product $(f, g)_L$ with norm $\|f\| = \sqrt{(f, f)_L} \geq 0$,

(iii) for each $F \in \mathcal{A} \times \mathcal{A} \times \mathcal{B}_0$, $|Q(F)| \leq L(F)$, a.e.

It may be observed that although Q is not positive it is positive definite since for each disjoint collection $A_1, \ldots, A_n$ from $\mathcal{A}$, and real numbers $a_1, \ldots, a_n$ one has

$$\sum_{i,j=1}^{n} a_i a_j Q(A_i, A_j, (s, t])$$

$$= [\sum_{i=1}^{n} (X_t(A_i) - X_s(A_i)), \sum_{i=1}^{n} a_i(X_t(A_i) - X_s(A_i)) \geq 0.$$

(However Q itself need not be a vector measure!)

One can verify that the controlledness property is present if X is a white noise measure, and more generally for martingale measures that are orthogonally valued, i.e., $X_t(A)$ and $X_t(B)$ satisfy the covariation relation $[X_t(A), X_t(B)] = 0$ for disjoint A, B. However, as Walsh (1984) detailed with an example, due to D. Bakry, that there exist martingale (hence $L^{2,2}$-bounded) measures that are not controlled. [He calls controlled martingales "worthy".] The purpose of these restrictions is to establish a stochastic Fubini type result on the product spaces such as $(A \times A \times [0, T], \mathcal{A} \otimes \mathcal{A} \otimes \mathcal{B}, L) \times (C, \mathcal{C}, \mu)$ where the last one is a given (finite) measure space so that several calculations are permissible by change of orders of the integrals in manipulating (11) to find solutions.

We now sketch the concept of the desired multiple stochastic integral relative to a controlled $L^{2,2}$-bounded measure X which gives a stochastic measure instead of a process (or a field). Thus let Q be the associated trimeasure of X, with L as its control (vector) measure. For a step function $f = u\chi_{(a,b]}\chi_B$, $0 \le a \le b$, $B \in \mathcal{A}$, and u as $\mathcal{F}_a$-measurable bounded random variable, define the integral of f relative to X as:

$$Y_t(C) = f \cdot X_t(C) = u(X_{t \wedge b}(B \cup C) - X_{t \wedge a}(B \cap C)), \qquad (14)$$

and extend it by linearity for all simple functions which are therefore finite linear combinations of step functions of the above type, to get a controlled $L^{2,2}$-bounded measure process $Y_t(\cdot)$. Letting $\mathcal{S}$ denote the class of all such simple functions and defining a norm $\| \cdot \|_Q$ on $\mathcal{S}$ by

$$\|f\|_Q^2 = E((|f|, |f|)_L), \qquad (15)$$

consider the set $\mathcal{P}_X = \{f : \|f\|_Q < \infty\}$. One verifies that $\mathcal{S} \subset \mathcal{P}_X = \{\mathcal{P}_X, \| \cdot \|_Q\}$ is dense and that the latter is a Banach space. Moreover the integral can be extended by continuity from $\mathcal{S}$ to $\mathcal{P}_X$ using a standard procedure. Following Walsh's work in a straightforward manner, one shows that for each $f \in \mathcal{P}_X$, $f \cdot X(\cdot)$ is again a controlled locally $L^{2,2}$-bounded measure process and its covariation and control vector measures can be calculated. To get the usual process from this, let $f \cdot X_t(B) = \int_{B \times [0,t]} f dX$, $B \in \mathcal{A}$, and $\int_{A \times \mathbb{R}^+} f dX = \lim_{t \to \infty} f \cdot X_t(A)$. After these preliminaries one can establish the following Fubini type result to use as a key technical ingredient in solving (11).

4. Theorem. *Let $(C, \mathcal{C}, \mu)$ be a σ-finite measure space and $(A \times A \times [0, T], \mathcal{A} \otimes \mathcal{A} \otimes \mathcal{B}, L)$ be the vector measure space introduced earlier. If $f : A \times [0, T] \times \Omega \times C \to \mathbb{R}$ is a jointly measurable function such that $f(\cdot, \cdot, \cdot, \lambda) \in \mathcal{P}_X$ for each $\lambda \in C$, suppose that*

$$E\left\{ \int_{A \times A \times [0,T] \times C} |f(x, s, \cdot, \lambda) f(y, s, \cdot \lambda)| L(dxdyds)\mu(d\lambda) \right\} < \infty.$$

Then one has for all $t > 0$,

$$\int_C [\int_{A \times [0,t]} f(x, s, \cdot, \lambda) X(dxds)] \mu(d\lambda)$$
$$= \int_{A \times [0,t]} [\int_C f(x, s, \cdot, \lambda) \mu(d\lambda)] X(dxds). \qquad (16)$$

with probability one.

This theorem is due to Walsh (1984), for worthy martingale measures, and the argument holds in our case without real changes. The result is needed in obtaining the following important assertion, due also to the same author, which is presented here for completeness.

5. Theorem. *Let X be a white noise measure (i.e., the $L^{2,2}$-bounded measure is now specialized), and f, g of (11) satisfy a Lipschitz condition of the form:*

$$|f(x,t) - f(y,t)| \leq K|x - y|; \quad |f(x,t)| \leq K(1+t)(1+|x|)$$

for some constant $K > 0$, and similarly for g, together with the initial and boundary conditions of (9). Then the stochastic wave equation, taken in the form (11), has a unique weak solution.

This theorem seems to be true for controlled $L^{2,2}$-bounded measures also, but the details have not been completed. The following further comments on these two results and related matters are of interest.

If X itself is a multimeasure, then one can consider some analogs of the Fubini type result from the available vector integration theory (as in, e.g., Dobrakov (1979 and 1987)). The latter approach may, however, be appealed for the associated trimeasure integration used here, and this connection should be explored for a possible simplification and then a generalization of the present case. It is also of interest to discuss the relation between the domination property for the stochastic measures and a similar problem appearing in the Banach space integration (cf., the preceding reference (1987)). In particular, a stronger "completely boundedness" property of the multimeasures (cf., e.g., Graham and Ylinen (1991)) and its use in the present work should be investigated. For bimeasures both these conditions coincide, and the weaker Morse-Transue integration, so vital for analysis of many nonstationary processes (cf., e.g., Chang and Rao (1986); Rao (1989)), was found to be appropriate. Their analogs, via the completely boundedness condition, should be analyzed in the context of the SPDEs, exemplified by Theorem 5 above. In fact, the application of real (= abstract) analysis for a study of SPDEs, especially appearing in Walsh's work, should be pursued further for other problems in this area.

Bibliography

[1] V. I. Arnold (1973), *Ordinary Differential Equations*, (translation), MIT Press, Cambridge, MA.

[2] S. Bochner (1955), *Harmonic Analysis and Probability Theory*, University of California Press.

[3] D. R. Borchers (1964), *Second Order Stochastic Differential Equations and Related Itô Processes*, Ph.D. thesis, Carnegie-Mellon University, Pittsburgh, PA.

[4] M. D. Brennan (1979), "Planar semimartingales", *J. Multivar. Anal.*, **9**, 465-486.

[5] R. Buckdahn (1991), "Linear Skorokhod stochastic differential equations," *Probab. Theor. Relat. Fields*, **90**, 223-240.

[6] R. Cairoli and J. B. Walsh (1975), "Stochastic integrals in the plane," *Acta Math.*, **134**, 111-183.

[7] S. Chandrasekhar (1943), "Stochastic problems in physics and astronomy," *Rev. Modern Phys.*, **15**, 1-89.

[8] D. K. Chang and M. M. Rao (1986), "Bimeasures and nonstationary processes," in *Real and Stochastic Analysis*, John Wiley & Sons, New York, 7-118.

[9] Ju. L. Dalecky and M. G. Krein (1974), *Stability of Solutions of Differential Equations in Banach Spaces*, Translation Series of the Am. Math. Soc., Providence, RI.

[10] I. Dobrakov (1979 and 1987), "Integration in Banach spaces-III, VIII" *Czech. Math. J.*, **29**, 478-499; **37**, 487-506.

[11] J. L. Doob (1942), "The Brownian movement and stochastic equations," *Ann. Math.*, **43**, 351-369.

[12] J. L. Doob (1953), *Stochastic Processes*, John Wiley & Sons, New York.

[13] J. L. Doob (1955), "Martingales and one-dimensional diffusion," *Trans. Am. Math. Soc.*, **78**, 168-208.

[14] M. Dozzi (1989), *Stochastic Processes with a Multidimensional Parameter*, Res. Notes in Math., Pitman, London.

[15] N. Dunford and J. T. Schwartz (1958), *Linear Operators, Part I: General Theory*, Wiley-Interscience, New York.

[16] H. Dym (1966), "Stationary measures for the flow of a linear differential equation driven by white noise," *Trans. Am. Math. Soc.*, **123**, 130-164.

[17] E. B. Dynkin (1961), *Theory of Markov Processes*, Prentice-Hall, Englewood Cliffs, NJ.

[18] W. Feller (1952), "The parabolic differential equations and the associated semi-group of transformations," *Ann. Math.* **55**, 468-519.

[19] J. A. Goldstein (1969), "Second order Itô process," *Nagoya Math. J.* **36**, 27-63.

[20] C. C. Graham and K. Ylinen (1991), "Classes of trimeasures: Applications of harmonic analysis," *Probability Measures on Groups-X*, Plenum Press, New York, 169-176.

[21] M. L. Green (1996), "Planar Stochastic Integrals Relative to Quasimartingales," *Chapter 2, this volume.*

[22] X. Guyon and B. Prum (1981), " Variations-produit et formule de Itô pour le semi-martingales representables à deux paramètres," *Z. Wahrs.*, **56**, 361-397.

[23] H. E. Hürzeller (1985), "Stochastic integration on partially ordered sets," *J. Multivar. Anal.*, **19**, 279-303.

[24] N. Ikeda and S. Watanabe (1989), *Stochastic Differential Equations and Diffusion Processes, (2nd ed.)*, North-Holland, Amsterdam.

[25] K. Itô and S. Watanabe (1978), "Introduction to stochastic differential equations," *Proc. Symp. SDE*, Kyoto, John Wiley & Sons, New York, i-xxx.

[26] H. Kunita (1980), "On the representation of solutions of stochastic differential equations," *Springer Lect. Notes in Math.* **784**, 282-304.

[27] H.-H. Kuo (1972), "Stochastic integrals in abstract Wiener space," *Pacific J. Math.* **41**, 469-483.

[28] T. L. Lai (1973), "Space-time processes, parabolic functions and one-dimensional diffusions," *Trans. Am. Math. Soc.* **175** ,409-438.

[29] P. Mandl (1968), *Analytical Treatment of One–dimensional Markov Processes*, Springer-Verlag, Berlin.

[30] G. Maruyama (1955), "Continuous Markov processes and stochastic equations," *Rend. Circ. Mat. Palermo*, **IV**, 48-90.

[31] O. A. Oleĭnik and E. V. Radkevič (1973), *Second Order Equations With Nonnegative Quadratic Form*, Am. Math. Soc. translation, Providence, RI.

[32] P. Protter (1990), *Stochastic Integration and Differential Equations, A New Approach*, Springer-Verlag, Berlin.

[33] M. M. Rao (1984), *Probability Theory with Applications*, Academic Press, New York.

[34] M. M. Rao (1989), "Bimeasures and harmonizable processes, (Analysis, classification and representation)," *Springer Lect. Notes in Math.* **1379**, 254-298.

[35] M. M. Rao (1993), *Conditional Measures and Applications*, Marcel Dekker, New York.

[36] M. M. Rao (1995), *Stochastic Processes: General Theory*, Kluwer Academic, Dordrecht, The Netherlands.

[37] M. M. Rao and Z. D. Ren (1991), *Theory of Orlicz Spaces*, Marcel Dekker, New York.

[38] F. Riesz and B. Sz.-Nagy (1955), *Functional Analysis*, F. Unger Publishing, New York.

[39] W. A. Rosencrantz (1974), "An application of Hille-Yosida theorem to the construction of martingales," *Indiana Univ. Math. J.* **24**, 527-532.

[40] W. A. Rosencrantz (1975), "A strong continuity theorem for a class of semi-groups of type Γ, with an application to martingales," *Indiana Univ. Math. J.*, **25**, 171-178.

[41] H. J. Sussman (1978), "On the gap between deterministic and stochastic differential equations," *Ann. Probab.*, **6**, 19-41.

[42] J. B. Walsh (1984), "An introduction to stochastic partial differential equations," *Springer Lect. Notes in Math.* **1180**, 265-439.

[43] E. Wong and M. Zakai (1974), "Martingales and stochastic integrals for processes with a multidimensional parameter," *Z. Warsh.*, **29**, 109-122.

[44] R. Wu (1985), *Stochastic Differential Equations*, Research Notes in Math. **130**, Pitman, Boston.

[45] J. Yeh (1986), " Two-parameter stochastic differential equations," in *Real and Stochastic Analysis*, Wiley, New York, 249-344.

Chapter 5

Some Aspects of Harmonizable Processes and Fields

Randall J. Swift

5.1 Introduction

Second-order stochastic processes and random fields play key roles in many areas of the applied and natural sciences. Several subclasses of such processes and fields are discussed in this chapter. The simplest and best understood is the stationary class. This is a second-order process whose covariance r is a continuous function which is invariant under time shifts. Covariances of stationary processes admit a Fourier transform by a classical theorem due to Bochner. This representation brings to bear the powerful Fourier analytic methods in the study of stationary processes and fields.

A motivation for the concept of harmonizability is to enlarge the applications of stationary processes and fields while retaining the Fourier analytic methods. As we will see in the course of this chapter, and can be observed throughout the development of representations for various nonstationary processes and fields, the stationary representations often extend. This "robustness" is characteristic of harmonizable processes and fields, though often new methods are required to obtain the desired results.

This chapter is divided into two distinct parts. Part I considers several new classes of nonstationary processes which are related to the harmonizable class. Representations for the covariances of these second-order processes are obtained, as well as spectral representations of the processes themselves.

These representations lead to structural and sample path analyses for these processes. Several applications are also considered.

Part II considers random fields. Random fields admit an additional property termed isotropy. The representation of harmonizable isotopic random fields is developed. The local behavior of several classes of fields is considered. Representations for these fields are obtained.

In addition, Parts I and II provide a unified treatment of recent developments in harmonizability. Several areas for additional research are indicated. The chapter may be studied as a whole, or each part may be considered individually.

5.2　Preliminaries

In the following work, let (Ω, Σ, P) be the underlying probability space.

Definition 5.2.1 *For $p \geq 1$, define $L_0^p(P)$ to be the set of all centered complex valued $f \in L^p(\Omega, \Sigma, P)$, that is $E(f) = 0$, where $E(f) = \int_\Omega f(\omega)dP(\omega)$ is the expectation.*

In this chapter, we will consider second order stochastic processes and random fields. More specifically, mappings $X : \mathbb{R} \to L_0^2(P)$ for processes and with X defined on $\mathbb{R}^n$ in the fields case. The classical results for $X(\cdot)$ are based upon the following assumption on the covariance $r(\cdot, \cdot)$:

Definition 5.2.2 *A stochastic process $X : \mathbb{R} \to L_0^2(P)$ is stationary (in the wide or Khintchine sense) if its covariance $r(s,t) = E(X(s)\overline{X(t)})$ is continuous and is a function of the difference of its arguments, so that*

$$r(s,t) = \tilde{r}(s - t).$$

An equivalent definition of a stationary process is one whose covariance function can be represented as

$$\tilde{r}(\tau) = \int_{\mathbb{R}} e^{i\lambda\tau}dF(\lambda), \tag{5.1}$$

for a unique non-negative bounded Borel measure $F(\cdot)$. This alternate definition is a consequence of a classical theorem due to Bochner (cf. Gihman and Skorohod [11]). Bochner's Theorem brings the powerful Fourier analytic methods to bear on stationary processes. In many applications, the assumption of stationarity is not always valid, and this provides a motivation for the following.

Suppose that the positive definite $r(\cdot, \cdot)$ is representable as

$$r(s,t) = \int_{\mathbb{R}} \int_{\mathbb{R}} e^{i\lambda s - i\lambda' t}dF(\lambda, \lambda') \tag{5.2}$$

where $F(\cdot, \cdot)$ is a positive definite function on $\mathbb{R} \times \mathbb{R}$ of bounded variation in the following sense: F is of bounded *Fréchet variation* if

$$\|F\|(\mathbb{R}, \mathbb{R}) = \sup\left\{ \left| \sum_{i=1}^{m} \sum_{j=1}^{m} a_i \overline{a_j} F(t_i, t_j) \right| : |a_i| \leq 1, t \in \mathbb{R}, i = 1, \ldots, m \right\}$$
$$< \infty.$$

and it is of bounded *Vitali variation* if

$$|F|(\mathbb{R}, \mathbb{R}) = \sup\left\{ \sum_{i=1}^{m} \sum_{j=1}^{m} |F(t_i, t_j)| : t \in \mathbb{R}, i = 1, \ldots, m \right\} < \infty. \quad (5.3)$$

It is clear that

$$\|F\|(\mathbb{R}, \mathbb{R}) \leq |F|(\mathbb{R}, \mathbb{R})$$

and

$$|F|(\mathbb{R}, \mathbb{R}) = +\infty$$

is possible. It can be shown that $\|F\|(\mathbb{R}, \mathbb{R})$ is always finite (cf., Rao [33]). Hence when $|F|(\mathbb{R}, \mathbb{R}) < \infty$, both variations are finite, and if the Vitali variation is finite, then the integrals in (5.2) are in the Lebesgue sense and all the standard results from Real Analysis apply. However, if $|F|(\mathbb{R}, \mathbb{R}) = +\infty$, then (5.2) has to be defined in a weaker form called the Morse-Transue (or MT-) integral for which the dominated convergence theorem is false. In this case some restriction has to be imposed. A restricted integral, still weaker then the Lebesgue integral but having a dominated convergence theorem is called a strict MT-integral, details of which can be found in Chang and Rao [6]. These strict MT-integrals will be used in what follows.

Definition 5.2.3 *A stochastic process* $X : \mathbb{R} \to L_0^2(P)$ *is weakly harmonizable if its covariance* $r(\cdot, \cdot)$ *is expressible as*

$$r(s, t) = \int_{\mathbb{R}} \int_{\mathbb{R}} e^{i\lambda s - i\lambda' t} dF(\lambda, \lambda') \quad (5.4)$$

where $F : \mathbb{R} \times \mathbb{R} \to \mathbb{C}$ *is a positive semi-definite bimeasure, hence of finite Fréchet variation.*

A stochastic process, $X(\cdot)$, is *strongly harmonizable* if the bimeasure $F(\cdot, \cdot)$ in (5.4) extends to a complex measure and hence is of bounded Vitali variation. In either case, $F(\cdot, \cdot)$ is termed the *spectral bi-measure* (or *spectral measure*) of the harmonizable process.

Comparison of Equation (5.4) with Equation (5.1) shows that when $F(\cdot, \cdot)$ concentrates on the diagonal $\lambda = \lambda'$, both the weak and strong harmonizability concepts reduce to the stationary concept. Harmonizable processes retain the powerful Fourier analytic methods inherent with stationary

processes, as seen in Bochner's Theorem, (5.1); but they relax the requirement of stationarity.

The structure and properties of harmonizable processes has been investigated and developed extensively by M.M. Rao and others. The following
sources are listed here to provide a partial summary of the literature. The
work of Rao [28] - [40], provides a basis and direction for the theory. Chang
and Rao [6] motivate and develop the necessary bi-measure theory. Some
moving average representations were obtained by Mehlman [23],[24]. The
structure of harmonizable isotropic random fields and some applications
have been considered by Swift [48] - [50], [54]. Asymptotic properties of
bispectral density estimators have recently been considered by H. Soedjak,
[47]. The forthcoming book by Kakihara [18] gives a general treatment of
multidimensional second order processes which include the harmonizable
class.

In the subsequent sections, the preceding work will be used as well
as the theory of almost periodic functions. For convenient reference the
classical definitions of the classes of u.a.p., $S^2 a.p$ and $B^2 a.p$ functions are
recalled here. The standard classical theory will be assumed throughout
and may be found in Besicovitch [3].

Let $\mathcal{A}$ be the class of all finite trigonometric polynomials

$$S(t) = \sum_{k=1}^{n} a_k e^{i\lambda_k t}. \tag{5.5}$$

The various forms of almost periodicity are obtained by considering the
following distances between two functions $f(t)$ and $\phi(t)$ from the class $\mathcal{A}$.

i) The uniform distance :

$$D_u[f(t), \phi(t)] = \sup_{-\infty < t < \infty} |f(t) - \phi(t)|. \tag{5.6}$$

ii) The S^2 (Stepanov) distance:

$$D_{S^2}[f(t), \phi(t)] = \sup_{-\infty < t < \infty} \left[\int_t^{t+1} | f(x) - \phi(x) |^2 \, dx \right]^{\frac{1}{2}}. \tag{5.7}$$

iii) The B^2 (Besicovitch) distance:

$$D_{B^2}[f(t), \phi(t)] = \left[\lim_{t \to \infty} \sup \frac{1}{t} \int_0^t | f(x) - \phi(x) |^2 \, dx \right]^{\frac{1}{2}}. \tag{5.8}$$

The classical results show that these are (semi) norms on $\mathcal{A}$, and that
the class of all uniformly almost periodic (u.a.p) functions is given by the
closure of $\mathcal{A}$ under $D_u[\cdot, \cdot]$. Similarly, the class of all S^2 almost periodic

($S^2 a.p$) functions, (respectively B^2 a.p.) is the closure of $\mathcal{A}$ under $D_{S^2}[\cdot, \cdot]$ (respectively $D_{B^2}[\cdot, \cdot]$).

The generalization of almost periodicity introduced by H. Weyl,(cf. Besicovitch [3]), as well as the further generalizations obtained by T. Hillmann [14], for classes of Besicovitch-Orlicz almost periodic functions, (B^Φ, where Φ is a Young function) could also be treated, but, for simplicity and space reasons, they will not be considered in the following work.

5.3 Part I: Second Order Nonstationary Processes Including the Harmonizable Class

5.3.1 (c,p)-Summable Processes

A general class of nonstationary processes which extends the ideas of the harmonizable class was first considered by Cramér in 1952 [8]. Rao has refined and studied [6], [35] these processes and the following definition is due to him.

Definition 5.3.1 *A second-order process* $X : T \to L^2(P)$ *is of Cramér class (or class (C)) if its covariance function* $r(\cdot, \cdot)$ *is representable as*

$$r(t_1, t_2) = \int_S \int_S g(t_1, \lambda)\overline{g(t_2, \lambda')} \, dF(\lambda, \lambda') \tag{5.9}$$

relative to a family $\{g(t, \cdot), t \in T\}$ *of Borel functions and a positive definite function* $F(\cdot, \cdot)$ *of locally bounded variation on* $S \times S$, *[S will be in the classical case* $\hat{T}$ *the dual of an LCA group* T, *and generally* (S, B) *is a measurable space] with each* g *satisfying the (Lebesgue) integrability condition:*

$$0 \le \int_S \int_S g(t_1, \lambda)\overline{g(t_2, \lambda')} \, dF(\lambda, \lambda') < \infty, \quad t \in T.$$

If $F(\cdot, \cdot)$ has a locally finite Fréchet variation, then the integrals in Equation (5.9) are in the sense of (strict) Morse-Transue and the corresponding concept is termed *weak class (C)*.

An integral representation of weak class (C) processes was obtained by Chang and Rao [6] and is given below.

Theorem 5.3.1 *If* $X : T \to L_0^2(P)$ *is of weak Cramér class relative to a family* $\{g(t, \cdot), t \in T\}$ *of Borel functions and a positive definite bimeasure* $F(\cdot, \cdot)$ *of locally bounded Fréchet variation on* $S \times S$, *then there exists a stochastic measure* $Z : B \to L_0^2(P)$, B *a* σ-algebra *of* S, *such that*

$$X(t) = \int_S g(t, \lambda)dZ(\lambda) \tag{5.10}$$

where

$$E(Z(A)\overline{Z(B)}) = F(A, B) \text{ for } (A, B) \in B \times B.$$

Conversely, if $X(\cdot)$ is a second-order process defined by (5.10) then it is a process of weak class (C).

Class (C) processes lend themselves to a finer analysis if some additional structure is imposed. The desired concept is introduced by considering $g(\cdot, \lambda)$ which satisfies a Cesàro summability condition.

Definition 5.3.2 *A second-order process $X : \mathbb{R} \to L_0^2(P)$ is (c, p)-summable weak Cramér class, $p \geq 1$ (or (c, p)-summable weak class (C)) if its covariance $r(\cdot, \cdot)$ has representation*

$$r(t_1, t_2) = \int_{\mathbb{R}} \int_{\mathbb{R}} g(t_1, \lambda)\overline{g(t_2, \lambda')} \, dF(\lambda, \lambda') \tag{5.11}$$

relative to a family $\{g(t, \cdot), t \in \mathbb{R}\}$ of Borel functions and a positive definite function $F(\cdot, \cdot)$ of locally bounded Fréchet variation on $\mathbb{R} \times \mathbb{R}$, with each g satisfying the condition that

$$\lim_{T \to \infty} a_T^{(p)}(|h|, \lambda, \lambda') \tag{5.12}$$

exists uniformly in h and is bounded for all h, $p \geq 1$, where

$$a_T^{(p)}(|h|, \lambda, \lambda') = \begin{cases} \frac{1}{T} \int_0^T a_\alpha^{(p-1)}(|h|, \lambda, \lambda') d\alpha & \text{for } p > 1, \\ \frac{1}{T} \int_0^{T-|h|} g(s, \lambda)\overline{g(s + |h|, \lambda')} ds & \text{for } p = 1. \end{cases} \tag{5.13}$$

If $F(\cdot, \cdot)$ has a locally finite Vitali variation, then $F(\cdot, \cdot)$ clearly determines a (Lebesgue-Stieltjes) measure in the plane. The corresponding processes will be termed class (c, p)-*summable Cramér* . The classes of (c, p)-summable weak Cramér processes are wide. Classical results on summability (cf. Hardy [13]) imply that if

$$\lim_{T \to \infty} a_T(|h|, \lambda, \lambda')$$

exists uniformly then

$$\lim_{T \to \infty} a_T^{(p)}(|h|, \lambda, \lambda')$$

exists uniformly for each integer $p \geq 1$. The converse implication is false. Hence, (c, p)-summable weak Cramér processes are contained in $(c, p + 1)$-summable weak Cramér class for $p \geq 1$. The inclusions are proper.

A subclass of $(c, 1)$-summable Cramér processes was introduced by Rao in 1978 [30]. This subclass requires the theory of uniformly almost periodic functions. An extension of the class introduced by Rao is given by the following definition.

Definition 5.3.3 *A second-order process $X : \mathbb{R} \to L^2(P)$ whose covariance is of weak class (C) is termed almost weakly harmonizable if $g(\cdot, \lambda)$ is a uniformly almost periodic function relative to $\mathbb{R}$.*

If the spectral bimeasure $F(\cdot, \cdot)$ admits a finite Vitali variation in the plane, the corresponding concept will be termed *almost strongly harmonizable*. The class of almost weakly harmonizable processes contains the class of weakly harmonizable processes. This can be immediately seen by setting $g(t, \lambda) = e^{i\lambda t}$. Further, if the spectral bimeasure $F(\cdot, \cdot)$ concentrates on the diagonal $\lambda = \lambda'$ the representation of the covariance would become

$$r(t_1, t_2) = \int_{\mathbb{R}} g(t_1, \lambda)\overline{g(t_2, \lambda)} \, d\tilde{F}(\lambda), \qquad (5.14)$$

where $g(\cdot, \lambda)$ is a uniformly almost periodic function. Processes with a covariance representable by (5.14) will be termed *almost stationary*.

These processes will be further studied in the ensuing work. We will first relate the class of almost weakly harmonizable processes to the (c, p)-summable Cramér class.

Proposition 5.3.1 *The class of almost weakly harmonizable processes is contained in the $(c, 1)$-summable weak Cramér class.*

Proof: If $X(\cdot)$ is an almost weakly harmonizable process, then $g(\cdot, \lambda)$ is a uniformly almost periodic function. Applying a classical result on almost periodic functions, (cf. Besicovitch, [3] p.15), one has that

$$\lim_{T \to \infty} \frac{1}{T} \int_0^{T-|h|} g(s, \lambda)\overline{g(s + |h|, \lambda')} ds$$

$$= \lim_{T \to \infty} \frac{T - h}{T} \lim_{T \to \infty} \frac{1}{T - h} \int_0^{T-|h|} g(s, \lambda)\overline{g(s + |h|, \lambda')} ds$$

exists uniformly in h. Hence $X(\cdot)$ satisfies Equation (5.13) with $p = 1$, and is of class $(c, 1)$-summable weak Cramér. $\square$

There are many conditions which will guarantee the existence of the limit in Equation (5.12). For instance, if g is locally square integrable, then $g(\cdot, \lambda)\overline{g(\cdot + h, \lambda)}$ will be $(c, 1)$-summable. Thus there are class $(c, 1)$-summable weak Cramér processes which are not almost weakly harmonizable.

The classes of (c, p)-summable Cramér processes may be further studied by considering a class of processes motivated by summability methods. These processes were first introduced by J. Kampé de Feriet and F.N. Frenkiel in [19]. They were also independently given by Yu. A. Rozanov [43], and E. Parzen [26], the latter under the name "asymptotic stationarity". These ideas are formalized in the following definition.

Definition 5.3.4 *A second-order process $X : \mathbb{R} \to L^2(P)$ with a continuous covariance r is of class (KF) if for each $h \in \mathbb{R}$ the following limit exists:*

$$\tilde{r}(h) = \lim_{T \to \infty} \frac{1}{T} \int_0^{T-|h|} r(s, s+|h|)ds. \tag{5.15}$$

A calculation verifies that the quantities on the right side of (5.15) before taking the limit are positive definite so that, when the limit exists, $\tilde{r}(\cdot)$ is also. The function $\tilde{r}(\cdot)$ is measurable even when it is not continuous, so that an application of the classical Herglotz-Bochner-Riesz theorem on a characterization of such functions (cf., Rao [34]), implies there is a unique positive bounded nondecreasing $G(\cdot)$ such that

$$\tilde{r}(h) = \int_{\mathbb{R}} e^{ih\lambda} G(d\lambda) \text{ a.a.}(h) \in \mathbb{R}. \tag{5.16}$$

The function $G(\cdot)$ is known as the *associated spectral function* of the process $X(t)$.

It is evident that if $X(\cdot)$ is real valued and stationary, so that its covariance $r(s, t)$ is a function of the difference $s - t$, and $\tilde{r}(s - t) = r(s, t)$. Thus the stationary processes are in class (KF). If $X = Y + Z$, where $Z(\cdot)$ is a zero mean stationary process and $Y(\cdot)$ a process with zero mean and periodic covariance [i.e., $r(s + k, t + k) = r(s, t)$ for some k, these processes are known as *periodically correlated*, a generalization of which will be presented in the next section] then it can be shown that X is in class (KF), but is not stationary. This example is in Rao [35] and further details may be found in that paper.

Every strongly harmonizable process is in class (KF). This was first noted by Rozanov [43] and later analyzed by Bhagavan [4]. Rao [30] extended this result to show that every almost strongly harmonizable process belongs to class (KF).

Examples can be constructed which show that the class of weakly harmonizable process is not properly contained in class (KF). One such example is given in Rao [35]. The relationship between the almost strongly harmonizable class and class (KF) suggests that a similar relationship exists for the (c, p)-summable Cramér classes.

Definition 5.3.5 *A second-order process $X : \mathbb{R} \to L^2(P)$ with a continuous covariance r is of class (KF,p), $p \geq 1$ if for each $h \in \mathbb{R}$ the following limit exists:*

$$\tilde{r}(h) = \lim_{T \to \infty} r_T^{(p)}(h) \tag{5.17}$$

where

$$r_T^{(p)}(h) = \begin{cases} \frac{1}{T} \int_0^T r_\alpha^{(p-1)}(h)d\alpha & \text{for } p > 1, \\ \frac{1}{T} \int_0^{T-|h|} r(s, s+|h|)ds & \text{for } p = 1 \end{cases} \tag{5.18}$$

The class $(KF, 1)$ corresponds to the original definition of class (KF) processes. Since in (5.18), $r_T(\cdot)$ is positive definite, it follows that $r_T^{(p)}(\cdot)$ is also positive definite. Thus $\tilde{r}(\cdot)$ satisfies the same hypothesis and (5.16) holds, so that the representing $G(\cdot)$ may now be called a *pth-order associated spectrum*. The classical summability result cited for the (c, p)-summable Cramér processes applies here so that

$$\text{class } (KF) \subset \text{class } (KF, p) \subset \text{class } (KF, p+1)$$

and the inclusions are proper. Thus one has an increasing sequence of classes of nonstationary processes each having an associated spectrum.

Rao first proposed the classes (KF, p) as a tool to show the breadth of the weakly harmonizable class. The structure of the class (KF, p) processes will shed further insight into the (c, p)-summable Cramér classes.

Theorem 5.3.2 *The class (c, p)-summable Cramér processes are contained in the class (KF, p) processes, $p \geq 1$.*

Proof: A class (c, p)-summable Cramér process $X(\cdot)$ has covariance $r(\cdot, \cdot)$ with representation

$$r(s, t) = \int_{\mathbb{R}} \int_{\mathbb{R}} g(s, \lambda) \overline{g(t, \lambda')} \, dF(\lambda, \lambda') \tag{5.19}$$

where $g(\cdot)$ satisfies (5.12). For $X(\cdot)$ to be of class (KF),

$$\lim_{T \to \infty} r_T^{(p)}(h) = \lim_{T \to \infty} \frac{1}{T} \int_0^T r_\alpha^{(p-1)}(h) d\alpha$$

must define a stationary covariance on $\mathbb{R}$. By symmetry it suffices to consider $h \geq 0$. Thus

$$R_T(h) = \frac{1}{T} \int_0^T r_{\alpha_1}^{(p-1)}(h) d\alpha_1$$

$$= \frac{1}{T} \int_0^T \frac{1}{\alpha_1} \int_0^{\alpha_1} \frac{1}{\alpha_2} \int_0^{\alpha_2} \cdots$$

$$\times \int_0^{\alpha_{p-1} - h} r(s, s+h) ds d\alpha_{p-2} \ldots d\alpha_1. \tag{5.20}$$

Replacing the representation (5.19) for $r(s, s+h)$ in Equation (5.20) and interchanging the integrals, (which is valid for class (C) processes) gives

$$R_T(h) = \int_{\mathbb{R}} \int_{\mathbb{R}} \frac{1}{T} \int_0^T \frac{1}{\alpha_1} \int_0^{\alpha_1} \frac{1}{\alpha_2} \int_0^{\alpha_2} \cdots$$

$$\times \int_0^{\alpha_{p-1} - h} g(s, \lambda) \overline{g(s, \lambda')} ds d\alpha_{p-2} \ldots d\alpha_1 \, dF(\lambda, \lambda').$$

The inner integrals are the same as those in the definition of class (c, p)-summable Cramér processes, Equation (5.13), that is

$$R_T(h) = \int_{\mathbf{R}} \int_{\mathbf{R}} a_T^{(p)}(h, \lambda, \lambda') dF(\lambda, \lambda')$$

where $a_T^{(p)}(\cdot, \cdot, \cdot)$ is given by Equation (5.13). Since $a_T^{(p)}(h, \lambda, \lambda')$ is bounded and exists uniformly for all h as $T \to \infty$, the dominated convergence theorem, valid for class (C) processes, can be applied. Hence, since for each T, $R_T(\cdot)$ is positive definite, then

$$\lim_{T \to \infty} R_T(h) = \int_{\mathbf{R}} \int_{\mathbf{R}} \lim_{T \to \infty} a_T^{(p)}(h, \lambda, \lambda') dF(\lambda, \lambda')$$

defines a continuous stationary covariance on $\mathbf{R}$ by the classical Bochner Theorem cited earlier. □

The structure of class (c, p)-summable Cramér processes is inherited by some other classes of processes. K. Karhunen [20] introduced the following extension of stationarity.

Definition 5.3.6 *A second-order process $X : G \to L^2(P)$ with covariance $r(\cdot, \cdot)$ is a Karhunen process if there is an auxiliary measure space (S, ς, ρ) and a set of complex functions $\{g(t, \cdot), t \in T\} \subset L^2(S, \varsigma, \rho)$ such that*

$$r(t_1, t_2) = \int_S g(t_1, \lambda)\overline{g(t_2, \lambda)} \, d\rho(\lambda) \tag{5.21}$$

In the following development let $G = \mathbf{R}$ so that $S = \hat{G} = \mathbf{R}$. Now, clearly, the Karhunen processes contain the class of stationary processes, but one can say much more. Consider a class (c, p)-summable Cramér process $X(\cdot)$, with spectral representation (5.10). That is,

$$X(t) = \int_{\mathbf{R}} g(t, \lambda) dZ(\lambda)$$

where $\{g(t, \cdot), t \in \mathbf{R}\}$ is a bounded Borel family which satisfies the limit condition (5.13). (The boundedness of the family $\{g(t, \cdot), t \in \mathbf{R}\}$ may be required, cf. Rao [32], Remark 6(a).) Using the dilation theorem for class (C) processes, (Corollary 6.2, Rao [33]), there is an enlargement of (Ω, Σ, P) to (S, ς, ρ) so that $L_0^2(P)$ may be identified as a closed subspace of $L_0^2(\rho)$ and there is an orthogonally scattered measure

$$Z_1 : B \to L_0^2(\rho)$$

so that

$$Z = \Pi \circ Z_1,$$

Π is the orthogonal projection on $L_0^2(\rho)$ onto $L_0^2(P)$. Further, $F(A, B) = E(Z(A)\overline{Z(B)})$ for $A, B \in \boldsymbol{B}$, $\boldsymbol{B}$ the Borel sets of $\boldsymbol{R}$. Now since each $g(t, \cdot)$ is bounded and is $\boldsymbol{B}$-measurable, $F(\cdot, \cdot)$ is positive definite. Thus $g(t, \cdot) \in L^2(F) \cap L^1(Z_1)$, where $L^1(Z_1)$ is the set of (Dunford-Schwartz)-integrable scalar functions for Z_1, (cf. Theorem 4.5, Chang and Rao [6]). Hence

$$X(t) = \int_{\boldsymbol{R}} g(t, \lambda) dZ_1(\lambda)$$

exists and is a Karhunen process, (by the integral representation for such processes, Rao [30]). But,

$$Y(t) = \Pi X(t) = \Pi \left(\int_{\boldsymbol{R}} g(t, \lambda) dZ_1(\lambda) \right)$$

$$= \int_{\boldsymbol{R}} g(t, \lambda) \Pi \circ Z_1(d\lambda)$$

(by Chapter 4, section 10 of Dunford and Schwartz [9]). Hence, $X(\cdot)$ is a Karhunen process that is the dilation in $L_0^2(\rho)$ of $Y(\cdot)$. The limit condition (5.13) upon $g(\cdot, \cdot)$ still applies. This is the motivation for the following definition.

Definition 5.3.7 *A second-order process $X : \boldsymbol{R} \to L^2(P)$ is of (c, p)-summable Karhunen class, $p \geq 1$ if its covariance $r(\cdot, \cdot)$ has representation*

$$r(t_1, t_2) = \int_{\boldsymbol{R}} g(t_1, \lambda)\overline{g(t_2, \lambda)} \, d\nu(\lambda) \tag{5.22}$$

relative to a family $\{g(t, \cdot), t \in \boldsymbol{R}\}$ of Borel functions and a positive function $\nu(\cdot)$ which is locally bounded on $\boldsymbol{R}$, with each g satisfying the condition that

$$\lim_{T \to \infty} b_T^{(p)}(|h|, \lambda) \tag{5.23}$$

exists uniformly in h and is bounded for all h, $p \geq 1$, where

$$b_T^{(p)}(|h|, \lambda) = \begin{cases} \frac{1}{T} \int_0^T b_\alpha^{(p-1)}(|h|, \lambda) d\alpha & \text{for } p > 1, \\ \frac{1}{T} \int_0^{T-|h|} g(s, \lambda)\overline{g(s + |h|, \lambda)} ds & \text{for } p = 1. \end{cases} \tag{5.24}$$

A similar set of inclusions for the (c, p)-summable Cramér processes is valid for the (c, p)-summable Karhunen processes, that is (c, p)-summable Karhunen processes are in $(c, p + 1)$-summable Karhunen class for $p \geq 1$, with the inclusions being proper.

Now that the (c, p)-summable Karhunen processes have been concretely defined, it is possible to obtain the converse of the previous dilation result.

If $X(t)$ is a (c, p)-summable Karhunen process, then there exists a measure $Z(\cdot)$ on the Borel sets B of R so that

$$E((Z(A)\overline{Z(B)}) = F(A \cap B)$$

and

$$X(t) = \int_{R} g(t, \lambda)dZ(\lambda)$$

with $g(\cdot, \cdot)$ satisfying the limit condition (5.24).

Now if $T : L_0^2(P) \to L_0^2(P)$ is a bounded linear operator and the integral representation for $X(\cdot)$ is in the Dunford-Schwartz sense, then there is a mapping

$$\tilde{Z} = T \circ Z : B \to L_0^2(P)$$

which is σ-additive and so that $L^1(Z) \subset L^1(\tilde{Z})$. Hence,

$$\tilde{X}(t) = TX(t) = \int_{R} g(t, \lambda)d\tilde{Z}(\lambda)$$

but $\tilde{Z}(\cdot)$ is not orthogonally scattered, thus $\tilde{X}(\cdot)$ is a (c, p)-summable weak class (C) process. The preceding arguments have shown that the *dilation procedure* is valid for these classes of processes, (see Rao [33]). The following result is stated here for reference.

Proposition 5.3.2 *Let* $X : R \to L_0^2(P)$ *be a* (c, p)-*summable Karhunen class,* $p \geq 1$, *process relative to a family* $\{g(t, \cdot), t \in R\}$ *of Borel functions satisfying (5.24) on a measure space* (S, ς, ρ). *If* $Q : L_0^2(P) \to L_0^2(P)$ *is a bounded linear operator, then the process*

$$Y(t) = QX(t), \ for \ t \in R$$

is a (c, p)-*summable weak class (C) process. Conversely, if* $Y : R \to L_0^2(P)$ *is a* (c, p)-*summable weak class (C) process relative to a family* $\{g(t, \cdot), t \in R\}$ *of bounded Borel functions satisfying (12) for a positive definite* $F(\cdot, \cdot)$, *then there is an enlargement of the basic* $(\Omega, \sum, P)$ *to* (S, ς, ρ) *such that* $L_0^2(P)$ *can be identified as a closed subspace of* $L_0^2(\rho)$ *and a Karhunen process* $X : R \to L_0^2(\rho)$ *relative to the same* $\{g(t, \cdot), t \in R\}$ *satisfying (11) on* (S, ς, ρ) *such that* $g(t, \cdot) \in L^2(\rho)$, $Y(t) = \Pi X(t)$ *where* Π *is the orthogonal projection on* $L_0^2(\rho)$ *with range* $L_0^2(P)$.

Rao showed, using a specialization of the dilation procedure, that every weakly harmonizable process belongs to the Karhunen class. An analogous argument shows that every almost weakly harmonizable process is of Karhunen class. This result is stated below. The proof is identical to Theorem 3.1, Rao [33].

Proposition 5.3.3 *Every almost weakly harmonizable process $X : \mathbb{R} \to L_0^2(P)$ is also a Karhunen process relative to a finite positive measure $\nu(\cdot)$ and a family $\{g(t, \cdot), t \in \mathbb{R}\}$ of Borel (uniformly almost periodic) functions.*

The following inclusion relations exist for the classes of processes introduced thus far. These inclusions are generally proper.

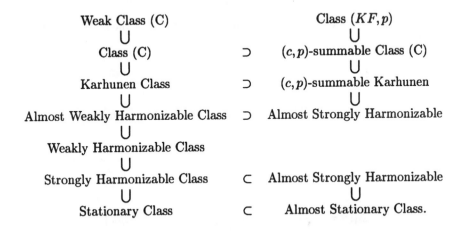

The example due to Rao, cited earlier, shows that there are weakly harmonizable (hence, weak class (C)) processes which do not belong to $\bigcup_{p \geq 1}$ Class (KF, p).

As was mentioned earlier (5.16), each of the classes (KF, p), $p \geq 1$ has an associated spectral function $G(\cdot)$. A key application for these classes of processes is in obtaining conditions for the weak or strong law of large numbers. This type of results has been obtained for the class of almost strongly harmonizable processes (Rao [30]) as well as for almost strongly harmonizable fields (Anh and Lunney [2]). The following result is an extension to the (c, p)-summable class (C) processes.

Theorem 5.3.3 *Let $X : \mathbb{R} \to L^2(P)$ be a (c, p)-summable Cramér process. Suppose that its mean function $m(\cdot)$ has the property that*

$$\alpha_0 = \lim_{T \to \infty} \frac{1}{T} \int_0^T m(t)dt$$

exists where $m(t) = E(X(t))$. Let

$$\hat{m}_T = \frac{1}{T} \int_0^T X(t)dt,$$

then

$$\lim_{T \to \infty} E(\hat{m}_T - \alpha_0)^2 = G(0+) - G(0-)$$

where $G(\cdot)$ is the associated spectral function [cf. (5.16)] of $X(\cdot)$. In particular, if $G(\cdot)$ is continuous at 0, then $\hat{m}_T$ is a strongly consistent estimator
of α_0.

Proof: Let

$$\alpha_T = \frac{1}{T} \int_0^T m(t)dt,$$

so that $\alpha_T \to \alpha_0$ as $T \to \infty$. Applying the definition of the covariance $r(\cdot, \cdot)$
with

$$E(\hat{m}_T) = \alpha_T$$

gives

$$
\begin{aligned}
E[(\hat{m}_T - \alpha_0)^2] &= E[((\hat{m}_T - \alpha_T) + (\alpha_T - \alpha_0))^2] \\
&= E[(\hat{m}_T - \alpha_T)^2] + (\alpha_T - \alpha_0)^2. \qquad (5.25)
\end{aligned}
$$

Now

$$
\begin{aligned}
E[(\hat{m}_T - \alpha_T)^2] &= E\left[\frac{1}{T} \int_0^T X(t) - m(t)dt\right]^2 \\
&= \frac{2}{T^2} \int_0^T \int_0^T r(s,t)dtds.
\end{aligned}
$$

This last expression follows from expanding the quadratic and interchanging
the order of integration. Substitution of this expression into (5.25) yields

$$
\begin{aligned}
E[(\hat{m}_T - \alpha_0)^2] &= \frac{2}{T^2} \int_0^T \int_0^T r(s,t)dtds + (\alpha_T - \alpha_0)^2 \\
&= \frac{2}{T} \int_{-T}^T r_T(h)dh + (\alpha_T - \alpha_0)^2, \qquad (5.26)
\end{aligned}
$$

where $r_T(\cdot)$ is given in the definition of class (KF, p) Equation (5.18).

Since $r(\cdot, \cdot)$ is of class (c, p)-summable Cramér (and hence of class
(KF, p)),

$$\tilde{r}(h) = \lim_{T \to \infty} r_T(h)$$

has an associated spectral distribution $G(\cdot)$. That is, since the limit in
(5.13) exists uniformly in h, it follows from Bochner's Theorem that

$$\tilde{r}(h) = \int_{\mathbf{R}} e^{i\lambda h} dG(\lambda).$$

Thus if G_T and G are bounded nondecreasing functions representing r_T and r as in (5.16), then $r_T(h) \to \tilde{r}(h)$ uniformly in h. But, these are Fourier transforms of G_T and G and hence must be uniformly bounded. Further, $G_T \to G$ at each continuity point of G.

Thus

$$\frac{1}{2T}\int_{-T}^{T} r_T(h)dh = \int_{\mathbf{R}} \frac{1}{2T}\int_{-T}^{T} e^{i\lambda h}dhdG_T(\lambda)$$

$$= \int_{\mathbf{R}} \frac{\sin(T\lambda)}{T\lambda}dG_T(\lambda).$$

But,

$$\frac{\sin(T\lambda)}{T\lambda} \to \begin{cases} 1 & \text{for } \lambda = 0, \\ 0 & \text{for } \lambda \neq 0, \end{cases}$$

as $T \to \infty$, hence

$$\frac{1}{2T}\int_{-T}^{T} r_T(h)dh \to G(0+) - G(0-).$$

So

$$\lim_{T\to\infty} E(\hat{m}_T - \alpha_0)^2 = G(0+) - G(0-)$$

completing the proof. $\square$

Since the limit $\tilde{r}$ and hence G are given only indirectly from r_T, it is desirable to have conditions on r_T that will ensure the continuity of G at 0 (or any given point). This will be true if $r_T(h) \to 0$ as $h \to \infty$ uniformly in T or if $\int_0^\infty |r_T(h)|dh$ is bounded as a function of T.

5.3.2 Almost Periodic Almost Harmonizable Processes

In this section a further analysis of the class of almost harmonizable processes will be considered. The results that will subsequently be developed follow from assumptions on the covariance $r(\cdot,\cdot)$ being a u.a.p. function of two variables. More precisely,

Definition 5.3.8 *A stochastic process* $X : \mathbb{R} \to L_0^2(P)$ *is quadratic mean uniformly almost periodic (q.m.u.a.p.) if for each* $\varepsilon > 0$,

$$\{\tau : \sup_{-\infty < t < \infty} E \mid X(t+\tau) - X(t) \mid^2 < \varepsilon\} \tag{5.27}$$

is relatively dense on $\mathbb{R}$, *(see Definition 3.10 below).*

Similarly, we may consider q.m. S^2 a.p. processes:

Definition 5.3.9 *A stochastic process* $X : \mathbb{R} \to L_0^2(P)$ *is quadratic mean* S^2 *almost periodic (q.m. S^2 a.p.) if for each $\varepsilon > 0$,*

$$\{\tau : \sup_{-\infty < t < \infty} \int_t^{t+1} E \mid X(t+\tau) - X(t) \mid^2 dt < \varepsilon\} \qquad (5.28)$$

is relatively dense on $\mathbb{R}$.

The concept of a relatively dense set may be found in Besicovitch [3], and is recalled here in the following definition.

Definition 5.3.10 *A set $E \subset \mathbb{R}$ is relatively dense if there exists a number $l > 0$ such that any interval of length l contains at least one number of E.*

Swift [51] showed that these concepts coincide for u.a.p. covariances $r(\cdot, \cdot)$ without any further assumptions on $r(\cdot, \cdot)$, that is;

Proposition 5.3.4 $X(\cdot)$ *is q.m.u.a.p. iff $r(\cdot, \cdot)$ is u.a.p.. $X(\cdot)$ is q.m. S^2 a.p. iff $r(\cdot, \cdot)$ is S^2 a.p..*

Since quadratic mean uniformly almost periodic processes have covariance functions which are uniformly almost periodic in two variables, it is natural to consider the class of processes with uniformly almost periodic covariances. A second-order process $X(\cdot)$ is almost periodically correlated in the sense of Gladyshev provided its covariance function $r(s,t)$ is uniformly continuous in s and t and if for every s and $t \in \mathbb{R}$, $r(s+\tau, t+\tau)$ is a uniformly almost periodic function of τ. An equivalent condition is that

$$B(t, \tau) = r(t+\tau, t) \qquad (5.29)$$

is uniformly almost periodic in t for each τ. Almost periodically correlated, termed *almost PC*, processes were first introduced by Gladyshev [12], and have been subsequently studied by Hurd [15], [16]. It is evident from the previous proposition that the class of almost PC processes contains the quadratic mean uniformly almost periodic processes. The previous proposition also motivates the introduction of an additional class of almost periodically correlated processes.

Definition 5.3.11 *A second-order process $X : \mathbb{R} \to L_0^2(P)$ will be Stepanov almost periodically correlated (S^2 almost PC) if $B(t, \tau) = r(t+\tau, t)$ is S^2 almost periodic in t for each τ where $r(\cdot, \cdot)$ is the covariance function.*

As for the almost (uniformly) PC process, the proposition shows that the class of S^2 almost PC process contains the quadratic mean S^2 almost periodic processes.

Since the generalization of the uniformly almost periodic concept afforded by Stepanov almost periodicity relaxes the continuity restrictions (see Besicovitch [3]), one notes here that the same relaxation is available for the S^2 almost PC processes. The definition of S^2 almost PC processes does not require that $r(s, t)$ be a uniformly continuous function in s and t.

Proposition 5.3.5 *If an S^2 almost PC process $X : \mathbb{R} \to L_0^2(P)$ has covariance $r(s, t)$ uniformly continuous in s, t then $X(\cdot)$ is a uniformly almost PC process.*

Proof: A classical result of Bochner's is that if $B(t, \tau) = r(t + \tau, t)$ is a uniformly continuous S^2 almost periodic function, it is then uniformly almost periodic. Hence $X(\cdot)$ is uniformly almost PC. $\square$

The class of S^2 almost PC processes hold promise to extend the works of Gladyshev and Hurd cited earlier. The current work focuses on the relationship between the S^2 and the uniformly almost PC classes and the harmonizable and almost harmonizable classes.

Theorem 5.3.4 *If the spectral measure $F(\cdot, \cdot)$ of an almost strongly harmonizable process has countable support then the covariance $r(\cdot, \cdot)$ is a uniformly almost periodic function of two variables.*

Proof: If $F(\cdot, \cdot)$ has countable support $\{(\lambda_j, \lambda_k')\}_{j,k=1}^\infty$, letting

$$a(\lambda, \lambda') = F(\lambda + 0, \lambda' + 0) - F(\lambda + 0, \lambda') - F(\lambda, \lambda' + 0) + F(\lambda, \lambda')$$

with $X(\cdot)$ almost strongly harmonizable implies

$$
\begin{aligned}
r(s, t) &= \int_{\mathbb{R}} \int_{\mathbb{R}} g(\lambda, s) \overline{g(\lambda', t)} dF(\lambda, \lambda') \\
&= \sum_{k=1}^\infty \sum_{j=1}^\infty a(\lambda_j, \lambda_k') g(\lambda_j, s) \overline{g(\lambda_k', t)} \quad (5.30)
\end{aligned}
$$

where $g(\lambda, \cdot)$ is a uniformly almost periodic function. Applying the classical approximation theorem for uniformly almost periodic functions to $g(\lambda, \cdot)$ gives

$$g(\lambda, t) = \sum_{n=1}^\infty b(\lambda_n) e^{i\lambda t},$$

where the series converges uniformly. Using this expression in the series representation (5.30) for $r(\cdot, \cdot)$ and rearranging the terms yields that $r(\cdot, \cdot)$ has a Fourier series representation and hence is u.a.p. in two variables. $\square$

Swift [51] showed that the converse is also true for strongly harmonizable processes with countable support. Thus every almost strongly harmonizable process with countable support is uniformly almost PC.

The previous Theorem is a key result for the study of the sample path behavior of almost harmonizable processes. The concept of almost periodic

stationary stochastic processes was first introduced and studied by Slutsky
[46], who obtained sufficient conditions for the sample paths of a stationary
process to be Besicovitch almost periodic. Later, Udagawa [55] gave conditions for the sample paths to be Stepanov almost periodic. Kawata [21]
extended these results to a very general setting and also gave conditions
for uniformly almost periodic sample paths. Swift [51] gave sufficient conditions for harmonizable processes to have almost periodic sample paths.
These conditions were similar to Kawata's in the stationary case. The sample path behavior for almost harmonizable processes will now be considered.

Suppose $X(\cdot)$ was an almost strongly harmonizable process with count-
able support $\{(\lambda_j, \lambda_k')\}_{j,k=1}^\infty$. Since the support may have limit points in the
plane, some regularity conditions on these limit points are needed and they
will now be introduced.

Condition 1: Let $\{(\mu_j, \mu_k')\}_{j,k=1}^\infty$ be the set of limit points of the support
$\{(\lambda_j, \lambda_k')\}_{j,k=1}^\infty$ of $F(\cdot, \cdot)$. It is required that

$$\inf_{k \neq j} \{|\,\mu_k - \mu_j\,|, |\,\mu_k' - \mu_j'\,|\} > 1. \tag{5.31}$$

The constant 1 on the right side here is chosen for simplicity; it may be
replaced by any positive constant. Condition (5.31) implies that each semi-
open square $(n, n+1] \times (m, m+1], n, m \in Z$, contains at most only one
limit point. Assume further that the limit points are enumerated so that
$\mu_k < \mu_{k+1}$ in each strip $(k, k+1] \times \{\mu_j'\}$ with the second coordinate μ_j'
ordered $\mu_j' < \mu_{j+1}'$.

Condition 2: When $\mu_{k_n} \neq n$, let $N_n(\alpha)$ be the number of discontinuities
between n and $\mu_{k_n} - \alpha$, where $0 < \alpha < \mu_{k_n} - n$, and when $\mu_{k_n} \neq n+1$, let
$M_n(\alpha)$ be the number of discontinuities between $\mu_{k_n} + \alpha$ and $n+1$, where
$0 < \alpha < n+1 - \mu_{k_n}$. Suppose there is a nondecreasing function $h(\cdot)$ on $I\!\!R^+$
such that $h(x) \nearrow \infty$ as $x \nearrow \infty$, with $h(x) \equiv C > 0$ for $0 \leq x < 2$ and

$$N_n(\alpha) \;\leq\; h(\frac{1}{\alpha}) \text{ for } 0 < \alpha < \mu_{k_n} - n$$

$$M_n(\alpha) \;\leq\; h(\frac{1}{\alpha}) \text{ for } 0 < \alpha < n+1 - \mu_{k_n}$$

if $[n, n+1] \times [m, m+1]$ had no limiting point, it will be assumed that the
set of discontinuities of $F(\cdot, \cdot)$ is bounded.

The following integral plays a key role in the analysis. Let

$$\Phi_{n,m} = \int_n^{n+1} \phi(|\,\lambda - \mu_{k_n}\,|)dF(\lambda, \lambda_m') \tag{5.32}$$

where

$$\phi(u) = \begin{cases} s(\frac{1}{u})h(\frac{2}{u}) & \text{if } u > 0 \\ 1 & \text{if } u = 0 \end{cases}$$

with $s(\cdot)$ a nondecreasing function on $\mathbb{R}^+$ such that $\int_1^\infty dx/(xs(x)) < \infty$ and $s(x) = 1$ for $0 \le x \le 1$.

The uniform almost periodicity of the sample paths of almost harmonizable processes may now be given.

Theorem 5.3.5 *If conditions 1 and 2 are satisfied and if*

$$\sum_{m=-\infty}^{\infty} \sum_{n=-\infty}^{\infty} \Phi_{n,m}^{\frac{1}{2}} < \infty \qquad (5.33)$$

then $X(\cdot)$ has almost all sample paths uniformly almost periodic.

Proof: The method of proof is similar to the strongly harmonizable case given in Swift [51]. A sketch of the proof will be given here. Since $X(\cdot)$ is an almost strongly harmonizable process, the process has integral representation

$$X(t) = \int_{\mathbb{R}} g(\lambda, t)dZ(\lambda) \qquad (5.34)$$

where $Z(\cdot)$ is the stochastic measure satisfying the condition $E(Z(B_1)\overline{Z(B_2)})$ $= F(B_1, B_2)$, $F(\cdot, \cdot)$ a function of bounded Vitali variation and $g(\lambda, \cdot)$ uniformly almost periodic. Let $A(\lambda) = Z(\lambda + 0) - Z(\lambda)$ and

$$a(\lambda, \lambda') = F(\lambda + 0, \lambda' + 0) - F(\lambda + 0, \lambda') - F(\lambda, \lambda' + 0) + F(\lambda, \lambda').$$

Swift [51] showed that if there were infinitely many $\lambda_{n,j}$ in $[n, \mu_{l_n})$ then

$$\sum_{m<\lambda'<m+1} \sum_{n<\lambda<n+1} |a|^{\frac{1}{2}}(\lambda, \lambda') < K\Phi_{n,m}^{\frac{1}{2}}, \qquad (5.35)$$

with $K > 0$ a constant.

Thus considering the spectral representation of the almost harmonizable process (5.34), with $F(\cdot, \cdot)$ having countable support, one has

$$X(t) = \sum_{j,k} A(\lambda_{k,j})g(\lambda_{k,j}, t),$$

then converges in $L^2(P)$.

For the uniform almost periodicity, $\sum_{j,k} |A(\lambda_{k,j})| < \infty$ with probability one must be shown. By the first Borel-Cantelli lemma, it suffices for this to show that for a given $\varepsilon > 0$,

$$\sum_{k,j} P(|A(\lambda_{k,j})| \ge \varepsilon) < \infty.$$

Now

$$P(|\,A(\lambda_{k,j})\,| \geq \varepsilon) \leq \frac{1}{\varepsilon}E(|\,A(\lambda_{k,j})\,|) \text{ by Markov's inequality}$$

$$\leq \frac{1}{\varepsilon}[E\,|\,A(\lambda_{k,j})\,|^2]^{\frac{1}{2}} = \frac{|a|^{\frac{1}{2}}(\lambda_k,\lambda_j')}{\varepsilon}.$$

But (5.35) implies $\sum_{k,j} P(|\,A(\lambda_{k,j})\,| \geq \varepsilon) \leq (1/\varepsilon)\sum_k\sum_j |a|^{\frac{1}{2}}(\lambda_k,\lambda_j') < \infty$. So

$$\sum_{k,j} |\,A(\lambda_{k,j})\,| < \infty \text{ with probability one.} \qquad (5.36)$$

Recalling that the definitions of $A(\lambda)$ and $a(\lambda,\lambda')$ one sets

$$S_{n,m}(t) = \sum_{n<\lambda<n+1} A(\lambda)g(\lambda,t).$$

By the spectral representation, converging in $L^2(P)$, this series exists in $L^2(P)$ as a sum representing $X(\cdot)$. Hence, Equation (5.36) implies that $S_{n,m}(\cdot)$ converges absolutely and uniformly with probability one.

Thus $S_{n,m}(t) = \sum_{n<\lambda<n+1} A(\lambda)g(\lambda,t)$ is u.a.p. with probability one. So there is a set $\Omega_{n,m}$ with $P(\Omega_{n,m}) = 1$ such that on $\Omega_{n,m}$, $S_{n,m}(t)$ is u.a.p. Now

$$\sum_{n=0}^{\infty}\sum_{n<\lambda<n+1} E\,|\,A(\lambda)\,| \leq \sum_{n=0}^{\infty}\sum_{n<\lambda<n+1} |a|^{\frac{1}{2}}(\lambda,\lambda')$$

$$\leq C'\sum_{m=0}^{\infty}\sum_{n=0}^{\infty}\Phi_{n,m}^{\frac{1}{2}} < \infty.$$

Thus $\sum_{n=0}^{\infty}\sum_{m=0}^{\infty} S_{n,m}(t)$ is uniformly and absolutely convergent with probability one.

A similar argument applies for $\sum_{n=-\infty}^{-1}\sum_{m=-\infty}^{-1} S_{n,m}(t)$. Hence $\sum_{n=-\infty}^{\infty}\sum_{m=-\infty}^{\infty} S_{n,m}(t)$ is uniformly and absolutely convergent, but each term of this series is u.a.p. for $\omega \in \bigcup_{m=-\infty}^{\infty}\bigcup_{n=-\infty}^{\infty} \Omega_{n,m}$ where

$$P\left(\bigcup_{n=-\infty}^{\infty}\bigcup_{m=-\infty}^{\infty} \Omega_{n,m}\right) = 1.$$

Thus $\sum_{n=-\infty}^{\infty}\sum_{m=-\infty}^{\infty} S_{n,m}(t)$ is u.a.p. with probability one. But, this series converges to $X(t)$ in $L^2(P)$, so $X(t)$ is u.a.p. with probability one. □

The proof of this theorem actually showed some further aspects; in particular, under the assumptions of the previous theorem, the uniformly almost periodic almost strongly harmonizable process $X(\cdot)$ has an absolutely convergent Fourier series and spectral measure satisfying

$$\int_{-\infty}^{\infty} \int_{-\infty}^{\infty} d|F|(\lambda, \lambda') \ < \infty.$$

The Stepanov and Besicovitch sample path behaviors will now be considered. The method of proof used to obtain Stepanov and Besicovitch almost periodicity of the sample paths of a harmonizable process relies upon the fact that for any sequence $\{v_j\}$ of real numbers such that $| v_j - v_k | \geq r$, $j \neq k$, the sequence $\{e^{iv_j t}\}$ is orthonormal with respect to

$$\sigma_r(E) = C_r \int_E \frac{\sin^{2r}(\frac{t}{2})}{(\frac{t}{2})^{2r}} dt$$

where E is any Borel measurable set of $\mathbb{R}$ and r is a positive integer. Since the spectral representation of an almost harmonizable process $X(\cdot)$ is of the form (5.34), an additional assumption on the process $X(\cdot)$ is required.

Theorem 5.3.6 *If $X : \mathbb{R} \to L_0^2(P)$ is an almost strongly harmonizable process with countable support and if the sequence $\{g(\lambda_j, t)\}$ is orthonormal with respect to $\sigma_r(E)$ and if conditions 1 and 2 are satisfied with*

$$\sum_{m=-\infty}^{\infty} \sum_{n=-\infty}^{\infty} \Phi_{n,m} < \infty \tag{5.37}$$

then $X(\cdot)$ has almost all sample paths S^2-almost periodic.

Proof: As was noted earlier, $S_{n,m}(t)$ is u.a.p. with probability one. Let $\Omega_{n,m}$ be a set such that $S_{n,m}(t,\omega)$ is u.a.p. for $\omega \in \Omega_{n,m}$ and let

$$\Omega' = \bigcap_{m=-\infty}^{\infty} \bigcap_{n=-\infty}^{\infty} \Omega_{n,m}.$$

Consider the sequence $\{S_{2n,2m}(t)\}$, since the limit points of discontinuity satisfy condition 1, the absolute convergence of $S_{n,m}(\cdot)$ and the lemma imply that $\{S_{2n,2m}(t)\}$ forms an orthogonal sequence with respect to $\sigma_1(E)$. Hence, for any y, $\{S_{2n,2m}(t + y)\}$ is also an orthonormal sequence with respect to $\sigma_1(E)$.

Following the proof of Theorem 4.1 of Swift [51], under the present assumptions one may show that

$$\sum_{n,m=-\infty}^{\infty} \left(\sum_{2n<\lambda<2n+1} | A(\lambda) | \right)^2 < \infty \text{ with probability one.}$$

So,

$$\lim_{M,N\to\pm\infty}\sup_y \int_{-\infty}^{\infty}\left|\sum_{n,m=N+1}^{M} S_{2n,2m}(t)\right|^2 d\sigma_1(t-y) = 0, \quad \text{a.e.,} \qquad (5.38)$$

from which it follows that

$$\lim_{M,N\to\pm\infty}\sup_y \int_{-\infty}^{\infty}\left|\sum_{n,m=N+1}^{M} S_{2n,2m}(t)\right|^2 dt = 0, \quad \text{a.e.,} \qquad (5.39)$$

so $\sum_{n,m=N+1}^{M} S_{2n,2m}(t)$ converges to an S^2 a.p. function in S^2 a.p.-norm with probability one. The same holds for $\sum_{n,m=N+1}^{M} S_{2n+1,2m+1}(t)$. Thus,

$$T_N(t) = \sum_{n,m=-N}^{N} S_{n,m}(t)$$

converges in L^2 to an S^2 a.p. function, say $X_1(t)$, with probability one. Thus, using the spectral representation of $X(\cdot)$,

$$\int_{-T}^{T} E\,|T_N(t) - X(t)|^2\, dt = \int_{-T}^{T} E\left|\sum_{-\infty}^{-N+1} A(\lambda_{k,j})g(\lambda_{k,j},t)\right.$$

$$+ \left.\sum_{N+1}^{\infty} A(\lambda_{k,j})g(\lambda_{k,j},t)\right|^2 dt. \qquad (5.40)$$

Interchanging the sum and expectation here, Equation (5.40) becomes

$$\int_{-T}^{T} E\,|T_N(t) - X(t)|^2\, dt = 2T\left(\int_{N+1}^{\infty}\int_{N+1}^{\infty}|\,dF(\lambda,\lambda')\,|\right.$$

$$+ \left.\int_{-\infty}^{-N}\int_{-\infty}^{-N}|\,dF(\lambda,\lambda')\,|\right).$$

So for fixed T as $N \nearrow \infty$, $\int_{-\infty}^{\infty} E|T_N(t) - X(t)|^2 dt \to 0$. Hence, $T_n(t) - X(t) \to 0$ in $L^2(P)$, so some subsequence $\{T_{n_k}(t)\}$ converges in $\Omega \times (-T,T)$ almost everywhere. That is, for $\omega \in \Omega$ fixed, $T_{n_k}(t,\omega) \to X(t,\omega)$, but this implies $X(t) = X_1(t)$ for almost all t with probability one. Hence, $X(\cdot)$ is S^2 a.p. with probability one. $\square$

B^2 a.p. sample paths will now be considered. This result extends Slutsky's theorem to almost harmonizable processes.

Theorem 5.3.7 *If $X : \mathbb{R} \to L_0^2(P)$ is an almost strongly harmonizable process with countable support and if the sequence $\{g(\lambda_j, t)\}$ is orthonormal with respect to $\sigma_r(E)$ and if conditions 1 and 2 are satisfied with $\Phi_{n,m} < \infty$, for every n, m, then $X(\cdot)$ has almost all sample paths B^2 almost periodic.*

Proof: Using the same argument for (5.39) with $S_{3n,3m}(\cdot), S_{3n+1,3m+1}(\cdot)$ and $S_{3n+2,3m+2}(\cdot)$, and $\sigma_2(\cdot)$, in place of $\sigma_1(\cdot)$ it follows that

$$\int_{y-\pi}^{y+\pi} \left| \sum_{|n|,|m|=k}^{M} S_{n,m}(t) \right|^2 dt$$

$$\leq K \int_{-\infty}^{\infty} \sum_{l=0}^{2} \left| \sum_{\substack{k \leq |3n+l| \leq M \\ k \leq |3m+l| \leq M}} S_{3n+l,3m+l}(t) \right|^2 d\sigma_2(t-y). \quad (5.41)$$

Let $A > 0$ be a fixed, large real number, and let $N \in \mathbb{Z}$ be the smallest integer so that $2N + 1 \geq A$. Writing $y = 2\nu\pi$, for $\nu = -N, \ldots, N$ in Equation (5.41) and adding, one obtains

$$\frac{1}{2A} \int_{-A}^{A} \left| \sum_{|n|,|m|=k}^{M} S_{n,m}(t) \right|^2 dt$$

$$\leq \frac{C}{2A} \sum_{\nu=-N}^{N} \int_{-\infty}^{\infty} \sum_{l=0}^{2} \left| \sum_{\substack{k \leq |3n+l| \leq M \\ k \leq |3m+l| \leq M}} S_{3n+l,3m+l}(t) \right|^2 d\sigma_2(t - 2\nu\pi). (5.42)$$

Now $\{S_{3n+l,3m+l}(\cdot)\}, n, m = 0, 1, 2, \ldots$, forms an orthogonal sequence of functions of t, for each l, with respect to $\sigma_2(\cdot)$.

Using the same computations as in the proof of the previous theorem one can show

$$\frac{1}{2A} \int_{-A}^{A} \left| \sum_{|n|,|m|=k}^{M} S_{n,m}(t) \right|^2 dt \leq C \int_{|t|>3N\pi} \frac{U_{k,m}(t)}{t^4} dt$$

$$+ \frac{C}{N} \int_{|t|>3N\pi} U_{k,m}(t) dt. \quad (5.43)$$

where

$$U_{k,m}(t) = \sum_{l=0}^{2} \sum_{\substack{k \leq |3n+l| \leq M \\ k \leq |3m+l| \leq M}} |S_{3n+l,3m+l}(t)|^2 d\sigma_2(t - 2\nu\pi).$$

Also Swift [51] showed that

$$\int_{|t|>3N\pi} U_{k,\infty}(t)\frac{dt}{t^4} \to 0, \text{ as } N \to \infty$$

and

$$\lim_{N\to\infty} \sup \frac{1}{N} \int_{|t|\leq 3N\pi} U_{k,\infty}(t)dt = 3\pi \sum_{\substack{k<|n| \\ k<|m|}} \sum_{n<\lambda\leq n+1} |A(\lambda)|^2. \qquad (5.44)$$

Thus

$$\lim_{A\to\infty} \sup \int_{-A}^{A} \left| X(t) - \sum_{\substack{|n|<k \\ |m|<k}} S_{n,m}(t) \right|^2 dt \leq 3\pi \sum_{\substack{|n|<k \\ |m|<k}} \sum_{n<\lambda\leq n+1} |A(\lambda)|^2.$$

$$(5.45)$$

Here

$$E\left(\sum_{n<\lambda\leq n+1} |A(\lambda)|^2 \right) = \int_{|\lambda'|>k} \int_{|\lambda|>k} |dF(\lambda,\lambda')|.$$

Choosing $k = k_j$ so that

$$\sum_{j=1}^{\infty} \int_{|\lambda'|>k_j} \int_{|\lambda|>k_j} |dF(\lambda,\lambda')| < \infty$$

one gets

$$\sum_{j=1}^{\infty} E\left(\sum_{n<\lambda\leq n+1} |A(\lambda)|^2 \right) < \infty.$$

So by the first Borel-Cantelli lemma, $\lim_{j\to\infty} \sum_{n<\lambda\leq n+1} |A(\lambda)|^2 = 0$ with probability one. This implies

$$\lim_{j\to\infty} \lim_{A\to\infty} \sup \int_{-A}^{A} \left| X(t) - \sum_{\substack{|n|<k \\ |m|<k}} S_{n,m}(t) \right|^2 dt = 0,$$

with probability one. Now since $S_n(\cdot)$ is u.a.p. with probability one, it follows that $X(\cdot)$ is B^2 a.p. with probability one. $\square$

5.4   Part II: Some Classes of Harmonizable Random Fields

In Part I of this chapter, several classes of nonstationary processes were studied. It is not difficult to extend most of the previous work to the fields case. This extension will be outlined for the harmonizable case. Random fields often satisfy an additional assumption termed *isotropy*. Harmonizable isotropic random fields play a special role in the subsequent material.

If the index set is $\mathbb{R}^n$, $n > 1$, then the mapping $X : \mathbb{R}^n \to L_0^2(P)$ is called a random field. In extending the definition of harmonizability to the fields case, one says that a random field $X : \mathbb{R}^n \to L_0^2(P)$ is weakly harmonizable if the covariance $r(\cdot, \cdot)$ is representable as

$$r(s, t) = \int_{\mathbb{R}^n} \int_{\mathbb{R}^n} e^{i\lambda \cdot s - i\lambda' \cdot t} dF(\lambda, \lambda') \qquad (5.46)$$

where $F(\cdot, \cdot)$ is a positive definite function on $\mathbb{R}^n \times \mathbb{R}^n$ of Fréchet bounded variation. If $F(\cdot, \cdot)$ has finite Vitali variation, the random field will be strongly harmonizable.

The spectral representation of a weakly harmonizable random field is

$$X(t) = \int_{\mathbb{R}^n} e^{i\lambda \cdot t} dZ(\lambda) \qquad (5.47)$$

where $Z(\cdot)$ is a stochastic measure which satisfies

$$E(Z(B_1)\overline{Z(B_2)}) = F(B_1, B_2)$$

where $B_1, B_2 \in B$, B the Borel sets of R^n and $F(\cdot, \cdot)$ is of finite Fréchet variation.

As can be seen from these definitions, much of the theory of harmonizable processes extends to random fields. A detailed treatment may be found in Rao [33] and Chang and Rao [6]. The next section details the consequences of isotropy.

5.4.1   Harmonizable Isotropic Random Fields

A subclass of random fields satisfies an additional condition called *isotropy*. Isotropic random fields $X(\cdot)$, have covariance $r(\cdot, \cdot)$ which are invariant under rotation and reflection. Isotropic fields play an important role in the statistical theory of turbulence, where direction in space is unimportant [59]. Swift [48] and recently, Rao [40], obtained the representation of a weakly harmonizable isotropic covariance as

$$r(s, t) = 2^\nu \Gamma\left(\frac{n}{2}\right) \int_0^\infty \int_0^\infty \frac{J_\nu(\|\lambda s - \lambda' t\|)}{\|\lambda s - \lambda' t\|^\nu} dF(\lambda, \lambda') \qquad (5.48)$$

where $J_\nu(\cdot)$ is the Bessel function (of the first kind) of order $\nu = \frac{n-2}{2}$ and $F(\cdot, \cdot)$ is a complex function of bounded Fréchet variation, with $||\cdot||$ denoting the vector norm.

Isotropic covariances $r(s, t)$ are functions of the lengths $||s||$, $||t||$ of the vectors s, t and of the angle θ between s and t. The following theorem gives a characterization in spherical-polar form for the covariances of harmonizable isotropic random fields, the proof may be found in Swift [48].

Theorem 5.4.1 *A random field* $X : \mathbb{R}^n \to L_0^2(P)$ *is weakly harmonizable isotropic if and only if the covariance function* $r(\cdot, \cdot)$ *is expressible for* $n \geq 2$ *as:*

$$r(s, t) = \alpha_n^2 \sum_{m=0}^{\infty} \sum_{l=1}^{h(m,n)} S_m^l(u) S_m^l(v) \int_0^\infty \int_0^\infty \frac{J_{m+\nu}(\lambda \tau_1) J_{m+\nu}(\lambda' \tau_2)}{(\lambda \tau_1)^\nu (\lambda' \tau_2)^\nu} dF(\lambda, \lambda')$$

(5.49)

where $\nu = \frac{n-2}{2}$ *and*
i) $s = (\tau_1, u), t = (\tau_2, v)$ *are the spherical polar coordinates of* s, t *in* $\mathbb{R}^n$, *here* $\tau_1 = || s ||, \tau_2 = || t ||$ *and* $u = \frac{s}{\tau_1}, v = \frac{t}{\tau_2}$ *are unit vectors.*
ii) $S_m^l(\cdot), 1 \leq l \leq h(m, n) = \frac{(2m+2\nu)(m+2\nu-1)!}{(2\nu)! m!}, m \geq 1, S_0^l(u) = 1$ *are the spherical harmonics on the unit* n-*sphere of order* m.
iii) $\alpha_n > 0, \alpha_n^2 = 2^{2\nu+1} \Gamma(\frac{n}{2}) \pi^{\frac{n}{2}}$
with $F(\cdot, \cdot)$ *as a complex function of bounded Fréchet variation.*

Representation (5.49) implies that a weakly harmonizable isotropic covariance $r(s, t)$ is a function of $||s||$, $||t||$ and θ.

Using (5.49) and Karhunen's Theorem, the spectral representation for a weakly harmonizable isotropic random field is given as

$$X(t) = \alpha_n \sum_{m=0}^{\infty} \sum_{l=1}^{h(m,n)} S_m^l(u) \int_0^\infty \frac{J_{m+\nu}(\lambda \tau)}{(\lambda \tau)^\nu} dZ_m^l(\lambda)$$

where $Z_m^l(\cdot)$ satisfies

$$E(Z_m^l(B_1) \overline{Z_{m'}^{l'}(B_2)}) = \delta_{mm'} \delta_{ll'} F(B_1, B_2)$$

with $F(\cdot, \cdot)$ a function of bounded Fréchet variation with the stochastic integral being in the Dunford-Schwartz sense, ([9], IV.10) and with $\delta_{mm'}$ the Kronecker delta.

In order to gain a deeper understanding of the structural properties of harmonizable isotropic fields, Swift [48] introduced the following classification of their covariance functions.

Let $\mathcal{G}_n$ be the class of all n-dimensional strongly harmonizable isotropic covariances on $\mathbb{R}^n \times \mathbb{R}^n$ and $\mathcal{G}_\infty$ the class of all covariances which belong to $\mathcal{G}_n$ for all $n \geq 1$.

If one identifies in a natural way, the random field $X : I\!\!R^n \to L_0^2(P)$ with a field $\tilde{X} : I\!\!R^{n+1} \to L_0^2(P)$ by taking

$$X(t_1, \ldots, t_n) \equiv \tilde{X}(t_1, \ldots, t_{n+1}), \tag{5.50}$$

with t_{n+1} fixed, then one has

$$\mathcal{G}_\infty \subset \ldots \subset \mathcal{G}_{n+1} \subset \mathcal{G}_n \subset \mathcal{G}_{n-1}$$

and that

$$\mathcal{G}_\infty = \bigcap_{n \geq 1} \mathcal{G}_n.$$

Similarly, let $\mathcal{D}_n$ be the class of all n-dimensional stationary isotropic covariances on $I\!\!R^n \times I\!\!R^n$ and $\mathcal{D}_\infty$ the class of all covariances which belong to $\mathcal{D}_n$ for all $n \geq 1$.

By the same natural identification mentioned above, one has

$$\mathcal{D}_\infty \subset \ldots \subset \mathcal{D}_{n+1} \subset \mathcal{D}_n \subset \mathcal{D}_{n-1}$$

and then

$$\mathcal{D}_\infty = \bigcap_{n \geq 1} \mathcal{D}_n.$$

This implies that

$$\mathcal{D}_n \subset \mathcal{G}_n \text{ and } \mathcal{D}_\infty \subset \mathcal{G}_\infty,$$

since the representation reduces to the form of the stationary case if and only if $F(\cdot, \cdot)$ concentrates on the diagonal $\lambda = \lambda'$. Clearly the classes $\mathcal{G}_n$ are not empty.

Examples can be constructed to show that these inclusions are proper (cf., Swift [48]). Swift also showed that as n increases, the covariance of a strongly harmonizable isotropic random field becomes smoother. More specifically:

Theorem 5.4.2 *The covariance $r : I\!\!R^n \times I\!\!R^n \to \mathcal{L}$ of a strongly harmonizable isotropic random field has at least $m = 1, 2, \ldots \lceil \frac{(n-1)}{2} \rceil$ partial derivatives with respect to τ_1, τ_2 and θ, where $\lceil \cdot \rceil$ is the greatest integer function, $\tau_1 = \|s\|, \tau_2 = \|t\|$, and $\theta = \arccos(s \cdot t)$.*

This result implies that members of the class $\mathcal{G}_\infty$ are infinitely differentiable. In fact, the following characterization can be given.

Theorem 5.4.3 *A covariance function $r(\cdot, \cdot)$ belongs to $\mathcal{G}_\infty$ if and only if*

$$r(s, t) = \int_0^\infty \int_0^\infty e^{-\frac{1}{2}\|\lambda s - \lambda' t\|^2} dF(\lambda, \lambda').$$

Swift obtained this result for strongly harmonizable isotropic fields. Rao [40] has recently extended it to the weakly harmonizable isotopic case. Using these smoothness properties of $r(\cdot, \cdot)$, it is now possible to consider the differentiability of the sample paths of $X(\cdot)$.

Analytic Random Fields

The concept of analyticity for the sample paths of harmonizable processes and processes with harmonizable increments of order k is considered in Swift [48], [52]. This concept for random fields is more difficult, since the theory of complex functions of several real variables plays a key role. Isotropy reduces some of these difficulties.

Theorem 5.4.4 *If the strongly harmonizable isotropic covariance $r(s,t)(= \tilde{\rho}(\tau_1, \tau_2, \theta)) \in \mathcal{G}_\infty$ is a holomorphic function in the variables τ_1 and τ_2 with θ fixed, $0 < \theta < \pi$, then almost all sample paths of the random field are entire.*

Proof: Fix θ with $0 < \theta < \pi$, so that the vectors s and t will not be collinear and consider the function $\tilde{\rho}(\tau_1, \tau_2)(= \rho(s, t))$. Here $\tau_1 = ||s||, \tau_2 = ||t||$, so $\tilde{\rho} : I\!\!R^+ \times I\!\!R^+ \to \mathbb{C}$. Suppose $\tilde{\rho}(\cdot, \cdot)$ is a class $\mathcal{G}_\infty$ covariance then since $\tilde{\rho}(\cdot, \cdot)$ is infinitely differentiable

$$\tilde{\rho}(\tau_1, \tau_2) = \sum_{k,l=0}^{\infty} \frac{\partial^{k+l} \tilde{\rho}(0,0)}{\partial \tau_1^k \partial \tau_2^l} \frac{\tau_1^k \tau_2^l}{k!l!} \qquad |\tau_1| < \infty, |\tau_2| < \infty.$$

Adler [[1], pp.26-27] shows that if the covariance $\rho(\cdot, \cdot)$ of a random field has partial derivatives up to order $2(n+1)$ then almost all sample functions of the field $X(\cdot)$ are mean square differentiable n times. This implies that the following

$$X^{(m)}(t) = D^{(m)} X(t)$$

where

$$D^{(m)} = \frac{\partial^{|m|}}{\partial t_1^{n_1} \dots \partial t_n^{m_n}},$$

and $t^k = t_1^{k_n} \cdot \ldots \cdot t_n^{k_n}$ exists in mean square, (using the standard multi-index notation, cf.[[27] pp. 4]). The multi-index notation will be conveniently used from now on. So, consider the random field

$$X_n(t) = \sum_{|k| \leq n} X^{(k)}(0) \frac{t^k}{k!}.$$

Since,

$$E \left| \sum_{|k| \geq 1} X^{(k)}(0) \frac{t^k}{k!} - X_n(t) \right|^2 = \sum_{|k|,|l| \geq n+1} \frac{\partial^{k+l} \tilde{\rho}(\tau_1, \tau_2)}{\partial \tau_1^k \partial \tau_2^l} \Big|_{\tau_1=\tau_2=0} \frac{t^{k+i}}{k!l!}$$

it follows that

$$\lim_{n \to \infty} E \left| \sum_{|k| \geq 0} X(k)(0) \frac{t^k}{k!} - X_n(t) \right|^2 = 0.$$

Thus,

$$X_n(t) \to \sum_k X^{(k)}(0)\frac{t^k}{k!}$$

in $L^2(P)$ so that

$$X_n(t) \to \sum_{k=0}^{\infty} X^{(k)}(0)\frac{t^k}{k!}$$

in $L^2(P)$ as $n \to \infty$.

To show the sample path analyticity, it must be shown that

$$X_n(t) \to \sum_k X^{(k)}(0)\frac{t^k}{k!}$$

holds with probability one. For this it suffices to show that

$$R_n(t) = \sum_{|k|\geq n} X^{(k)}(0)\frac{t^k}{k!} \to 0$$

with probability one. Let $\varepsilon > 0$ be given, and consider $A_n = [|\,R_n(t)\,| > \varepsilon]$. If $\sum_n P(A_n) < \infty$ then by the first Borel-Canteli lemma

$$P(\limsup_n A_n) = 0,$$

that is $R_n(t) \to 0$ with probability one. Now $\sum_n P(A_n) < \infty$, since

$$\sum_n P(A_n) \leq \frac{1}{\varepsilon^2}\sum_n E\,|\,R_n(t)\,|^2$$

$$= \sum_{|k|,|l|\geq n} \frac{\partial^{k+l}\tilde{\rho}(0,0)}{\partial \tau_1^k \partial \tau_2^l}\frac{t^{k+l}}{k!l!}.$$

Expanding and collecting appropriate terms gives:

$$\sum_n E\,|\,R_n(t)\,|^2 \leq \sum_{|k|,|l|\geq 1} \left|\frac{\partial^{k+l}\tilde{\rho}(0,0)}{\partial \tau_1^k \partial \tau_2^l}\right|\frac{||t||^{k+l}}{k!l!} < \infty$$

since the convergence of a power series implies its absolute convergence in the circle of convergence. Hence $\lim_n X_n(t)$ exists with probability one. Let

$$\tilde{X}(t) = \lim_{n\to\infty} X_n(t)$$

so $\tilde{X}(t)$ is an analytic random field for $||t|| < \infty$.

But

$$X_n(t) \to X(t) = \sum_k X^{(k)}(0)\frac{t^k}{k!}$$

in $L^2(P)$. Thus

$$\tilde{X}(t) = X(t) \quad \text{a.e.}$$

So for $0 < \theta < \pi$, $\rho \in \mathcal{G}_\infty$, almost all sample paths of the field are entire, completing the proof. $\square$

Using this result, one may obtain the following sampling result for harmonizable isotropic random fields of class $\mathcal{G}_\infty$:

Theorem 5.4.5 *If $X : \mathbb{R}^n \to L^2_0(P)$ is a strongly harmonizable isotropic random field in class $\mathcal{G}_\infty$ so that $r(\tau_1, \tau_2, \theta)$ is holomorphic in τ_1 and τ_2, for θ fixed and $0 < \theta < \pi$, then for any bounded infinite set of distinct points $\{t_n, |n| \geq 1\} \subset \mathbb{R}^n$, the sample $\{X(t_n), |n| \geq 1\}$ determines the field on $\mathbb{R}^n$.*

Proof: Since $\{t_n, |n| \geq 1\} \subset \mathbb{R}^n$ is bounded and infinite it has a convergent subsequence with limit t_0. Thus $X(t)$ is analytic in a neighborhood D of t_0 and has series representation (using multi-index notation)

$$X(t) = \sum_{|n| \geq 1} \frac{(t - t_0)^n}{n!} X^{(n)}(t_0). \tag{5.51}$$

Now since t_0 is a limit point of $\{t_{n_k}, |k| \geq 1\}$, it follows that $X(t)$ and $\{X(t_{n_k}), |k| \geq 1\}$ agree in this neighborhood D as well as the function defined by Equation (5.51). Since the domain of these functions is $\mathbb{R}^n$, a connected set, the field given by (5.51) is analytic; by analytic continuation, the field $X(t), t \in \mathbb{R}^n$ is the only one which agrees with $\{X(t_{n_k}), |k| \geq 1\}$ proving the assertion. $\square$

A result of Rao [28] follows from this:

Corollary 5.4.1 *If $X : \mathbb{R} \to L^2_0(P)$ is a strongly harmonizable process with an entire covariance $r(\cdot, \cdot)$ then for any bounded infinite set of distinct points $\{t_n, n \geq 1\} \subset \mathbb{R}$ the sample $\{X(t_n), n \geq 1\}$ determines the given process on $\mathbb{R}$.*

Continuity

Sample path continuity for harmonizable isotropic random fields has not yet been thoroughly investigated; the following is one such result and provides a direction for further investigation.

Definition 5.4.1 *A random field $X : \mathbb{R}^n \to L^2_0(P)$ is stochastically continuous at a point t_0 if*

$$\lim_{t \to t_0} P(\omega : |X(\omega, t) - X(\omega, t_0)| \geq \varepsilon) = 0 \tag{5.52}$$

A random field is continuous a.e. at t_0 if the limit can be moved inside $P(\cdot)$. These notions are said to hold on a set if they hold at every point of the set.

A sufficient condition for a harmonizable isotropic random field to be continuous a.e. at the origin is given in the following.

Theorem 5.4.6 *A strongly harmonizable isotropic random field $X : \mathbb{R}^n \to L_0^2(P)$ is continuous a.e. in a neighborhood of the origin provided*

$$\int_0^\infty \int_0^\infty \lambda'(|\lambda - \lambda'| + \lambda + \lambda')\, d\,|\,F\,|\,(\lambda, \lambda') < \infty. \tag{5.53}$$

Proof: Since

$$P(|\,X(h) - X(0)\,| > \varepsilon) \quad = \quad P(|\,X(h) - X(0)\,|^2 > \varepsilon^2)$$

$$\leq \quad \frac{1}{\varepsilon^2} E(|\,X(h) - X(0)\,|^2)$$

$$= \quad \frac{1}{\varepsilon^2}(r(h,h) - r(h,0) - r(0,h) + r(0,0)).$$

By symmetry, it suffices to consider just the first two terms. Now $X(\cdot)$ harmonizable isotropic implies that

$$|\,r(h,h) - r(h,0)\,|$$
$$\leq \quad 2^\nu \Gamma\left(\frac{n}{2}\right) \int_0^\infty \int_0^\infty \left| \frac{J_\nu(\|\lambda h - \lambda' h\|)}{\|\lambda h - \lambda' h\|^\nu} - \frac{J_\nu(\|\lambda h\|)}{\|\lambda h\|^\nu} \right| d\,|\,F\,|\,(\lambda, \lambda').$$

The function $J_\nu(x)/x^\nu$ is uniformly continuous and differentiable on $\mathbb{R}^+$, [56], so by the mean value theorem there exists an element z_h,

$$z_h \in (|\lambda - \lambda'|\|h\|, \lambda\|h\|)$$

so that

$$\frac{J_\nu(|\lambda - \lambda'|\|h\|)}{(|\lambda - \lambda'|\|h\|)^\nu} - \frac{J_\nu(\lambda\|h\|)}{(\lambda\|h\|)^\nu} = \left(\frac{J_\nu(z_h)}{z_h^\nu}\right)' \|h\|(|\lambda - \lambda'| - \lambda)$$

but

$$\frac{d}{dx} \frac{J_\nu}{x^\nu} = \frac{-J_{\nu+1}}{x^\nu},$$

hence

$$\frac{J_\nu(|\lambda - \lambda'|\|h\|)}{(|\lambda - \lambda'|\|h\|)^\nu} - \frac{J_\nu(\lambda\|h\|)}{(\lambda\|h\|)^\nu} = \left(\frac{-J_{\nu+1}(z_h)}{z_h^\nu}\right) \|h\|(|\lambda - \lambda'| - \lambda).$$

So

$$|r(h,h) - r(h,0)|$$

$$\leq\ 2^\nu \Gamma\left(\frac{n}{2}\right) \int_0^\infty \int_0^\infty \left|\frac{J_{\nu+1}(z_h)}{z_h^\nu}\right| \mid\ ||h||(|\lambda - \lambda'| - \lambda)\ \mid\ d\mid F\mid(\lambda, \lambda').$$

Using the Poisson integral representation for Bessel functions, [45], yields

$$\left|2^\nu \Gamma\left(\frac{n}{2}\right) \frac{J_{\nu+1}(z_h)}{z_h^\nu}\right|\ =\ \left|\frac{\Gamma\left(\frac{n}{2}\right) z_h}{\sqrt{\pi}\Gamma\left(\frac{n+1}{2}\right)} \int_0^{\frac{\pi}{2}} \cos(z_h \sin\phi) \cos^{2(\nu+1)}\phi d\phi\right|$$

$$\leq\ c_n z_h \leq c_n ||h||(|\lambda - \lambda'| + \lambda),$$

where c_n is a constant depending only on n. Now noting

$$\mid\ |\lambda - \lambda'| - \lambda\ \mid\ \leq \lambda'$$

gives

$$\mid r(h,h) - r(h,0)\mid\ \leq c_n ||h||^2 \int_0^\infty \int_0^\infty \lambda'(|\lambda - \lambda'| + \lambda)\ d\mid F\mid(\lambda, \lambda').$$

Similar analysis also applies for $\mid r(0, h) - r(0,0)\mid$, so that

$$\mid r(h,h) - r(h,0)\mid\ +\ \mid r(0,h) - r(0,0)\mid$$

$$\leq\ c_n' ||h||^2 \int_0^\infty \int_0^\infty \lambda'(|\lambda - \lambda'| + \lambda + \lambda')\ d\mid F\mid(\lambda, \lambda').$$

Thus

$$P(\mid X(h) - X(0)\mid > \varepsilon) \leq \frac{K}{\varepsilon^2}||h||^2$$

where K is a finite constant. Hence, by Kolmogorov's sufficiency Theorem for sample path continuity, (cf. Rao, [31]) $X(\cdot)$ is continuous a.e. in a neighborhood of the origin. $\square$

The following result for the class $\mathcal{G}_\infty$ is a consequence of the last Theorem and is given here for reference.

Corollary 5.4.2 *A strongly harmonizable isotropic random field $X : \mathbb{R}^n \to L_0^2(P)$ with covariance from $\mathcal{G}_\infty$ is continuous a.e. in a neighborhood of the origin provided*

$$\int_0^\infty \int_0^\infty \lambda'(|\lambda - \lambda'| + \lambda + \lambda')\ d\mid F\mid(\lambda, \lambda') < \infty. \tag{5.54}$$

Theorem 4.6 implies almost all sample paths are continuous for stationary isotropic random fields, since the covariance $r(s, t)$ depends only on the difference $s - t$. Moreover, since a harmonizable random field is a Fourier transform of a vector measure, this result and the Lévy continuity Theorem imply that the field is weakly uniformly continuous, although the pointwise case is the desirable one.

A Moving Average Representation

Moving average representations for harmonizable processes have been considered extensively by Mehlman [23], [24]. Chang and Rao [7] gave conditions upon Karhunen class processes to have moving average representations. Moving average representations will now be considered for harmonizable isotropic random fields. Extending the results of Chang and Rao to the fields case, one can introduce:

Definition 5.4.2 *A Karhunen type random field $X : \mathbb{R}^n \to L_0^2(P)$ has a moving average representation if $X(\cdot)$ can be represented in the form*

$$X(t) = \int_{\mathbb{R}^n} g(t - u)dZ(u), \ for \ t \in \mathbb{R}^n \tag{5.55}$$

where $Z(\cdot)$ is a stochastic measure which satisfies

$$F_Y(A, B) = E(Z_Y(A)\overline{Z_Y(B)}),$$

and $g(t - \cdot) \in L_2(F)$.

Using this definition, the following extension of the result of Chang and Rao may be given.

Theorem 5.4.7 *Let $X : \mathbb{R}^n \to L_0^2(P)$ be a field with representation (5.55), $g = \hat{f}$, where $f \in L^1(\mathbb{R}^n)$, and Z is a stochastic measure with values in $L_0^2(P)$. Then the field is strongly harmonizable and its spectral measure is absolutely continuous relative to the planar Lebesgue measure.*

Proof: The proof of this result is virtually identical to that given by Chang and Rao. A brief sketch is given here so that essential details may be referred to in the subsequent development.

Since $g = \hat{f}$ so that g is bounded, one has

$$X(t) = \int_{\mathbb{R}^n} \hat{f}(t - u)dZ(u)$$

$$= \int_{\mathbb{R}^n} dZ(u) \left(\int_{\mathbb{R}^n} e^{i(t-u)\cdot\lambda} f(\lambda)d\lambda \right)$$

$$= \int_{R^n} e^{i\lambda \cdot t} f(\lambda) \left(\int_{R^n} e^{-iu \cdot \lambda} dZ(u) \right) d\lambda,$$

by a form of Fubini's Theorem,

$$= \int_{R^n} e^{i\lambda \cdot t} f(\lambda) \overline{Y}(\lambda) d\lambda, \tag{5.56}$$

where $\{\overline{Y}(\lambda), \lambda \in R^n\}$ is a weakly harmonizable field. Let

$$\tilde{Z} : A \mapsto \int_A \overline{Y}(t) f(t) dt \text{ for } A \in B$$

so that $\tilde{Z}$ is a stochastic measure on the Borel sets of R^n, the integral being in Bochner's sense. Now

$$\nu(A, B) = E(\tilde{Z}(A) \overline{\tilde{Z}}(B))$$

can be shown to have finite Vitali variation, so that using properties of the Bochner integral, the field has representation

$$X(t) = \int_{R^n} e^{it \cdot \lambda} d\tilde{Z}(\lambda),$$

hence the field $X(\cdot)$ is strongly harmonizable. □

If the field $X(\cdot)$ satisfies (5.55) then the covariance $r(\cdot, \cdot)$ can be expressed as

$$r(s, t) = \int_{R^n} \int_{R^n} e^{i\lambda \cdot s - i\lambda' \cdot t} d\nu(\lambda, \lambda') \tag{5.57}$$

$$= \int_{R^n} \int_{R^n} e^{i\lambda \cdot s - i\lambda' \cdot t} r_Y(\lambda, \lambda') \overline{f}(\lambda) f(\lambda') d\lambda d\lambda'$$

where $r_Y(\cdot, \cdot)$ is the covariance of the weakly harmonizable field $Y(\cdot)$, (5.56) given above. If the field is also isotropic, so that $r(\cdot, \cdot)$ is invariant under $\varrho \in SO(n)$, then by a theorem of Rao [29], $\nu(\cdot, \cdot)$ in (5.57) is also invariant under ϱ. But this implies $r_Y(\cdot, \cdot)$ and $f(\cdot)$ are also invariant, and these observations are summarized in the following:

Theorem 5.4.8 *Let* $X : R^n \to L_0^2(P)$ *be an isotropic random field with moving average representation (5.55) with* $g = \hat{f}$, *where* $f \in L^1(R^n)$ *and* Z *is a stochastic measure with values in* $L_0^2(P)$. *Then* f *is invariant under* $\varrho \in SO(n)$, *the field is strongly harmonizable isotropic with spectral representation*

$$X(t) = \int_{R^n} e^{i\lambda \cdot t} f(\lambda) Y(\lambda) d\lambda$$

where $Y(\cdot)$ is a weakly harmonizable isotropic random field and the spectral measure $F(\cdot, \cdot)$ is absolutely continuous relative to the planar Lebesgue measure.

Fields with absolutely continuous spectral measures are also known as fields with an absolutely continuous (bi) spectra.

Harmonizable Spatially Isotropic Random Fields

An interesting addition to the theory of random fields, often useful in applications, is given by considering a random field $X : \mathbb{R}^k \times \mathbb{R}^n \to L_0^2(P)$. The random field $X(t, x)$ is both a function of a spatial variable x and a time variable t. These fields are often useful in applications such as the theory of turbulence, cf. Yaglom [59], and meteorology, cf. Jones [17]. They have been previously considered in the stationary isotropic case, and some results for these fields may be found in papers by Jones, [17] and Roy [42], as well as the books by Adler [1], Yadrenko [58], and Yaglom [59]. Recently, Swift [49] considered these fields in the harmonizable case. The spectral representation of these fields, termed *weakly harmonizable spatially isotropic*, is

$$X(t, x) = \alpha_n \sum_{m=0}^{\infty} \sum_{l=0}^{h(m,n)} S_m^l(u) \int_{\mathbb{R}^k} \int_0^\infty e^{i\omega \cdot t} \frac{J_{\nu+k}(\lambda ||x||)}{(\lambda ||x||)^\nu} dZ_m^l(\omega, \lambda).$$

(5.58)

The covariance of a weakly harmonizable spatially isotropic random field is given by

$$r(s, t, x, y) = 2^\nu \Gamma\left(\frac{n}{2}\right) \int_{\mathbb{R}^k} \int_{\mathbb{R}^k} \int_0^\infty \int_0^\infty e^{i\omega \cdot s - i\omega' \cdot t}$$

$$\times \frac{J_\nu(||\lambda x - \lambda' y||)}{||\lambda x - \lambda' y||^\nu} dF(\omega, \omega,' \lambda, \lambda')$$

(5.59)

where $F(\cdot, \cdot, \cdot, \cdot)$ is a function of bounded Fréchet variation.

Observe that when $k = 1$, the parameter t is a scalar, so that it is natural to term these representations as *time-varying*.

If $F(\cdot, \cdot, \cdot, \cdot)$ concentrates on the diagonals $\lambda = \lambda'$ and $\omega = \omega'$, the representation (5.59) reduces to the stationary representation

$$r(s, t, x, y) = 2^\nu \Gamma\left(\frac{n}{2}\right) \int_{-\infty}^{\infty} \int_0^\infty e^{i\omega(s-t)} \frac{J_\nu(\lambda || x - y ||)}{(\lambda || x - y ||)^\nu} d\Phi(\omega, \lambda). \quad (5.60)$$

Further it should be noted that in representation (5.59), with additional restrictions placed upon $F(\cdot, \cdot, \cdot, \cdot)$, includes the case of stationary in time, harmonizable and isotropic in space, as well as the case harmonizable in time, stationary and isotropic in space. Swift also obtained the representation of a spatially isotropic random fields which need not be harmonizable as

$$X(t, x) = \sum_{m=0}^{\infty} \sum_{l=0}^{h(m,n)} S_m^l(u) Y_m^l(t, \|x\|), \qquad (5.61)$$

where

$$Y_m^l(\cdot, \cdot), m = 0, 1, \ldots, \ l = 1, \ldots, h(m, n)$$

are a sequence of random fields such that

$$E(Y_m^l(t, \| x \|) \overline{Y_{m'}^l(t, \| x \|)}) = \delta_{mm'} \delta_{ll'} b_m(t, t, \| x \|, \| x \|)$$

and

$$\sum_{m=0}^{\infty} h(m, n) b_m(t, t, \| x \|, \| x \|) < \infty.$$

Time-varying representations play a key role in the studies of turbulence and meteorology. It will be useful to provide further representations of time-varying fields.

The previous section showed that the differentiability of the covariance of a strongly harmonizable isotropic random field provides insight into the structural properties of these fields. Similar investigation into the structure of harmonizable spatially isotropic random fields is possible. We shall consider one such application after briefly outlining the related results for harmonizable isotropic fields. Let us introduce the following concept.

Definition 5.4.3 *A random field* $X : \mathbb{R}^n \to L_0^2(P)$ *is harmonic if*

$$P[(\Delta X)(t) = 0] = 1$$

where Δ *is the Laplace operator and* ΔX *is defined in the sense of* L^2-*mean.*

Swift [48] recently proved the following result.

Theorem 5.4.9 *A strongly harmonizable isotropic random field* $X(t)$ *is harmonic iff its spectral measure* $F(\cdot, \cdot)$ *satisfies:*

$$\left| \int_0^{\infty} \int_0^{\infty} \lambda^2 (\lambda')^2 \frac{J_\nu(\tau \mid \lambda - \lambda' \mid)}{(\tau \mid \lambda - \lambda' \mid)^\nu} dF(\lambda, \lambda') \right| < \infty.$$

This result is specially interesting since there are nontrivial fields $X(t)$, as above satisfying the inequality, whereas for stationary isotropic random fields Yadrenko (cf., [[58] pp.16]) gave the following:

Theorem 5.4.10 *Every stationary isotropic harmonic random field is a constant with probability one.*

Thus, Swift's result shows that the natural relaxation of stationarity provided by the harmonizable concept gives further classes of fields useful in applications. This work provides motivation for considering Dirichlet type problems for time-varying spatial random fields.

Theorem 5.4.11 *If for each $t \in \mathbb{R}$ a random field $X(t, x)$ satisfies*

$$\Delta X(t, x) = 0 \qquad a.e. \; \|x\| < 1$$

$$X(t, x) = Y(t, x) \quad a.e. \; \|x\| = 1$$

where $Y(\cdot, \cdot)$ is a prescribed random field and the Laplacian Δ is in terms of the spatial variable x, then $X(\cdot, \cdot)$ has spectral representation

$$X(t, x) = \sum_{m=0}^{\infty} \sum_{l=1}^{h(m,n)} \tau^n S_m^l(u) \int_{S_n} S_m^l(v) Y(t, v) d\mu_n(v), \qquad (5.62)$$

where $\mu_n(\cdot)$ is the surface measure of the n-sphere S_n.

Proof: For each $t \in \mathbb{R}$, consider the Poisson integral formula as a vector integral,

$$X(t, x) = \frac{1 - \tau^2}{\omega_n} \int_{S_n} \frac{Y(t, v)}{[1 + \tau^2 - 2\tau \cos \theta]^{\frac{n}{2}}} d\mu_n(v),$$

where $\tau = \|x\|$. Using the identity

$$\sum_{m=0}^{\infty} \frac{(m + \nu)}{\nu} C_m^\nu(x) \tau^m = \frac{1 - \tau^2}{[1 - 2\tau x + \tau^2]^{\frac{n}{2}}} \;, | \tau | < 1, n > 2.$$

and the addition formula for spherical harmonics, [22], this representation becomes

$$X(t, x) = \int_{S_n} Y(t, v) \left(\sum_{m=0}^{\infty} \sum_{l=0}^{h(m,n)} \tau^m S_m^l(u) S_m^l(v) \right) d\mu_n(v)$$

$$= \sum_{m=0}^{\infty} \sum_{l=1}^{h(m,n)} \tau^m S_m^l(u) \int_{S_n} S_m^l(v) Y(t, v) d\mu(v) \qquad (5.63)$$

as desired. $\square$

The previous result gives the spectral representation (5.63) of a time-varying field which is spatially harmonic. Specializing this yields the following:

Theorem 5.4.12 *If a random field* $X : \mathbb{R} \times \mathbb{R}^n \to L_0^2(P)$ *is spatially harmonic inside the unit sphere* S_n *for each* $t \in \mathbb{R}$ *and is strongly harmonizable spatially isotropic on the surface then it is spatially isotropic inside* S_n *with a spectral representation*

$$X(t, x) = \alpha_n \sum_{m=0}^{\infty} \sum_{l=1}^{h(m,n)} \tau^m S_m^l(u) \int_{-\infty}^{\infty} \int_0^{\infty} e^{i\lambda t} \frac{J_{m+\nu}(\omega)}{\omega^\nu} dZ_m^l(\lambda, \omega) \quad (5.64)$$

where $\tau = \|x\|$ *and* $u = x/\tau$ *and covariance*

$$r(s, t) = \alpha_n^2 \sum_{m=0}^{\infty} \sum_{l=1}^{h(m,n)} \tau_1^m \tau_2^m S_m^l(u) S_m^l(v) \int_0^{\infty} \int_0^{\infty}$$

$$\times \quad \int_{-\infty}^{\infty} \int_{-\infty}^{\infty} e^{i\lambda s - i\lambda' t} \frac{J_{m+\nu}(\omega) J_{m+\nu}(\omega')}{\omega^\nu (\omega')^\nu} dF(\lambda, \lambda', \omega, \omega') \quad (5.65)$$

Proof: Since for each $t \in \mathbb{R}$, $Y(t, v) (= X(t, x)$ on the sphere) is a strongly harmonizable spatially isotropic random field on the surface of S_n i.e., $\|v\| = 1$, it follows that

$$Y(t, v) = \alpha_n \sum_{j=0}^{\infty} \sum_{k=1}^{h(j,n)} S_j^k(v) \int_{S_n} S_m^l(v) \alpha_n \sum_{j=0}^{\infty} \sum_{k=0}^{h(j,n)} S_j^k(v)$$

$$\times \quad \int_0^{\infty} \int_{-\infty}^{\infty} e^{i\lambda t} \frac{J_{j+\nu}(\omega)}{\omega^\nu} dZ_j^k(\lambda, \omega) d\mu_n(v)$$

$$= \sum_{m=0}^{\infty} \sum_{l=1}^{h(m,n)} \tau^m S_m^l(u) \alpha_n \sum_{j=0}^{\infty} \sum_{k=1}^{h(j,n)} \left(\int_{S_n} S_m^l(v) S_j^k(v) d\mu_n(v) \right)$$

$$\times \quad \int_0^{\infty} \int_{-\infty}^{\infty} e^{i\lambda t} \frac{J_{j+\nu}(\omega)}{\omega^\nu} dZ_j^k(\lambda, \omega)$$

$$= \alpha_n \sum_{m=0}^{\infty} \sum_{l=1}^{h(m,n)} \tau^m S_m^l(u) \int_0^{\infty} \int_{-\infty}^{\infty} e^{i\lambda t} \frac{J_{m+\nu}(\omega)}{\omega^\nu} dZ_m^l(\lambda, \omega).$$

Since

$$\int_{S_n} S_m^l(v) S_j^k(v) d\mu_n(v) = \delta_{mj} \delta_{lk},$$

this gives the covariance of $X(t, x)$ as (5.65). Thus the field is spatially isotropic, but not harmonizable. This finishes the proof. $\square$

One notes here that the previous result is similar to that obtained in the strongly harmonizable isotropic case, Swift [48], but (5.64) provides a representation of a time-varying harmonic field.

5.4.2 Local Classes of Fields

In the modern statistical theory of turbulence, random fields with certain local properties are often considered. A useful addition to this theory is given by considering a random field $X(t)$ which is not necessarily of class (C), but whose increment field

$$I_\tau X(t) = X(t + \tau) - X(t)$$

is of class (C). Rao [29], obtained the spectral representations for these locally class (C) random fields. Rao showed that the representations are obtained by considering generalized (in the sense of Gel'fand and Vilenkin, [10]) random fields, since they provide the required differentiability structure.

Generalized Harmonizable Random Fields

The following is a brief outline of the development of generalized fields from Gel'fand and Vilenkin, [10] and Yaglom, [59] to make the exposition clear, and it will be used below.

Consider the space $\mathcal{K}$ of infinitely differentiable functions $h(t)$ having compact supports, which with compact convergence becomes a locally convex linear topological space. A generalized random field $\tilde{X}$ is a linear functional $\tilde{X} : \mathcal{K} \to L$ such that if $\{\phi_n\}_{n=1}^\infty \subset \mathcal{K}, \phi_n \to 0$ in the topology of $\mathcal{K}$, then $\tilde{X}(\phi_n) \to 0$ in probability, as $n \to \infty$.

The mean of a generalized field is the linear functional

$$m(h) = E(\tilde{X}(h)), \quad h \in \mathcal{K}$$

and similarly its covariance is the bilinear (conjugate linear in the complex case) functional

$$r(h_1, h_2) = E(\tilde{X}(h_1)\overline{\tilde{X}(h_2)}), \quad h_i \in \mathcal{K}, i = 1, 2.$$

Ordinary fields generate the corresponding generalized fields by the relation

$$\tilde{X}(h) = \int_{R^n} X(t)h(t)dt \quad \text{for } h \in \mathcal{K},$$

The converse is not true unless an additional condition is assumed. That is, if a generalized field $\tilde{X}(\cdot)$ has point values (also called "of function space type") then the reverse implication holds.

Using this, and results from the theory of generalized functions, one defines the derivative $\tilde{X}^{(m_1,\ldots,m_n)}(h)$ of a generalized field $\tilde{X}(h)$ as

$$\tilde{X}^{(m_1,\ldots,m_n)}(h) = (-1)^M \tilde{X}(h^{(m_1,\ldots,m_n)}), \quad M = m_1 + \ldots + m_n.$$

Then it follows that if $\tilde{X}(h)$ has point values $X(t)$ and if $X(t)$ is harmonizable then $\tilde{X}^{(m_1,\ldots,m_n)}(h)$ also has point values coinciding with $X^{(m_1,\ldots,m_n)}(t)$. Hence, an ordinary field $X(t)$ may be regarded as having a weak derivative, in the above sense, i.e., by allowing this derivative to be a generalized field. This condition will be used in the subsequent development of fields with harmonizable increments.

Definition 5.4.4 *A generalized field $X(\cdot)$ is weakly harmonizable if*

$$E(\tilde{X}(h)) = 0$$

and $r(\cdot,\cdot)$ admits a representation

$$r(h_1, h_2) = \int_{R^n} \int_{R^n} \hat{h}_1(\lambda)\overline{\hat{h}_2(\lambda')}dF(\lambda,\lambda'), \tag{5.66}$$

where $F(\cdot,\cdot)$ is a positive definite function which satisfies

$$\int_{R^n} \int_{R^n} \frac{|dF(\lambda,\lambda')|}{((1+\|\lambda\|^2)\,(1+\|\lambda'\|^2))^{\frac{p}{2}}} < \infty \tag{5.67}$$

where $p > 0$, $\|\cdot\|$ is the Euclidean length. Further the integrals relative to F are in the strict Morse-Transue sense.

Spectral bi-measures $F(\cdot,\cdot)$ which satisfy Equation (5.67) are known as *tempered*. It may be shown that such an $\tilde{X}(\cdot)$ admits a representation

$$\tilde{X}(h) = \int_{R^n} \hat{h}(\lambda)dZ(\lambda)$$

where $Z : \mathcal{B} \to L^2(P)$ is a vector measure such that

$$E(Z(A)\overline{Z(B)}) = \int_A \int_B dF(\lambda,\lambda').$$

The theory of generalized fields will be used throughout the remaining sections of this chapter.

Generalized Fields with Class (C) Increments

The definition of generalized class (C) fields given by Rao [29] is extended in the following manner.

Definition 5.4.5 *A generalized random field* $\tilde{X} : \mathcal{K} \to \mathcal{L}$ *with zero mean and covariance functional* $r(\cdot, \cdot)$ *is of weak class (C) if it can be expressed as*

$$r(h_1, h_2) = \int_{\mathbb{R}^n} \int_{\mathbb{R}^n} \hat{h}_1(\lambda)\overline{\hat{h}_2(\lambda')}dF(\lambda, \lambda') \tag{5.68}$$

where $F(\cdot, \cdot)$ *is a function of locally bounded Fréchet variation satisfying Equation (5.67). Further, the integrals relative to F are in the strict Morse-Transue sense and $\hat{h}_i$ are the g-transforms of h_i, $i = 1, 2$*

$$\hat{h}_i(\lambda) = \int_{\mathbb{R}^n} h_i(t)g(t, \lambda)dt. \tag{5.69}$$

If in the representation (5.68) $g(t, \lambda) = e^{i\lambda \cdot t}$, then the generalized random field $\tilde{X}(\cdot)$ will have the representation of a generalized weakly harmonizable random field.

A useful extension of the classes thus far considered is given by the class of fields $X(\cdot)$ for which the increments of order M are of class (C). More specifically, if $\tilde{X}(h)$, is an arbitrary generalized random field, since its partial derivatives $\tilde{X}^{(m_1, m_2, \ldots, m_n)}(h)$, where $m_1 + m_2 + \ldots + m_n = M$ always exists, the following definition makes sense.

Definition 5.4.6 *A mapping* $X : \mathbb{R}^n \to L^2(P)$ *is a random field with class (C) increments of order M if the partial derivatives $\tilde{X}^{(m_1, m_2, \ldots, m_n)}(h)$, where $m_1 + m_2 + \ldots + m_n = M$ are of class (C).*

Some results concerning the covariance of generalized random fields with stationary increments of order M are given by Gel'fand and Vilenkin [10]. The representation for processes with harmonizable increments of order M was recently given by Swift [52]. It is now possible to obtain the representation of generalized random fields with class (C) increments of order M. Rao obtained the representation of a generalized random field with class (C) increments. As Rao showed, for a generalized random field to have class (C) increments, g must satisfy further conditions. Specifically for the case of class (C) increments of order M, it is required that g satisfies $g(t, \lambda - \lambda') = g(t, \lambda)\overline{\beta(t, \lambda')}$ for all $t, \lambda, \lambda' \in \mathbb{R}^n$ where

$$\frac{\partial^M g(t, 0)}{\partial \lambda_1^{m_1} \partial \lambda_2^{m_2} \ldots \partial \lambda_n^{m_n}} \neq 0 \text{ for } t_k \neq 0 \tag{5.70}$$

with

$$\frac{\partial^M \beta(t, \lambda)}{\partial^M t_k} = \alpha_k(t, \lambda)\lambda_k, \tag{5.71}$$

and $\alpha_k(0,\lambda) = \alpha_k \neq 0$. One notes that these restrictions are satisfied if $g(t,\lambda) = e^{i\lambda \cdot t}$.

Theorem 5.4.13 *A generalized random field $\tilde{X}(h)$ with Mth order class (C) increments has spectral representation:*

$$\tilde{X}(h) = \int_{\mathbb{R}^n - \{0\}} \hat{h}(\lambda)dZ_Y(\lambda) + (\alpha, \hat{\nabla}_M \hat{h}(0))$$

where $Z_Y(\cdot)$ is the spectral measure associated with its class(C) Mth order partial derivative field $Y(\cdot) = \tilde{X}^{(m_1,m_2,\ldots,m_n)}(h)(\cdot)$ and $\hat{h}$ is the g-transform (5.69) of h with g satisfying $g(t, \lambda - \lambda') = g(t, \lambda)\beta(t, \lambda')$ for all $t, \lambda, \lambda' \in \mathbb{R}^n$ and (5.70) and (5.71). More specifically,

$$Z_Y : \mathcal{B}(\mathbb{R}^n - \{0\}) \to L_0^2(P)$$

is a measure such that

$$F_Y(A, B) = E(Z_Y(A)\overline{Z_Y(B)}),$$

which is of finite Vitali variation, where $\mathcal{B}(\mathbb{R}^n - \{0\})$ is the Borel σ-algebra of $\mathbb{R}^n - \{0\}$. Further, $(\cdot, \cdot)$ is the inner product and the "Mth order gradient" is defined as:

$$\hat{\nabla}_M = (-1)^M \alpha \cdot (\partial^{m_1}/\partial\lambda_1^{m_1}, \partial^{m_2}/\partial\lambda_2^{m_2}, \ldots, \partial^{m_n}/\partial\lambda_n^{m_n}).$$

The covariance functional of $\tilde{X}(\cdot)$ is given by

$$r(h_1, h_2) = \int_{\mathbb{R}^n - \{0\}} \int_{\mathbb{R}^n - \{0\}} \hat{h}_1(\lambda)\overline{\hat{h}_2(\lambda')}dF(\lambda, \lambda') + (A\hat{\nabla}_M \hat{h}_1(0), \hat{\nabla}_M \hat{h}_2(0))$$
$$(5.72)$$

with A a positive definite matrix.

Proof: The proof here is, in essence, similar to the proofs given by Rao for fields with class (C) increments and for processes with harmonizable increments of order M given by Swift. A modification is needed only to use the covariance functional (5.68) and we sketch the proof below.

Using the relationship between $\tilde{X}$ and $\tilde{X}^{(m_1,m_2,\ldots,m_n)}$ for Mth order partial derivatives, it follows that since the measure $F(\cdot, \cdot)$ is tempered,

$$\tilde{X}^{(m_1,m_2,\ldots,m_n)}(h) = (-1)^M \tilde{X}(h^{(m_1,m_2,\ldots,m_n)}), \quad M = m_1 + m_2 + \ldots + m_n$$

$$= (-1)^M \int_{\mathbb{R}^n} \hat{h}(\lambda)dZ_Y(\lambda). \qquad (5.73)$$

where $\hat{h}$ is the g-transform (5.69) of h. Swift showed that the behavior of the stochastic measure $Z_Y(\cdot)$ near zero must be considered. Define $\tilde{Z}_Y(\cdot)$ by

$$\tilde{Z}_Y(A) = \begin{cases} Z_Y(A) & \text{if } A \not\supset \{0\} \\ 0 & \text{if } A = \{0\} \end{cases}$$

which is a stochastic measure. Thus integrating (5.69) by parts repeatedly and substituting this representation in (5.73) one obtains

$$\tilde{X}^{(m_1, m_2, \ldots, m_n)}(h) = (-1)^M \int_{\mathbb{R}^n - \{0\}} \hat{h}(\lambda) d\tilde{Z}_Y(\lambda)$$

$$+ \lim_{\varepsilon \to 0} \int_{-\varepsilon}^{\varepsilon} \hat{\nabla}_M \hat{h}(\lambda) dZ_Y(\lambda),$$

Now $g(t, \lambda - \lambda') = g(t, \lambda)\overline{\beta(t, \lambda')}$ for all $t, \lambda, \lambda' \in \mathbb{R}^n$ where

$$\frac{\partial^M g(t, 0)}{\partial \lambda_1^{m_1} \partial \lambda_2^{m_2} \ldots \partial \lambda_n^{m_n}} \neq 0 \text{ for } t_k \neq 0$$

with g satisfying (5.71) and $\alpha_k(0, \lambda) = \alpha_k \neq 0$. Thus α is a second-order random vector representing the "jump" of $Z_Y(\cdot)$ at $\lambda = 0$. Hence

$$\tilde{X}(h) = \int_{\mathbb{R}^n - \{0\}} \hat{h}(\lambda) dZ_Y(\lambda) + (\alpha, \hat{\nabla}_M \hat{h}(0)). \tag{5.74}$$

Using this representation and the temperedness of $F(\cdot, \cdot)$, the covariance functional (5.72) follows, completing the proof. $\square$

As noted before, the conditions upon g are satisfied when $g(t, \lambda) = e^{i\lambda \cdot t}$, in which case the previous result reduces to the following:

Theorem 5.4.14 *A generalized random field $\tilde{X}(h)$ with Mth order strongly harmonizable increments has spectral representation:*

$$\tilde{X}(h) = \int_{\mathbb{R}^n - \{0\}} \hat{h}(\lambda) dZ_Y(\lambda) + (-1)^M (\alpha, \hat{\nabla}_M \hat{h}(0))$$

where $Z_Y(\cdot)$ is the spectral measure associated with its strongly harmonizable Mth order partial derivative field $Y(\cdot) = \tilde{X}^{(m_1, m_2, \ldots, m_n)}(h)(\cdot)$ and $\hat{h}$ is the Fourier transform of h and Z_Y, α a second-order random vector and $\hat{\nabla}_M$ as defined above. The covariance functional of $\tilde{X}(\cdot)$ is given by

$$r(h_1, h_2) = \int_{\mathbb{R}^n - \{0\}} \int_{\mathbb{R}^n - \{0\}} \hat{h}_1(\lambda)\overline{\hat{h}_2(\lambda')} dF(\lambda, \lambda') + (A \hat{\nabla}_M \hat{h}_1(0), \hat{\nabla}_M \hat{h}_2(0))$$

$$\tag{5.75}$$

with A a positive definite matrix.

If $F(\cdot, \cdot)$ in (5.75) concentrates on the diagonal $\lambda = \lambda'$ the representation becomes that obtained by Gel'fand and Vilenkin for the covariance of a generalized random field with Mth order stationary increments. Thus (5.75) is an extension of their result.

Fields with Harmonizable Isotropic Increments of Order M

As we have seen in the course of this chapter, and can be observed throughout the development of representations for various nonstationary processes and fields, the stationary formulations often extend with different proofs. This "robustness" is one of the underlying motivations for the forms we consider. Let us now obtain the representation of a generalized random field with harmonizable isotropic increments of order M.

Definition 5.4.7 *A mapping $X : \mathbb{R}^n \to L^2(P)$ is a random field with strongly harmonizable isotropic increments of order M if its generalized partial derivatives $\hat{X}^{(m_1, m_2, \ldots, m_n)}(h)$, where $m_1 + m_2 + \ldots + m_n = M$ are strongly harmonizable isotropic.*

The covariance $r(\cdot, \cdot)$ of a generalized random field with strongly harmonizable (not necessarily isotropic) increments of order M is given by (5.75). Isotropy yields that $r(\cdot, \cdot)$ is invariant under rotations $\varrho \in SO(n)$, where $SO(n)$ is the group of orthogonal matrices on $\mathbb{R}^n$. That is,

$$r(\varrho h_1, \varrho h_2) = r(h_1, h_2) \text{ for } \varrho \in SO(n).$$

One can easily show that the bi-measure $F(\cdot, \cdot)$ in (5.75) is invariant under $\varrho \in SO(n)$. Further, the bilinear form

$$(A \hat{\nabla}_M \hat{h}_1(0), \ \hat{\nabla}_M \hat{h}_2(0)) \tag{5.76}$$

in (5.75) must satisfy

$$(A \hat{\nabla}_M \varrho \hat{h}_1(0), \ \hat{\nabla}_M \varrho \hat{h}_2(0)) = (A \hat{\nabla}_M \hat{h}_1(0), \ \hat{\nabla}_M \hat{h}_2(0)), \tag{5.77}$$

where A is a positive definite matrix. The relation (5.76) upon simplification becomes

$$(A \hat{\nabla}_M \hat{h}_1(0), \ \hat{\nabla}_M \hat{h}_2(0)) = \sum_{|j|=|k|=M} a_{jk} \mu_j \overline{\theta_k} \tag{5.78}$$

where μ_j and θ_k denote the moments of h_1 and h_2, so that (5.77) is equivalent to

$$\sum_{|j|=|k|=M} a_{jk} \mu_j^\varrho \overline{\theta_k^\varrho} = \sum_{|j|=|k|=M} a_{jk} \mu_j \overline{\theta_k} \tag{5.79}$$

where μ_j^ϱ and θ_k^ϱ are the moments of ϱh_1 and ϱh_2.

Using the invariance of $F(\cdot, \cdot)$ under ϱ we can show that the first term of (5.75) can be expressed as

$$2^{\nu}\Gamma(\frac{n}{2})\int_{+0}^{\infty}\int_{+0}^{\infty}\int_{\mathbf{R}^n}\int_{\mathbf{R}^n}h_1(s)\overline{h_2(t)}\frac{J_{\nu}(\|\lambda s-\lambda' t\|)}{(\|\lambda s-\lambda' t\|)^{\nu}}ds\,dt\,dF(\lambda,\lambda').$$
(5.80)

If the bi-measure $F(\cdot,\cdot)$ is tempered so that Equation (5.67) is satisfied, the derivation of (5.80) is identical to the derivation of the covariance of a harmonizable isotropic random field and may be found in the papers of Rao [40] and Swift [49].

Since μ_j and θ_k are the moments of h_1 and h_2 Equation (5.77) can be written as

$$\sum_{|j|=|k|=M} a_{jk}\mu_j\overline{\theta_k} = \int_{\mathbf{R}^n}\int_{\mathbf{R}^n} P(s,t)h_1(s)\overline{h_2(t)}ds\,dt \qquad (5.81)$$

where $P(s,t)$ is the polynomial

$$P(s,t) = \sum_{|j|=|k|=M} a_{jk}s^j t^k.$$

Hence, Equation (5.79) implies that $P(s,t)$ is invariant under ϱ, that is

$$P(\varrho s, \varrho t) = P(s,t) \text{ for } \varrho \in SO(n).$$

Using a classical result from group theory (cf. Weyl [57] pp. 31-32), any polynomial which is invariant under ϱ can be expressed in the form of a polynomial in (s,s), (s,t), and (t,t) where $(\cdot, \cdot)$ is the inner product. Thus

$$P(s,t) = \sum_{i,j,k} b_{ijk}(s,s)^i(s,t)^j(t,t)^k.$$

Since every term of the polynomial $P(s,t)$ is of degree $2M$ then $i+j+k = M$, further, since every term of $P(s,t)$ has the same degree M in s and t then $i = k$. Hence, $P(s,t)$ has the following form

$$P(s,t) = \sum_{k=0}^{\lfloor\frac{M}{2}\rfloor} b_k(s,s)^k(s,t)^{M-2k}(t,t)^k.$$

Substituting this expression into (5.81) gives

$$\sum_{|j|=|k|=M} a_{jk}\mu_j\overline{\theta_k} = \sum_{k=0}^{\lfloor\frac{M}{2}\rfloor} b_k\int_{\mathbf{R}^n}\int_{\mathbf{R}^n}(s,s)^k(s,t)^{M-2k}(t,t)^k h_1(s)\overline{h_2(t)}ds\,dt.$$
(5.82)

Finally, observe that the coefficients b_k are nonnegative. This follows from (5.78) being positive-definite so that

$$\sum_{k=0}^{\lfloor \frac{M}{2} \rfloor} b_k \int_{R^n} \int_{R^n} (s,s)^k (s,t)^{M-2k} (t,t)^k h(s) \overline{h(t)} ds dt \geq 0 \qquad (5.83)$$

for any $h \in \mathcal{K}$. But, the inequality (5.83) is valid if and only if each b_k is nonnegative. The preceding is summarized as follows:

Theorem 5.4.15 *A generalized random field $X : \mathcal{K} \to L_0^2(P)$ with strongly harmonizable isotropic increments of order M has a covariance functional representable as*

$$r(h_1, h_2) = 2^\nu \Gamma(\frac{n}{2}) \int_{+0}^{\infty} \int_{+0}^{\infty} \int_{R^n} \int_{R^n}$$

$$\times h_1(s) \overline{h_2(t)} \frac{J_\nu(\|\lambda s - \lambda' t\|)}{(\|\lambda s - \lambda' t\|)^\nu} ds \, dt \, dF(\lambda, \lambda') + \sum_{k=0}^{\lfloor \frac{M}{2} \rfloor} b_k$$

$$\times \int_{R^n} \int_{R^n} (s,s)^k (s,t)^{M-2k} (t,t)^k h_1(s) \overline{h_2(t)} ds dt. \quad (5.84)$$

where $F(\cdot, \cdot)$ is a tempered function of bounded Vitali variation, and the b_k are nonnegative constants.

When $M = 1$, the representation (5.84) reduces to the representation obtained by Swift [51] for a generalized random field with strongly harmonizable isotropic increments. Further, if $F(\cdot, \cdot)$ concentrates on the diagonal $\lambda = \lambda'$, (5.84) reduces to the representation of a generalized random field with stationary isotropic increments of order M obtained by Gel'fand and Vilenkin.

The following theorem is a useful characterization of the covariance functional of a generalized random field with harmonizable isotropic increments of order M.

Theorem 5.4.16 *A generalized random field $X : \mathcal{K} \to L_0^2(P)$ has strongly harmonizable isotropic increments of order M if and only if its covariance functional is representable as:*

$$r(h_1, h_2) = \alpha_n^2 \int_{R^n} \int_{R^n} \int_{+0}^{\infty} \int_{+0}^{\infty} \sum_{m=0}^{\infty} \sum_{l=1}^{h(m,n)} h_1(s) \overline{h_2(t)}$$

$$\times S_m^l(u) S_m^l(v) \frac{J_{m+\nu}(\lambda \|s\|) J_{m+\nu}(\lambda' \|t\|)}{(\lambda \|s\|)^\nu (\lambda' \|t\|)^\nu} dF(\lambda, \lambda') ds dt$$

$$+ \sum_{k=0}^{\lfloor \frac{M}{2} \rfloor} b_k \int_{R^n} \int_{R^n} (s,s)^k (s,t)^{M-2k} (t,t)^k h_1(s) \overline{h_2(t)} ds dt,$$

$$(5.85)$$

where $\nu = \frac{n-2}{2}$ and $S_m^l(\cdot), 1 \le l \le h(m,n) = \frac{(2m+2\nu)(m+2\nu-1)!}{(2\nu)!m!}, m \ge 1, S_0^l(u) = 1$ are the spherical harmonics on the unit n-sphere of order m, and $\alpha_n > 0, \alpha_n^2 = 2^{2\nu+1} \Gamma(\frac{n}{2}) \pi^{\frac{n}{2}}$ with $F(\cdot,\cdot)$ a tempered function of bounded Vitali variation, b_k nonnegative constants, and the series converges absolutely.

The proof of this result depends upon the following lemma.

Lemma 5.4.1 *With the above notation, one has*

$$\alpha_n^2 \sum_{m=0}^{\infty} \sum_{l=1}^{h(m,n)} S_m^l(u) S_m^l(v) \frac{J_{m+\nu}(\lambda r_1) J_{m+\nu}(\lambda' r_2)}{(\lambda r_1)^\nu (\lambda' r_2)^\nu} = 2^\nu \Gamma \left(\frac{n}{2} \right) \frac{J_\nu(\lambda R(\lambda,\lambda'))}{(\lambda R(\lambda,\lambda'))^\nu}$$

$$(5.86)$$

where $R(\lambda,\lambda') = (r_1^2 + (\frac{\lambda'}{\lambda})^2 r_2^2 - 2(\frac{\lambda'}{\lambda}) r_1 r_2 \cos\theta)^{\frac{1}{2}}$ and $\cos\theta = <u,v>$

The proof of this lemma follows from standard arguments and may be found in Swift [48].

Proof of Theorem: In one direction, if $X(\cdot)$ is a generalized random field with harmonizable isotropic increments of order M, then the covariance is representable as

$$r(h_1, h_2) = 2^\nu \Gamma(\frac{n}{2}) \int_{+0}^{\infty} \int_{+0}^{\infty} \int_{R^n} \int_{R^n} h_1(s) \overline{h_2(t)}$$

$$\times \frac{J_\nu(\|\lambda s - \lambda' t\|)}{(\|\lambda s - \lambda' t\|)^\nu} ds\, dt\, dF(\lambda, \lambda')$$

$$+ \sum_{k=0}^{\lfloor \frac{M}{2} \rfloor} b_k \int_{R^n} \int_{R^n} (s,s)^k (s,t)^{M-2k} (t,t)^k h_1(s) \overline{h_2(t)} ds dt.$$

But, by the law of cosines

$$\|\lambda s - \lambda' t\| = (\|\lambda s\|^2 + \|\lambda' t\|^2 - 2 <\lambda s, \lambda' t>)^{\frac{1}{2}}$$

so that by lemma

$$r(h_1, h_2) = \alpha_n^2 \int_{R^n} \int_{R^n} \int_{+0}^{\infty} \int_{+0}^{\infty} \sum_{m=0}^{\infty} \sum_{l=1}^{h(m,n)} h_1(s) \overline{h_2(t)}$$

$$\times \; S_m^l(u) S_m^l(v) \frac{J_{m+\nu}(\lambda \|s\|) J_{m+\nu}(\lambda'\|t\|)}{(\lambda\|s\|)^\nu (\lambda'|t|)^\nu} dF(\lambda, \lambda') ds dt$$

$$+ \; \sum_{k=0}^{\lfloor \frac{M}{2} \rfloor} b_k \int_{R^n} \int_{R^n} (s,s)^k (s,t)^{M-2k} (t,t)^k h_1(s)\overline{h_2(t)} ds dt.$$

since the series is absolutely convergent and the interchange of the integrals and summations is easily justified. This establishes one implication. For the converse, it is only necessary to reverse the above steps. This completes the proof of the theorem. $\square$

This theorem will be used to obtain the spectral representations of the generalized and ordinary fields.

Theorem 5.4.17 *A generalized random field $\tilde{X} : \mathcal{K} \to L_0^2(P)$ has strongly harmonizable isotropic increments of order M if and only if it has a spectral representation (with a mean convergence series):*

$$\tilde{X}(h) \;=\; \alpha_n \int_{R^n} h(t) \sum_{m=0}^{\infty} \sum_{l=1}^{h(m,n)} S_m^l(u)$$

$$\times \; \int_{+0}^{\infty} \frac{J_{m+\nu}(\lambda\|t\|)}{(\lambda\|t\|)^\nu} dZ_m^l(\lambda) \, dt$$

$$+ \; X_M \cdot \sum_{|j|=M} \mu_j, \tag{5.87}$$

where $E(Z_m^l(B_1) Z_{m'}^{l'}(B_2)) = \delta_{mm'} \delta_{ll'} F(B_1, B_2)$, with $F(\cdot, \cdot)$ as a tempered function of bounded Vitali variation, $X_M = (X_{M1}, X_{M2}, \ldots, X_{Mn})$ is a random vector which satisfies

$$E(X_{Mk} Z_m^l(B)) = 0, k = 1, \ldots, n$$

and

$$E(X_{Mk}\overline{M_{1j}}) = \begin{cases} 0 & \text{for } k \neq j \\ b & \text{for } k = j, \end{cases}$$

and the μ_j denote the moments of h.

Proof: Suppose $\tilde{X}(\cdot)$ is a generalized random field with strongly harmonizable isotropic increments of order M, then the covariance functional is expressible as Equation (5.85),

$$r(h_1, h_2) \;=\; \alpha_n^2 \int_{R^n} \int_{R^n} \int_{+0}^{\infty} \int_{+0}^{\infty} \sum_{m=0}^{\infty} \sum_{l=1}^{h(m,n)} h_1(s)\overline{h_2(t)}$$

$$\times \quad S_m^l(u)S_m^l(v)\frac{J_{m+\nu}(\lambda\|s\|)J_{m+\nu}(\lambda'\|t\|)}{(\lambda\|s\|)^\nu(\lambda'\|t\|)^\nu}dF(\lambda,\lambda')dsdt$$

$$+ \quad \sum_{k=0}^{\lfloor\frac{M}{2}\rfloor} b_k \int_{\mathbf{R}^n}\int_{\mathbf{R}^n}(s,s)^k(s,t)^{M-2k}(t,t)^k h_1(s)\overline{h_2(t)}dsdt,$$

where the series converges absolutely. Now by (5.82)

$$\sum_{|j|=|k|=M} a_{jk}\mu_j\overline{\theta_k} = \sum_{k=0}^{\lfloor\frac{M}{2}\rfloor} b_k \int_{\mathbf{R}^n}\int_{\mathbf{R}^n}(s,s)^k(s,t)^{M-2k}(t,t)^k h_1(s)\overline{h_2(t)}dsdt,$$

(5.88)

so that the representation of the covariance functional becomes

$$r(h_1,h_2) \quad = \quad \alpha_n^2\int_{\mathbf{R}^n}\int_{\mathbf{R}^n}\int_{+0}^\infty\int_{+0}^\infty\sum_{m=0}^\infty\sum_{l=1}^{h(m,n)} h_1(s)\overline{h_2(t)}$$

$$\times \quad S_m^l(u)S_m^l(v)\frac{J_{m+\nu}(\lambda\|s\|)J_{m+\nu}(\lambda'\|t\|)}{(\lambda\|s\|)^\nu(\lambda'\|t\|)^\nu}dF(\lambda,\lambda')dsdt$$

$$+ \quad \sum_{|j|=|k|=M} a_{jk}\mu_j\overline{\theta_k}.$$

Using the classical Karhunen type theorem for such a series, the spectral representation is obtained as

$$\tilde{X}(h) \quad = \quad \alpha_n\int_{\mathbf{R}^n} h(t)\sum_{m=0}^\infty\sum_{l=1}^{h(m,n)} S_m^l(u)$$

$$\times \quad \int_{+0}^\infty\frac{J_{m+\nu}(\lambda\|t\|)}{(\lambda\|t\|)^\nu}dZ_m^l(\lambda)\,dt$$

(5.89)

$$+ \quad X_M\cdot\sum_{|j|=M}\mu_j$$

where $X_M = (X_{M1}, X_{M2}, \ldots, X_{Mn})$ is a random vector which satisfies

$$E(X_{Mk}Z_m^l(B)) = 0, k = 1,\ldots,n$$

and

$$E(X_{Mk}\overline{X_{Mj}}) = \begin{cases} 0 & \text{for } k \neq j \\ b & \text{for } k = j \end{cases} .$$

The converse statement follows directly from the definition of a covariance. The proof is complete. □

Using this theorem, the spectral representation of the ordinary field $X(\cdot)$ can be obtained.

Theorem 5.4.18 *A random field $X : \mathbf{R}^n \to L^2_0(P)$ with strongly harmonizable isotropic increments of order M has spectral representation (convergence of series in mean)*

$$X(t) = \alpha_n \sum_{m=0}^{\infty} \sum_{l=1}^{h(m,n)} S_m^l(u) \int_{+0}^{\infty} \frac{J_{m+\nu}(\lambda\|t\|)}{(\lambda\|t\|)^{\nu}} dZ_m^l(\lambda)$$

$$+ \ \mathbf{X}_M \cdot t^M, \tag{5.90}$$

where $F(\cdot, \cdot)$ is a tempered function of bounded Vitali variation, related to the $Z_m^l(\cdot)$-measure, $\mathbf{X}_M = (X_{M1}, X_{M2}, \ldots, X_{Mn})$ is a random vector which satisfies

$$E(X_{Mk} Z_m^l(B)) = 0, k = 1, \ldots, n, \forall B \in \Sigma$$

and

$$E(X_{Mk}\overline{X_{Mj}}) = \begin{cases} 0 & \text{for } k \neq j \\ b & \text{for } k = j \end{cases}$$

b a nonnegative constant and $t^M = t_1^{m_1} t_2^{m_2} \ldots t_n^{m_n}$ with $M = m_1 + m_2 + \ldots + m_n$.

Proof: Using the relationship

$$\tilde{X}(h) = \int_{\mathbf{R}^n} h(t) X(t) \, dt$$

with the spectral representation of the previous theorem, one has (since $X(\cdot)$ is point valued),

$$X(t) = \alpha_n \sum_{m=0}^{\infty} \sum_{l=1}^{h(m,n)} S_m^l(u) \int_{+0}^{\infty} \frac{J_{m+\nu}(\lambda\|t\|)}{(\lambda\|t\|)^{\nu}} dZ_m^l(\lambda)$$

$$+ \ \mathbf{X}_M \cdot t^M,$$

which completes the proof. □

As was done for harmonizable isotropic fields, Swift [48], it is possible to obtain the spectral representation of a random field with isotropic increments of order M which are not necessarily harmonizable.

Theorem 5.4.19 *A random field* $X : \mathbb{R}^n \to L_0^2(P)$ *with isotropic increments of order* M *has spectral representation (convergence of series in mean)*

$$X(t) = \sum_{m=0}^{\infty} \sum_{l=1}^{h(m,n)} S_m^l(u) Y_m^l(\|t\|) + X_M \cdot t^M, \qquad (5.91)$$

where $Y_m^l(\cdot)$ *is a sequence of stochastic processes which satisfy*

$$E(Y_m^l(s)\overline{Y_{m'}^{l'}(t)}) = \delta_{mm'} \delta_{ll'} \gamma_m(s,t)$$

with $\gamma_m(\cdot, \cdot)$ *satisfying*

$$\left| \sum_{m=0}^{\infty} h(m,n)\gamma_m(\tau_1, \tau_2) \right| < \infty$$

and

$$\gamma_m(0, \tau) = 0 \text{ for } m \neq 0.$$

Further, X_M *is a fixed random vector satisfying*

$$E(X_{Mk} Z_m^l(B)) = 0, k = 1, \dots, n$$

and

$$E(X_{Mk}\overline{X_{Mj}}) = \begin{cases} 0 & \text{for } k \neq j \\ b & \text{for } k = j \end{cases},$$

with b *a non-negative constant;* $t^M = t_1^{m_1} t_2^{m_2} \dots t_n^{m_n}$ *and* $M = m_1 + m_2 + \dots + m_n$.

Proof: Letting

$$Y_m^l(r) = \alpha_n \int_0^{\infty} \frac{J_{m+\nu}(\lambda r)}{(\lambda r)^\nu} dZ_m^l(\lambda)$$

one has

$$E(Y_m^l(r)) = 0$$

and

$$E(Y_m^l(r_1)\overline{Y_{m'}^{l'}(r_2)}) = \delta_{mm'} \delta_{ll'} F(r_1, r_2)$$

using a form of Fubini's theorem. In fact, first apply $x^* \in (L_0^2(P))^*$ to both sides, then moving x^* inside the integral which is permissible and since $x^* Z_m^l(\cdot)$ is a scalar measure, a form of the classical Fubini theorem applies, cf., Dunford and Schwartz, [9]. ☐

Thus the above representation can be extended for all random fields with isotropic increments of order M which need not be harmonizable.

It is now possible to study the structure of fields with strongly harmonizable isotropic increments of order M. As was done for strongly harmonizable isotropic fields, let $\mathcal{G}_n^{(M)}$ be the class of all n-dimensional covariance functions $r(\cdot, \cdot)$ of a field with strongly harmonizable isotropic increments of order M. Let $\mathcal{G}_\infty^{(M)}$ be the class of all covariance functions belonging to $\mathcal{G}_n^M$ for all n. Using the natural identification mentioned above (see Equation (5.50)), one has

$$\mathcal{G}_{n+1}^{(M)} \subset \mathcal{G}_n^{(M)} \text{ and } \mathcal{G}_\infty^{(M)} = \bigcap_{n \geq 1} \mathcal{G}_n^{(M)}.$$

Observe further that any field with strongly harmonizable isotropic increments of order M is also a field with strongly harmonizable isotropic increments of order $M - 1$, so $\mathcal{G}_n \supset \mathcal{G}_n^{(M-1)} \supset \mathcal{G}_n^{(M)}$. An analysis of the classes $\mathcal{G}_n^{(M)}$ similar to Swift [48] is now possible. However it will not be considered here as it is a subject which should be investigated in detail in the future.

It is also possible to investigate the behavior of fields $X(t, x)$ with various assumptions upon the increments. In light of the preceding, one can formulate the following:

Definition 5.4.8 *A mapping $X : \mathbb{R} \times \mathbb{R}^n \to L_0^2(P)$ is a strongly harmonizable spatially isotropic random field with strongly harmonizable time increments of order k if its generalized kth partial derivative $\partial^k \tilde{X}(h(t, x))/\partial t^k$ is a generalized strongly harmonizable spatially isotropic random field.*

Using this definition, one can obtain in a similar manner the representation of a strongly harmonizable spatially isotropic random field $X(t, x)$ with strongly harmonizable time increments of order k. This representation was recently obtained by Swift [54]. The following is stated here for reference.

Theorem 5.4.20 *The spectral representation of an ordinary strongly harmonizable spatially isotropic random field with strongly harmonizable time increments of order k is given by a mean (or $L^2(P)$-) convergent series:*

$$X(t, x) = \int_{\mathbb{R}-\{0\}} \int_0^\infty \frac{e^{i\omega t} - \Delta_k(\omega, t)}{(i\omega)^k}$$

$$\times \; \alpha_n \sum_{m=0}^\infty \sum_{l=0}^{h(m,n)} S_m^l(u) \frac{J_{\nu+m}(\lambda \parallel x \parallel)}{(\lambda \parallel x \parallel)^\nu} d\tilde{Z}_m^l(\omega, \lambda)$$

$$+ \; \sum_{\eta=0}^{k-1} A_\eta(x) t^\eta \tag{5.92}$$

where $\Delta_k(\cdot, \cdot)$ are the jumps at the origin, defined by

$$\Delta_k(\omega, t) = \begin{cases} \sum_{j=0}^{k-1} \frac{(i\omega t)^j}{j!} & \text{for } |\omega| \leq 1, \\ 0 & \text{for } |\omega| > 1, \end{cases} \tag{5.93}$$

and where $Z_m^l(\cdot, \cdot)$ is the spectral measure associated with its strongly harmonizable spatial isotropic partial derivative random field

$$Y(t, x) = \frac{\partial^k \tilde{X}(t, x)}{\partial t^k}.$$

More specifically,

$$Z_m^l : \mathcal{B}(\mathbb{R} - \{0\} \times \mathbb{R}^n) \to L_0^2(P)$$

is a measure such that $F(\cdot, \cdot, \cdot, \cdot)$ is of finite Vitali variation, where $\mathcal{B}(\mathbb{R} - \{0\} \times \mathbb{R}^n)$ is the Borel σ-algebra of $(\mathbb{R} - \{0\}) \times \mathbb{R}^n$. Further, $A : \mathbb{R}^n \to L_0^2$ is a strongly harmonizable isotropic random field.

The technique used to obtain representation (5.90) can be employed to obtain the representation of a field $X(t, x)$ that is strongly harmonizable in the time parameter t and has strongly harmonizable spatially isotropic increments of order M in the spatial parameter x. This representation as well as the representation of $X(t, x)$ with strongly harmonizable time increments of order k and strongly harmonizable spatially isotropic increments of order M will not be given here. Instead, we will consider the following application.

A Prediction Problem For Local Fields

Let S_r denote a sphere in $\mathbb{R}^n$, of radius r, centered at the origin. Suppose that one observes $X(t)$, a locally harmonizable isotropic random field, at a countable number $\|t\| = r_i$, $i \geq 1$, of points. These are some "instances", and could be taken as $0 < r_1 < r_2 < \ldots$.

In this formulation, the problem is to obtain the best linear predictor in $L_0^2(P)$ of $X(t_0)$ for $\|t\| = r(\neq r_i, i \geq 1)$. In the following work, it is assumed that the tempered spectral bimeasure, $F(\cdot, \cdot)$ of $X(\cdot)$ is strictly positive definite.

Rao [38] solved the linear prediction problem for harmonizable isotropic fields and Swift [50] extended the result to fields with harmonizable isotropic increments. The following work is an extension of these works to fields with harmonizable isotropic increments of order M.

Let $G = \overline{sp}\{X(t) : \|t\| = r_i, i \geq 1\}$, then $G \subset L_0^2(P)$ and for $\|t_0\| = r > 0$, $X(t_0) \notin G$. The linear prediction problem is to find a unique Y in G such that

$$|X(t_0) - Y|_2 = \inf\{|X(t_0) - W|_2 : W \in G\}.$$

Using a classical result due to F. Riesz [41], there is a unique Y in G satisfying

$$(X(t_0 - Y)) \perp G.$$

That is, Y is a solution of the infinite system of linear equations

$$E(X(t_0)\overline{X(t)}) = E(Y\overline{X(t)}) \text{ for all } ||t|| = r_i, i \geq 1.$$

This produces the desired solution once a representation of elements of G is obtained. This representation can be obtained from an isometric identification of G with an appropriate sequence space ℓ_F^2 with respect to a positive definite weight function $F(\cdot,\cdot)$. This is given by

Proposition 5.4.1 *Each element Y of G, the subspace of $L_0^2(P)$ introduced above is representable as a mean convergent series:*

$$Y = \sum_{k=1}^{\infty} \sum_{m=0}^{\infty} \sum_{l=1}^{h(m,n)} a(l,m,k)v(k,t)$$

where

$$v(k,t) = \int_{+0}^{\infty} \frac{J_{m+\nu}(\lambda||t||)}{(\lambda||t||)^{\nu}} dZ_m^l(\lambda_k) + \boldsymbol{X}_M \cdot t^M$$

and $Z_m^l(\cdot)$ is the stochastic measure representing the random field $X(\cdot)$, and the sequence $\{a(l,m,k) : m \geq 0, 1 \leq l \leq h(m,n), k \geq 1\}$ satisfies the condition:

$$0 \leq \sum_{m=0}^{\infty} \sum_{l=1}^{h(m,n)} \sum_{k=1}^{\infty} \sum_{k'=1}^{\infty} a(l,m,k)\overline{a(l,m,k')}v(k,s_k)\overline{v(k',t_{k'})} < \infty$$

with $E(Z_m^l(A)\overline{Z}_{m'}^{l'}(B)) = F(A,B)\delta_{mm'}\delta_{ll'}$ and $\nu = \frac{n-2}{2}$

Proof: The proof of this proposition follows from standard arguments and is similar to that of Rao [38]; the details are outlined here. Let $\ell_{J_\nu,F}$ be the vector space of scalar sequences $\boldsymbol{a} = \{a(l,m,k) : m \geq 0, 1 \leq l \leq h(m,n), k \geq 1\}$ such that $|a|_{2,J_{\nu,F}^2}$ is given by

$$0 \leq |a|_{2,J_{\nu,F}^2} = \alpha_n^2 \sum_{m=0}^{\infty} \sum_{l=1}^{h(m,n)} \sum_{k=1}^{\infty} \sum_{k'=1}^{\infty} a(l,m,k)\overline{a(l,m,k')}v(k,s)\overline{v(k',t)} < \infty$$

so that $|\cdot|_{2,J_{\nu,F}}$ is a norm. For $\boldsymbol{a}, \boldsymbol{b} \in \ell_{J_\nu,F}$ the inner product is defined by polarization. Using this, it follows that $(\ell_{J_\nu,F}, |\cdot|_{2,J_{\nu,F}})$ is a Hilbert sequence space and taking $a(l,m,k) = S_m^l(\boldsymbol{u})$, the spherical harmonic, for $k = ||t_k||$ and $= 0$ for $k \neq ||t_k||$, one gets the mapping $i : X(t) \to \boldsymbol{a}$ with $t = (||t||, \boldsymbol{u})$ which satisfies $E(|X(t)|^2) = |a|_{2,J_{\nu,F}^2}^2 = E(|iX(t)|^2)$. Thus i is an isometry and can be extended by linearity and polarization to all of G onto $\ell_{J_\nu,F}$. It thus follows from standard arguments that there is a unique $\boldsymbol{a} \in \ell_{J_\nu,F}$ so that $Y = i\boldsymbol{a}$, yielding the desired result in explicit form. $\square$

Let $Y_0 \in G$ be the unique element that is closest to $X(t_0)$. Since Y_0 satisfies a system of equations as mentioned above, the desired solution is obtained if the vector a is found such that

$$E(X(t_0)\overline{X(t)}) = E(Y_0\overline{X(t)})$$

Using the proposition, along with the spectral representation of $X(\cdot)$ in spherical-polar coordinates, one multiplies by $S_m^l(v)$ then integrates over S_n, the unit sphere in $\mathbb{R}^n$, which gives

$$\sum_{k=1}^{\infty} a(l,m,k) \left(\frac{\|t\|}{\|t_k\|}\right)^{\nu} \int_{+0}^{\infty} \int_{+0}^{\infty} \frac{J_{m+\nu}(\lambda\|t_k\|)J_{m+\nu}(\lambda'\|t_l\|)}{(\lambda\lambda')^{\nu}} dF(\lambda,\lambda')$$

$$= \alpha_n S_m^l(u) \int_{+0}^{\infty} \int_{+0}^{\infty} \frac{J_{m+\nu}(\lambda\|t\|)J_{m+\nu}(\lambda'\|t_l\|)}{(\lambda\lambda')^{\nu}} dF(\lambda,\lambda'),$$

by the orthonormality of the $S_m^l(v)$ relative to the surface measure on S_n. Further simplification gives

Theorem 5.4.21 *Let* $\{X(\|t_k\|, u_k) : k = 1, 2, \ldots\}$ *be an observed sequence of a locally weak harmonizable isotropic random field* $X : \mathbb{R}^n \to L_0^2(P)$. *Then the best linear least squares predictor* Y_0 *of* $X(t_0)$, $t_0 = (\|t_0\|, u_0)$, $0 < \|t_0\| \neq \|t_k\|$, $k = 1, 2, \ldots$, *is given by*

$$Y = \sum_{k=1}^{\infty} \sum_{m=0}^{\infty} \sum_{l=1}^{h(m,n)} a(l,m,k)v(k,t_k)$$

with coefficients

$$a = \{a(l,m,k) : 1 \leq l \leq h(m,n), m \geq 0, k \geq 1\}$$

as a unique solution to the system of equations:

$$\sum_{k=1}^{\infty} a(l,m,k) \left(\frac{\|t\|}{\|t_k\|}\right)^{\nu} \int_{+0}^{\infty} \int_{+0}^{\infty} \frac{J_{m+\nu}(\lambda\|t_k\|)J_{m+\nu}(\lambda'\|t_l\|)}{(\lambda\lambda')^{\nu}} dF(\lambda,\lambda')$$

$$= \alpha_n S_m^l(u) \int_{+0}^{\infty} \int_{+0}^{\infty} \frac{J_{m+\nu}(\lambda\|t\|)J_{m+\nu}(\lambda'\|t_l\|)}{(\lambda\lambda')^{\nu}} dF(\lambda,\lambda').$$

Here the random variables $v(k,t_k)$ *are as in the preceding proposition and the series for* Y *is in the* $L^2(P)$ *sense.*

Some Further Directions

From the point of view of applications as discussed in Yaglom [59], it is important to have a multidimensional version of the preceding work. The concept of isotropy on $I\!\!R^n$ is susceptible to two different (multidimensional) generalizations. The first and simplest method is given by:

Definition 5.4.9 *A mapping $X = (X_1, X_2, \ldots, X_k) : I\!\!R^n \to L_0^2(P, \mathcal{L}^k)$ is a multidimensional weakly harmonizable isotropic random field if the mean values of all components X_j, $j = 1, \ldots, k$ are constant (assumed to be zero for convenience) and all the elements of the covariance matrix*

$$\mathcal{R}(s, t) = \|r_{jl}(s, t)\|$$

depend upon the lengths $\|s\|$, $\|t\|$ and θ the angle between s and t.

This definition is equivalent (as shown by Rao [38]) to requiring that for each complex k-vector, $\alpha = (\alpha_1, \alpha_2, \ldots, \alpha_k)$ the scalar

$$Y_\alpha = \alpha \cdot X = \sum_{i=1}^{k} \alpha_i X_i : I\!\!R^n \to L_0^2(P) \tag{5.94}$$

is a corresponding harmonizable random field. One notes here that the components $X_j(t)$, $j = 1 \ldots k$ of X are one-dimensional weakly harmonizable isotropic random fields. This is given in the following.

Proposition 5.4.2 *The elements $r_{jl}(s, t)$ of the covariance matrix $\mathcal{R}(s, t)$ of a multidimensional weakly harmonizable isotropic random field can be represented as*

$$r_{jl}(s, t) = 2^\nu \Gamma\left(\frac{n}{2}\right) \int_0^\infty \int_0^\infty \frac{J_\nu(\|\lambda s - \lambda' t\|)}{\|\lambda s - \lambda' t\|^\nu} dF_{jl}(\lambda, \lambda') \tag{5.95}$$

where $J_\nu(\cdot)$ is the Bessel function (of the first kind) of order $\nu = \frac{n-2}{2}$ and $F_{jl}(\cdot, \cdot)$ are complex functions of bounded Fréchet variation.

    The spectral representation of $X_j(\cdot)$ is then given as a mean convergent series:

$$X_j(t) = \alpha_n \sum_{m=0}^{\infty} \sum_{l=1}^{h(m,n)} S_m^l(u) \int_0^\infty \frac{J_{m+\nu}(\lambda\tau)}{(\lambda\tau)^\nu} dZ_{jm}^l(\lambda)$$

where $Z_{jm}^l(\cdot)$ satisfies

$$E(Z_{jm}^l(B_1)\overline{Z_{j'm'}^{l'}(B_2)}) = \delta_{mm'}\delta_{ll'} F_{jj'}(B_1, B_2)$$

with $F_{jj'}(\cdot, \cdot)$ a function of bounded Fréchet variation.

These results are to be expected and are natural multidimensional extensions. It is possible to extend much of the above work as well as the results contained in the papers of Rao [38], [40] and Swift [48] - [50]. There is, however, a more intricate generalization of isotropy which is often encountered in the study of turbulence. This second extension requires that for each $k \times k$ permutation matrix T, the multidimensional random field TX is weakly harmonizable isotropic. The basic structure theory of such random fields requires a deeper study, as it depends upon certain aspects of group representations. This awaits a serious investigation.

Acknowledgments

The author expresses his thanks to Professor M.M. Rao for his continuing advice, encouragement, and guidance during the work of this project. The author also expresses his gratitude to the Mathematics Department at Western Kentucky University for release time during the spring 1995 and 1996 semesters, during which this work was completed.

Bibliography

[1] Robert J. Adler, *The Geometry of Random Fields,* John Wiley and Sons, New York, 1981.

[2] V.V. Anh and K.E. Lunney, Covariance Function and Ergodicity of Asymptotically Stationary Random Fields, *Bull. Aust. Math. Soc.,* **44**, 49-62, 1991.

[3] A. S. Besicovitch, *Almost Periodic Functions.* Dover Publications, New York, 1954.

[4] C.S.K. Bhagavan, *Non-stationary Processes, Spectra and Some Ergodic Theorems,* Andhra University Press, 1974.

[5] C.S.K. Bhagavan, On Nonstationary Time Series, *Handbook of Statistics,* Vol. 5, p. 311-320, E.J. Hannan, P.R. Krishnaiah, M.M. Rao, Eds., 1985.

[6] D. K. Chang and M. M. Rao. Bimeasures and Nonstationary Processes, *Real and Stochastic Analysis,* John Wiley & Sons, New York, 1986, 7-118.

[7] D. K. Chang and M. M. Rao, Special Representations of Weakly Harmonizable Processes, *Stochastic Anal. Appl.,* **6**, 169-190, 1988.

[8] H. Cramér, A Contribution to the Theory of Stochastic Processes, *Proc. Second Berkeley Symp. Math. Stat. Probab.,*2, 55-77, 1952.

[9] N. Dunford and J. T. Schwartz, *Linear Operators,* Part I, Interscience, New York, 1957.

[10] I. M. Gel'fand and N. Ya Vilenkin, *Generalized Functions, Volume 4, Applications of Harmonic Analysis,* Academic Press, New York, 1964.

[11] I. I. Gihman and A. V. Skorohod, *The Theory of Stochastic Processes I,* Springer-Verlag, New York, 1974.

[12] E.G.Gladyshev, Periodically Correlated Random Sequences, *Soviet Math.,* **2**, 383-388, 1961.

[13] G. H. Hardy, *Divergent Series,* Oxford University Press, London, 1949.

[14] T. R. Hillman, Besicovitch-Orlicz Spaces of Almost Periodic Functions, *Real and Stochastic Analysis,* John Wiley & Sons, New York, 1986, 119-167.

[15] H. L. Hurd, Correlation Theory of Almost Periodically Correlated Processes, *J. Multivar. Anal.,* **37**, (1), 24-45, 1991.

[16] H. L. Hurd, Almost Periodically Unitary Stochastic Processes, *Stochastic Processes Appl.,* **43**, 99-113, 1992.

[17] R. H. Jones, Stochastic Processes on a Sphere, *Ann. Math. Stat.,* **34**, 213-218, 1963.

[18] Y. Kakihara, Multidimensional Second Order Stochastic Processes, In preparation.

[19] J. Kampé de Feriet and F.N. Frenkiel, Correlations and Spectra of Nonstationary Random Functions, *Math. Comput.,* **10**, 1-21, 1962.

[20] K. Karhunen, Über lineare Methoden in der Wahrscheinlichkeitsrechnung, *Acad. Sci. Fenn. Ser. AI, Math,* **37**, 3-79, 1947.

[21] T. Kawata, Almost Periodic Weakly Stationary Processes, *Statistics and Probability: Essays in Honor of C. R. Rao,* Kallianpur, Krishnaiah, and Ghosh, Eds., North Holland, New York, 1982, 383-396.

[22] N. N. Lebedev, *Special Functions and Their Applications,* Dover Publications, New York, 1972.

[23] M. H. Mehlman, Structure and Moving Average Representation for Multidimensional Strongly Harmonizable Processes, *Stochastic Anal. Appl.,* **9**, 323-361, 1991.

[24] M. H. Mehlman, Prediction and Fundamental Moving Averages for Discrete Multidimensional Harmonizable Processes, *J. Multivar. Anal.,* **43**, 147-170, 1992.

[25] C. Müller, *Spherical Harmonics,* Lecture Notes in Mathematics, Vol. 17, Springer-Verlag, New York, 1966.

[26] E. Parzen, Spectral Analysis of Asymptotically Stationary Time Series. *Bull. Int. Stat. Inst.,* **39**, 87-103, 1962.

[27] R. M. Range, *Holomorphic Functions and Integral Representations in Several Complex Variables,* Springer-Verlag, New York, 1986.

[28] M. M. Rao, Inference in Stochastic Processes - III (Nonlinear Prediction, Filtering, and Sampling Theorems), *Z. Wahrs.,* **8**, 49-72, 1967.

[29] M.M. Rao, Representation Theory of Multidimensional Generalized Random Fields, *Proc. Symp. Multivariate Analysis,* Vol. 2, Academic Press, New York, 1969, 411-435.

[30] M.M. Rao, Covariance Analysis of Non Stationary Time Series, *Devel. Stat.,* Vol. 1, 171-275, 1978.

[31] M.M. Rao, *Stochastic Processes and Integration,* Sijthoff and Noordhoff, Alphen aan den Rijn, The Netherlands, 1979.

[32] M.M. Rao, Domination Problem for Vector Measures and Applications to Nonstationary Processes, *Springer Lecture Notes in Math.,* **945**, Springer-Verlag, New York, 1982, 296-313.

[33] M.M. Rao, Harmonizable Processes: Structure Theory, *L'Enseign Math,,* 28, 295-351, 1984.

[34] M.M. Rao, *Probability Theory with Applications,* Academic Press, New York, 1984.

[35] M. M. Rao, Harmonizable, Cramér, and Karhunen Classes of Processes, *Handbook of Statistics,* Vol. 5, E.J. Hannan, P.R. Krishnaiah, and M.M. Rao, Eds., 279-310, 1985.

[36] M.M. Rao, Harmonizable Signal Extraction, Filtering and Sampling, *Topics in Non-Gaussian Signal Processing,* E.J. Wegman, S.C. Schwartz, J.B. Thomas, Eds., Springer-Verlag, New York, 98-117, 1989.

[37] M. M. Rao, A View of Harmonizable Processses, *Statistical Data Analysis and Inference,* Y. Dodge, Ed., 597-615, 1989.

[38] M.M. Rao, Sampling and Prediction for Harmonizable Isotropic Random Fields. *J. Combinatorics, Inf. Sys. Sci.,* **16**, 207-220, 1991.

[39] M. M. Rao, Harmonizable processes and inference: unbiased prediction for stochastic flows, *J. Stat. Plan. Infer.,* **39**, 187-209.

[40] M. M. Rao, Characterization of Isotropic Harmonizable Covariances and Related Representations (preprint)

[41] F. Riesz and B. Sz-Nagy, *Functional Analysis,* Dover, New York.

[42] R. Roy, Spectral Analysis of a Random Process on the Circle, *J. Appl. Probab.*, **9**, 745-757, 1972.

[43] Yu. Rozanov, Spectral Analysis of Abstract Functions, *Theor. Probab. Appl.*, **4**, 271-287, 1959.

[44] Yu. Rozanov, *Stationary Random Processes*, Holden-Day, San Francisco, 1967.

[45] I.M. Ryshik and I.S. Gradstein. *Tables of Series, Products and Intergrals*, VEB Deutscher Verlag Der Wissenschaften, Berlin, 1963.

[46] E. Slutsky, Sur les Functions Aléatiores Presque Periodques et Sur la Decomposition des Functions Aléatiores Stationaries en Compusantes, *Actual. Sci. Ind.*, 738, 33-55, 1938.

[47] H. Soedjak, *Asymptotic Properties of Bispectral Density Estimators of Harmonizable Processes*, Ph.D. Thesis, University of California, Riverside, 1996.

[48] R. Swift, The Structure of Harmonizable Isotropic Random Fields, *Stochastic Anal. Appl.*, **12**, 583-616, 1994.

[49] R. Swift, A Class of Harmonizable Isotropic Random Fields, *J. Combinatorics, Inf. Sys. Sci.*, **20**, 111-127, 1995.

[50] R. Swift, Representation and Prediction for Locally Harmonizable Isotropic Random Fields, *J. Appl. Math. Stochastic Anal.*, **8**, 101-114, 1995.

[51] R. Swift, Almost Periodic Harmonizable Processes, *Georgian Math. J.*, 3, 275-292, 1996.

[52] R. Swift, Stochastic Processes with Harmonizable Increments, *J. Combinatorics, Inf. Sys. Sci.*, **21**, 47-60, 1996.

[53] R. Swift, An Operator Characterization of Oscillatory Harmonizable Processes, Proceedings of the Probability and Modern Analysis Conference in honor of M.M. Rao, (to appear).

[54] R. Swift, Locally Time-Varying Harmonizable Spatially Isotropic Random Fields, *Indian J. Pure Appl. Math.*, (to appear).

[55] M. Udagawa, Asmptotic Properties of Distributions of Some Functionals of Random Variables, *Rep. Stat. Appl. Res. JUSE*, 2, No. 2 and 3, 1-66, 1952.

[56] G.N. Watson, *A Treatise on the Theory of Bessel Functions*. Cambridge University Press, London, 1962.

[57] H. Weyl, *The Classical Groups,* Princeton University Press, Princeton, New Jersey, 1946.

[58] M.I. Yadrenko, *Spectral Theory of Random Fields,* Optimization Software Inc., New York (English Translation), 1983.

[59] A.M. Yaglom, *Correlation Theory of Stationary and Related Random Functions,* Vol. 1 and 2, Springer-Verlag, New York, 1987.

[60] A.M. Yaglom, Some Classes of Random Fields in n-Dimensional Space Related to Stationary Random Processes. *Theory Probab. Appl.,* **2,** 273-320, (English Translation), 1957.

Chapter 6

On Singularity and Equivalence of Gaussian Measures

N. Vakhania and V. Tarieladze

6.1 Introduction

In the remarkable paper [11] it was shown that for the countable products of pairwise equivalent probability measures the following alternative holds: the products are either mutually equivalent or singular. Later on in [7] and [5] it was stated that the similar alternative also takes place for any two general Gaussian measures. After these papers, many works have appeared in which different proofs of the alternative were presented, conditions for equivalence were studied, and the form of density (Radon-Nikodym derivative) for equivalent measures were found. A rich bibliography about these questions is contained in [3],[4], [13]. These questions are considered also in books [6],[9],[13],[15],[16],[17],[22]. Let us also mention the papers [8], [14],[20], [24] where short proofs of Gaussian dichotomy are given. Especially should be mentioned the works [14], [19], [18], written by the specialists in Constructive field theory; for the first time they have underlined that the density of a Gaussian measure with respect to another such measure is always integrable in some power strictly greater than 1.

In all works which contain the detailed study of conditions for equivalence of two arbitrary Gaussian measures, the exposition usually is too long and complicated, they use assertions elaborated especially for this aim and have little general interest. In the present chapter we give the proofs, using only the basic notions from [5] and [11] and, in case of equivalent measures, we find the greatest power of integrability for the corresponding densities. It is worth to note that in our proofs we do not apply the concept of conditional expectations, and use rather the Hilbert space technique.

6.2   Definitions and Auxiliary Results

Let X be an infinite set, $\mathcal{B}$ be a σ-algebra of its subsets, μ and ν be positive measures on $\mathcal{B}$. We write $\mu \perp \nu$ if μ and ν are (mutually) singular, and $\nu << \mu$ if ν is absolutely continuous with respect to μ; if ν is finite and μ is σ-finite and $\nu << \mu$, then $\frac{d\nu}{d\mu}$ denotes the corresponding Radon-Nikodym derivative or density. If $\nu << \mu$ and $\mu << \nu$ then we say that ν and μ are equivalent and write $\nu \sim \mu$.

Suppose $(X, \mathcal{B})$ and $(Y, \mathcal{C})$ are measurable spaces and $\eta : X \to Y$ is a measurable mapping. Then for an arbitrary measure μ on $\mathcal{B}$ we denote by μ_η the image of μ under η, i.e., $\mu_\eta(C) = \mu(\eta^{-1}(C))$ for all $C \in \mathcal{C}$. If, moreover, $\mathcal{B} = \eta^{-1}(\mathcal{C}) := \{\eta^{-1}(C) : C \in \mathcal{C}\}$ and μ and ν are probability measures on $\mathcal{B}$, then $\nu << \mu$ if and only if $\nu_\eta << \mu_\eta$ and in case of absolute continuity the equality

$$\frac{d\nu}{d\mu} = \frac{d\nu_\eta}{d\mu_\eta} \circ \eta \quad \mu\text{-a.e.}$$

holds,where $\circ$ is the composition sign. This simple assertion will be used later without any reference to it.

Following [11], for the study of questions of absolute continuity of measures we shall use the notion of Hellinger integral. Let $(X, \mathcal{B})$ be a measurable space, μ and ν be probability measures on $\mathcal{B}$, κ be a positive σ-finite measure on $\mathcal{B}$ such that $\mu << \kappa$ and $\nu << \kappa$. Put

$$\rho(\mu, \nu) = \int_X \left(\frac{d\mu}{d\kappa}\right)^{\frac{1}{2}} \left(\frac{d\nu}{d\kappa}\right)^{\frac{1}{2}} d\kappa.$$

It is easy to see that the value of $\rho(\mu, \nu)$ does not depend on the choice of measure κ. This implies in particular that if $\nu << \mu$ then we have

$$\rho(\mu, \nu) = \int_X \left(\frac{d\nu}{d\mu}\right)^{\frac{1}{2}} d\mu.$$

The number $\rho(\mu, \nu)$ is called the Hellinger integral of the pair of measures (μ, ν). It is not difficult to prove the following formula (see [28], p.43; see also [2] for a generalization):

$$\rho(\mu, \nu) = \inf \sum_k \sqrt{\mu(B_k)\nu(B_k)}.$$

Here infimum is taken over all $\mathcal{B}$-measurable countable partitions of X. In the following easy assertion we collect all necessary properties related to this concept.

Proposition 1.1. *Let $(X, \mathcal{B})$ be a measurable space, μ and ν be probability measures on $\mathcal{B}$. Then the following assertions are valid:*

(a) $0 \le \rho(\mu, \nu) \le 1$;

(b) $\rho(\mu, \nu) = 1 \iff \mu = \nu$;

(c) $\rho(\mu, \nu) = 0 \iff \mu \perp \nu$;

(d) If $\mathcal{B}_\infty \subset \mathcal{B}$ is a σ-algebra, μ_1, ν_1 are restrictions of μ, ν to $\mathcal{B}_\infty$, then $\rho(\mu, \nu) \le \rho(\mu_1, \nu_1)$;

(e) If $(Y, \mathcal{C})$ is another measurable space, $\eta : X \to Y$ is a measurable mapping and $\eta^{-1}(\mathcal{C}) = \mathcal{B}$, then $\rho(\mu, \nu) = \rho(\mu_\eta, \nu_\eta)$;

(f) If n is a natural number, $(X_k, \mathcal{B}_k, \mu_k), (X_k, \mathcal{B}_k, \nu_k), k = 1, ..., n$ are probability spaces,

$$\mu = \prod_{k=1}^n \mu_k, \qquad \nu = \prod_{k=1}^n \nu_k,$$

then

$$\rho(\mu, \nu) = \prod_{k=1}^n \rho(\mu_k, \nu_k).$$

Remark. Though we shall not need it, we note that the function $(\mu, \nu) \to 1 - \rho(\mu, \nu)$ is a metric on the set of all probability measures defined on $\mathcal{B}$, and convergence with respect to this metric is equivalent to convergence in variation; this and related facts can be found in [10].

Let Λ be an infinite set. Denote by $\mathcal{F}(*)$ the set of all finite nonempty subsets of Λ. Evidently $\mathcal{F}(*)$ is a directed set with respect to the set-theoretic inclusion $\subset$. If now $(y_\Delta)_{\Delta \in \mathcal{F}(*)}$ is a net in a metric space (Y, d), then the meaning of expressions: this net is convergent or is a Cauchy net should be clear. Suppose that $\xi_\lambda, \lambda \in \Lambda$, is a family of measurable functions on a probability space $(X, \mathcal{B}, \mu)$ and $0 \le p \le \infty$; let us say that this family is (μ, p)-*productable* (= product representable) if the net

$$\left(\prod_{\lambda \in \Delta} \xi_\lambda\right)_{\Delta \in \mathcal{F}(*)}$$

is a convergent net in $L_p(X, \mathcal{B}, \mu)$. For a (μ, p)-productable family $\xi_\lambda, \lambda \in \Lambda$, we put

$$\prod_{\lambda \in \Lambda} \xi_\lambda := \lim_\Delta \prod_{\lambda \in \Delta} \xi_\lambda.$$

The following assertion is a slight refinement of a similar statement in [11].

Proposition 1.2. *Let* $\xi_\lambda, \lambda \in \Lambda$, *be a family of independent nonnegative random variables on a probability space* $(X, \mathcal{B}, \mu)$ *such that*

$$\mathbf{E}_\mu \xi_\lambda := \int_X \xi_\lambda d\mu = 1 \qquad \forall \lambda \in \Lambda,$$

and

$$\prod_{\lambda \in \Lambda} \mathbf{E}_\mu \xi_\lambda^{\frac{1}{2}} > 0. \tag{1.1}$$

Then $\xi_\lambda, \lambda \in \Lambda$, *is* $(\mu, 1)$-*productable and* $\prod_{\lambda \in \Lambda} \xi_\lambda$ *is nonzero on a set of a positive* μ-*measure. If, moreover,* $\xi_\lambda > 0$ μ-*a.e. for all* λ, *then*

$$\prod_{\lambda \in \Lambda} \xi_\lambda > 0$$

μ-*a.e.*

Proof. Relation (1.1) implies that the set $\{\lambda \in \Lambda : \mathbf{E}_\mu \xi_\lambda^{1/2} \neq 1\}$ is at most countable. So we can suppose without loss of generality that $\Lambda = \mathcal{B}$. Put $g_n = \prod_{k=1}^n \xi_k^{\frac{1}{2}}$ for any $n \in \mathcal{B}$. For the verification of the first statement it is enough to show that (g_n) is a Cauchy sequence in L_2. Fix $\varepsilon > 0$. It follows easily from (1.1) that for some $n_\varepsilon \in \mathcal{B}$ and all $n \geq n_\varepsilon$ we have

$$\prod_{k=n_\varepsilon}^n \mathbf{E}_\mu \xi_k^{\frac{1}{2}} > 1 - \frac{\varepsilon}{2}. \tag{1.2}$$

Then the independence and relation (1.2) imply that for all n, m with $min(n, m) \geq n_\varepsilon$ we have $\mathbf{E}_\mu (g_n - g_m)^2 < \varepsilon$, i.e. our sequence is a Cauchy sequence. For the proof of the second statement note that again by (1.1) we have

$$\lim_{n,m} \mathbf{E}_\mu \left(1 - \prod_{k=n}^m \xi_k^{\frac{1}{2}} \right)^2 = 0.$$

This relation implies $\lim_{n,m} \prod_{k=n}^m \xi_k = 1$ in measure μ, so

$$\lim_{n,m} \sum_{k=n}^m \ln \xi_k = 0$$

again in measure μ; consequently, the series $\sum_k \ln \xi_k$ is convergent in measure μ. Therefore

$$\prod_k \xi_k = \exp(\sum_k \ln \xi_k) > 0$$

μ-a.e. and the proof is finished.

Remark. It is easy to see that if in notations of Proposition 1.2 the family $\xi_\lambda, \lambda \in \Lambda$, is $(\mu, 1)$-productable (or even $(\mu, 0)$-productable and $\prod_{\lambda \in \Lambda} \xi_\lambda$ is nonzero on a set of a positive μ-measure, then the relation (1.1) also holds. More information about the convergence of the infinite products of nonnegative independent random variables can be found in [18].

In the next section we shall see that these two propositions are sufficient for the proof of Kakutani's dichotomy theorem. For the proof of a similar result for Gaussian measures we need a result on Hilbert-Schmidt operators. Let us recall that if H_1 and H_2 are real, not necessarily separable, Hilbert spaces, $D : H_1 \to H_2$ is a Hilbert-Schmidt operator and $(e_\lambda)_{\lambda \in \Lambda}$ is an orthonormal basis of H_1, then the number

$$\|D\|_2 = \left(\sum_\lambda \|De_\lambda\|^2 \right)^{\frac{1}{2}}$$

does not depend on a particular choice of (e_λ) and the functional $\|.\|_2$ is a norm on the vector space of all Hilbert-Schmidt operators. $\|D\|_2$ is called the *Hilbert-Schmidt norm* of D.

Proposition 1.3. *Let H be an infinite-dimensional Hilbert space, $D : H \to H$ be a continuous linear operator, and $H_0 \subset H$ be a dense vector subspace. Suppose that there is a finite constant $c > 0$ such that for any finite-dimensional vector subspace $F \subset H_0$ the inequality $\|P_F D P_F\|_2 \leq c$ holds (here and in the sequel $P_F : H \to F$ denotes the operator of orthogonal projection).*
Then D is a Hilbert-Schmidt operator with $\|D\|_2 \leq c$.

Proof. Denote by $\mathcal{FD}(\mathcal{H}_l)$ the family of all finite-dimensional vector subspaces of H_0. Evidently $\mathcal{FD}(\mathcal{H}_l)$ is a directed set with respect to the set-theoretic inclusion $\subset$. Therefore $(P_F)_{F \in \mathcal{FD}(\mathcal{H}_l)}$ is a net of finite rank orthogonal projectors. Since H_0 is dense in H, we can conclude easily that the equality

$$\lim_{F \in \mathcal{FD}(\mathcal{H}_l)} P_F h = h \qquad \forall h \in H \tag{1.3}$$

holds in the sense of convergence in H. Now continuity of D and (1.3) imply

$$\lim_{F \in \mathcal{FD}(\mathcal{H}_l)} P_F D P_F h = Dh \qquad \forall h \in H. \tag{1.4}$$

Recall that the space $\mathcal{HS}(\mathcal{H})$ of all Hilbert-Shmidt operators itself is a Hilbert space with respect to the inner product

$$(A|B) = tr A^* B, \qquad \forall A, B \in \mathcal{HS}(\mathcal{H}).$$

This implies that it is a closed ball $\mathcal{K}_c$ with the center at zero and with the radius c is a weakly compact subset of $\mathcal{HS}(\mathcal{H})$. Since this weak topology is coarser than the weak operator topology, we obtain that $\mathcal{K}_c$ is also compact, and hence, it is a closed subset in the space of all continuous linear operators with respect to the weak operator topology. Now, since according to our assumption we have

$$P_F D P_F \in \mathcal{K}_c \qquad \forall F \in \mathcal{FD}(\mathcal{H}_l),$$

the equality (1.4) implies $D \in \mathcal{K}_c$, i.e., D is a Hilbert-Schmidt operator and $\|D\|_2 \leq c$. QED

6.3 Kakutani's Theorem

We give the formulation of Kakutani's theorem in a slightly different form.

Theorem 2.1. *Let $(X, \mathcal{B})$ be a measurable space and μ and ν be probability measures on $\mathcal{B}$. Suppose that there are measurable spaces $(Y_\lambda, \mathcal{D}_\lambda)$, and measurable mappings $\eta_\lambda : X \to Y_\lambda, \lambda \in \Lambda$, with the following properties:*

(1) The mappings $\eta_\lambda, \lambda \in \Lambda$, generate the σ-algebra $\mathcal{B}$;

(2) The mappings $\eta_\lambda, \lambda \in \Lambda$, are independent with respect to both of measures μ and ν;

(3) $\nu_\lambda := \nu_{\eta_\lambda} << \mu_\lambda := \mu_{\eta_\lambda}$ for all $\lambda \in \Lambda$.

Then either $\nu \perp \mu$ or $\nu << \mu$.

Moreover, we have

$$\nu \perp \mu \iff \prod_{\lambda \in \Lambda} \rho(\mu_\lambda, \nu_\lambda) = 0$$

and

$$\nu << \mu \iff \prod_{\lambda \in \Lambda} \rho(\mu_\lambda, \nu_\lambda) > 0.$$

If instead of (3) the condition

(4) $\nu_\lambda \sim \mu_\lambda$ for all $\lambda \in \Lambda$ holds, then either $\nu \perp \mu$ or $\nu \sim \mu$ and

$$\nu \sim \mu \iff \prod_{\lambda \in \Lambda} \rho(\mu_\lambda, \nu_\lambda) > 0.$$

If $\nu \ll \mu$ then the functions

$$\frac{d\nu_\lambda}{d\mu_\lambda} \circ \eta_\lambda, \lambda \in \Lambda,$$

are $(\mu, 1)$-productable and

$$\frac{d\nu}{d\mu} = \prod_{\lambda \in \Lambda} \frac{d\nu_\lambda}{d\mu_\lambda} \circ \eta_\lambda.$$

Proof. Let $\Delta \subset \Lambda$ be any finite subset. Then Proposition 1.1(d,e) implies that

$$\rho(\mu, \nu) \le \prod_{\lambda \in \Delta} \rho(\mu_\lambda, \nu_\lambda),$$

and so,

$$\rho(\mu, \nu) \le \inf_\Delta \prod_{\lambda \in \Delta} \rho(\mu_\lambda, \nu_\lambda) = \prod_{\lambda \in \Lambda} \rho(\mu_\lambda, \nu_\lambda). \tag{2.1}$$

If now we assume that

$$\prod_{\lambda \in \Lambda} \rho(\mu_\lambda, \nu_\lambda) = 0,$$

then relation (2.1) will imply the equality $\rho(\mu, \nu) = 0$. Consequently, by Proposition 1.1(c), we obtain $\mu \perp \nu$.

Suppose now that $\mu \not\perp \nu$. Then according to relation (2.1) and Proposition 1.1(c) we have

$$\prod_{\lambda \in \Lambda} \rho(\mu_\lambda, \nu_\lambda) > 0. \tag{2.2}$$

Fix some $\lambda \in \Lambda$ and denote

$$\xi_\lambda = \frac{d\nu_\lambda}{d\mu_\lambda} \circ \eta_\lambda.$$

Then we can write

$$\rho(\mu_\lambda, \nu_\lambda) = \mathbf{E}_\mu \xi_\lambda^{\frac{1}{2}}.$$

Using this equality and relation (2.2) we obtain

$$\prod_{\lambda \in \Lambda} \mathbf{E}_\mu \xi_\lambda^{\frac{1}{2}} > 0.$$

This inequality, since the random variables are independent, implies by Proposition 1.2 that $\xi_\lambda, \lambda \in \Lambda$, are $(\mu, 1)$-productable. Put $\xi = \prod_{\lambda \in \Lambda} \xi_\lambda$. Now it is easy to see that $\nu(B) = \int_B \xi d\mu$ for all $B \in \mathcal{B}$ and so $\nu \ll \mu$. If $\nu_\lambda \sim \mu_\lambda$, then $\xi_\lambda > 0$ μ-a.e., and this fact by Proposition 1.2 implies $\xi > 0$ μ-a.e. Evidently this gives that $\nu \sim \mu$. QED

It is easy to reformulate Theorem 2.1 in ordinary form by using the product measures.

Theorem 2.2. *Let $(Y_\lambda, \mathcal{D}_\lambda, \mu_\lambda)$, $(Y_\lambda, \mathcal{D}_\lambda, \nu_\lambda)$, $\lambda \in \Lambda$, be infinite families of probability spaces. Suppose that the following conditions are satisfied:*

(1) $X = \prod_{\lambda \in \Lambda} Y_\lambda$ and $\mathcal{B} = \prod_{\lambda \in \Lambda} \mathcal{D}_*$ is the product σ-algebra.
(2) $\mu = \prod_{\lambda \in \Lambda} \mu_\lambda$ and $\nu = \prod_{\lambda \in \Lambda} \nu_\lambda$ are the product measures.
(3) $\nu_\lambda << \mu_\lambda$ for all $\lambda \in \Lambda$.
Then either $\nu \perp \mu$ or $\nu << \mu$.
Moreover, we have

$$\nu \perp \mu \Longleftrightarrow \prod_{\lambda \in \Lambda} \rho(\mu_\lambda, \nu_\lambda) = 0$$

and

$$\nu << \mu \Longleftrightarrow \prod_{\lambda \in \Lambda} \rho(\mu_\lambda, \nu_\lambda) > 0.$$

If instead of (3) the condition
(4) $\nu_\lambda \sim \mu_\lambda$ for all $\lambda \in \Lambda$ holds, then either $\nu \perp \mu$ or $\nu \sim \mu$ and

$$\nu \sim \mu \Longleftrightarrow \prod_{\lambda \in \Lambda} \rho(\mu_\lambda, \nu_\lambda) > 0.$$

If $\nu << \mu$ and η_λ, $\lambda \in \Lambda$ are natural projection mappings then the functions

$$\frac{d\nu_\lambda}{d\mu_\lambda} \circ \eta_\lambda, \ \lambda \in \Lambda,$$

are $(\mu, 1)$-productable and

$$\frac{d\nu}{d\mu} = \prod_{\lambda \in \Lambda} \frac{d\nu_\lambda}{d\mu_\lambda} \circ \eta_\lambda.$$

Proof. By the definition, the family of mappings η_λ, $\lambda \in \Lambda$ generates the σ-algebra $\mathcal{B}$. Evidently we also have $\nu_\lambda := \nu_{\eta_\lambda}$ and $\mu_\lambda := \mu_{\eta_\lambda}$ for all $\lambda \in \Lambda$. Therefore we can apply Theorem 2.1. QED

This theorem easily implies that if $\mu_\lambda \neq \nu_\lambda$ for uncountably many indices λ then $\mu \perp \nu$ even if μ_λ and ν_λ are equivalent measures for all $\lambda \in \Lambda$.

In Section 4 we shall show that an analog of Theorem 2.1 for Gaussian measures can be proved essentially on the same lines. We start the next section with several simple assertions concerning Gaussian measures on the real line.

6.4 Gaussian Measures on $\mathbb{R}$

A probability measure μ on Borel σ-algebra $\mathcal{B}(\mathbb{R})$ of the real line $\mathbb{R}$ is called Gaussian if it is absolutely continuous with respect to the Lebesgue measure and the density is of the following form

$$g_{a,t}(x) = \frac{1}{\sqrt{2\pi t}} \exp\{-\frac{(x-a)^2}{2t}\}, \qquad x \in \mathbb{R},$$

where a and $t > 0$ are real numbers, which are called the parameters of μ. It is convenient to denote by $\gamma_{a,t}$ the Gaussian measure with parameters (a, t). The measure $\gamma_{0,1}$ is called *the standard Gaussian measure* and is denoted by γ. In what follows it will be agreed that for any $a \in \mathbb{R}$ the Dirac measure δ_a is a Gaussian measure and put $\gamma_{a,0} = \delta_a$.

Let us recall also that if $(X, \mathcal{B}, \mu)$ is a probability space and $\xi : X \to \mathbb{R}$ is a random variable, then ξ is said to be *a Gaussian random variable* if its distribution $\mu_\xi := \mu \circ \xi^{-1}$ is a Gaussian measure on $\mathbb{R}$.

It is clear that any two Gaussian measures $\gamma_{a,t}, \gamma_{b,s}$ are either singular or equivalent; they are equivalent if and only if either $ts > 0$ or $t + s = 0$ and $a = b$. In case of equivalence we have

$$\frac{d\gamma_{b,s}}{d\gamma_{a,t}}(x) = \exp\left\{ \frac{1}{2}\left[\frac{(x-b)^2}{s} - \frac{(x-a)^2}{t} + \ln\frac{s}{t} \right] \right\}, \qquad x \in \mathbb{R}. \tag{3.1}$$

It is easy to see using (3.1) that in case of equivalence for any $p \in \mathbb{R}$ we have

$$\int_{\mathbb{R}} \left(\frac{d\gamma_{b,s}}{d\gamma_{a,t}} \right)^p d\gamma_{a,t} = \left(p\left(\frac{s}{t}\right)^{p-1} + (1-p)\left(\frac{s}{t}\right)^p \right)^{-\frac{1}{2}} \times$$

$$\exp\left\{ -\frac{p(1-p)}{2} \frac{(a-b)^2}{pt + (1-p)s} \right\} < \infty \tag{3.2}$$

if $pt + (1-p)s > 0$, and this integral is infinite otherwise. In particular, this integral is finite for all p with $1 \leq p < (\frac{s}{t})'$, where, as usual, for a number $r > 0$, r' denotes the conjugate number in the following sense: $r' = \frac{r}{r-1}$ if $r > 1$ and $r' = \infty$ if $0 < r \leq 1$. Below we shall see that a similar assertion also remains valid in the case of general Gaussian measures.

If we put $p = \frac{1}{2}$ in (3.2) we obtain the following relation that is usefull for the sequel:

$$\rho(\gamma_{a,t}, \gamma_{b,s}) = \left(\frac{1}{2}\left(\frac{s}{t}\right)^{-\frac{1}{2}} + \frac{1}{2}\left(\frac{s}{t}\right)^{\frac{1}{2}} \right)^{-\frac{1}{2}} \exp\left\{ -\frac{1}{4}\frac{(a-b)^2}{t+s} \right\}. \tag{3.3}$$

6.5   General Gaussian Measures

We consider general Gaussian measures in the sense of [5]. Let X be a set and G be a nonempty family of real functions defined on X. Denote by $\hat{\mathcal{C}}(X, G)$ the smallest σ-algebra of subsets of X with respect to which all functions $f \in G$ are measurable. Suppose now that $G \subset \mathbb{R}^X$ is a vector subspace and μ is a probability measure on $\hat{\mathcal{C}}(X, G)$. The measure μ is said to be *a Gaussian measure with respect to G* if the measures $\mu \circ f^{-1}, f \in G$, are Gaussian measures on $\mathbb{R}$. When it is clear which vector space G is considered, the words "...with respect to G" will be omitted. When X is a topological vector space, the role of G will be played by the topological dual space X^* or some of its vector subspaces.

Let μ be a Gaussian measure on $\hat{\mathcal{C}}(X, G)$. Then its *mean functional* $m_\mu : G \to \mathbb{R}$ and *covariance* $r_\mu : G \times G \to \mathbb{R}$ are defined, respectively, by the equalities

$$m_\mu(f) = \int_X f d\mu$$

and

$$r_\mu(f_1, f_2) = \int_X f_1 f_2 d\mu - m_\mu(f_1) m_\mu(f_2).$$

Evidently m_μ is a linear functional and r_μ is a symmetric and positive $(r_\mu(f, f) \geq 0$ for all $f \in G)$ bilinear functional and μ is uniquely defined by its parameters m_μ, r_μ; this follows from the equality

$$\hat{\mu}(f) := \int_X \exp(if) d\mu = \exp[im_\mu(f) - \frac{1}{2} r_\mu(f, f)] \qquad \forall f \in G.$$

and the uniqueness theorem for characteristic functionals (see [27], p. 200).

Denote also for a Gaussian measure μ on $\hat{\mathcal{C}}(X, G)$ by A_μ the linear operator from G to $L_2(X, \mu)$ defined by the condition: $A_\mu f$ is the equivalence class in $L_2(X, \mu)$ which contains the function $x \to f(x) - m_\mu(f)$. If now $(.|.)_\mu$ denotes the inner product in $L_2(X, \mu)$, then the equality

$$r_\mu(f_1, f_2) = (A_\mu f_1 | A_\mu f_2)_\mu, \qquad f_1, f_2 \in G$$

holds. In what follows H_μ always will denote the closure of $A_\mu(G)$ in $L_2(X, \mu)$; this set with the induced Hilbert space structure is a Hilbert space consisting of symmetric Gaussian random variables. Note that H_μ is isometric to the Hilbert space associated with the pre-Hilbert space (G, r_μ).

Now we formulate our form of the Feldman-Hajek theorem.

Theorem 4.1. *Let X be a set, G be a vector space of real functions defined on X, μ and ν be Gaussian measures on $\hat{\mathcal{C}}(X, G)$ with respect to G. Then μ and ν are either singular or equivalent.*

Moreover, $\nu << \mu$ and therefore $\nu \sim \mu$ if and only if the following conditions are satisfied:

(1) The linear functional $m = m_\nu - m_\mu$ is continuous on (G, r_μ) or, equivalently, there exists an element $\eta_0 \in H_\mu$ such that

$$m(f) = (A_\mu f | \eta_0)_\mu \qquad \forall f \in G;$$

(2) The bilinear functional r_ν is continuous on (G, r_μ) or, equivalently, there is a continuous linear operator $S : H_\mu \to H_\mu$ such that

$$r_\nu(f_1, f_2) = (S A_\mu f_1 | A_\mu f_2)_\mu \qquad \forall f_1, f_2 \in G;$$

(3) The operator S in condition (2) is injective;

(4) The operator S in condition (2) is such that the difference $D = S - I$ is a Hilbert-Schmidt operator in H_μ (here I denotes the identity operator on H_μ).

Suppose that these conditions are satisfied. Consider (finite or infinite) sequence of nonzero eigenvalues $s_1 - 1, \dots$ of the symmertic Hilbert-Schmidt operator D listed according to their multiplicities and the (finite or infinite) complete orthonormal sequence of corresponding eigenvectors $e_1, \dots$ in H_μ. Then $s_n > 0$ for all possible indices n, the sequence

$$\exp\left\{ -\frac{1}{2}[(\frac{1}{s_n} - 1)e_n^2 + \ln s_n] \right\}, \qquad n = 1, \dots$$

is $(\mu, 1)$-productable, the corresponding product is strictly positive μ-a.e. and

$$\frac{d\nu}{d\mu} = \left(\prod_n \exp\left\{ -\frac{1}{2}[(\frac{1}{s_n} - 1)e_n^2 + \ln s_n] \right\} \right) \exp\left(h_0 - \frac{1}{2}(h_0 | h_0)_\mu \right), \quad (4.1)$$

where $h_0 = S^{-1}\eta_0$.

We have also that for $r = \sup_n s_n (= \|S\|)$ the relation

$$\int_X \left(\frac{d\nu}{d\mu} \right)^p d\mu < \infty \quad \forall p \in]1, r'[\qquad (4.2)$$

holds, where r' is the number conjugate to r.

Proof. Assume

$$\nu \not\perp \mu. \qquad (4.3)$$

We shall derive from this assumption that all conditions (1),...,(4) are satisfied, these conditions imply $\nu << \mu$, the equality (4.1) holds, $\frac{d\nu}{d\mu} > 0$ μ-a.e. and therefore $\nu \sim \mu$.

According to Proposition 1.1(c) relation (4.3) implies

$$\rho(\mu, \nu) > 0. \qquad (4.4)$$

Fix arbitrarily $f \in G$. Denote by ν_f, μ_f the measures $\nu \circ f^{-1}, \mu \circ f^{-1}$. Then these measures are Gaussian measures on $\mathbb{R}$ with parameters $(m_\nu(f), r_\nu(f, f))$ and $(m_\mu(f), r_\mu(f, f))$, respectively, and (4.3) implies that $\nu_f \not\perp \mu_f$. It is clear that if $r_\mu(f, f) = 0$ then $r_\nu(f, f) = 0$ and $m(f) = 0$. Suppose now that $r_\mu(f, f) = 1$. Then relation (4.4), Proposition 1.1(d), and equality (3.3) imply that

$$0 < \rho(\mu, \nu) \leq \rho(\mu_f, \nu_f) = \left(\frac{1}{2} r_\nu(f, f)^{-\frac{1}{2}} + \frac{1}{2} r_\nu(f, f)^{\frac{1}{2}} \right)^{-\frac{1}{2}} \times$$
$$\exp\left[-\frac{m^2(f)}{4(1 + r_\nu(f, f))} \right].$$

This relation gives the inequalities

$$\frac{1}{4} \rho^4(\mu, \nu) \leq r_\nu(f, f) \leq 4\rho^{-4}(\mu, \nu) \tag{4.5}$$

and

$$\frac{m^2(f)}{1 + r_\nu(f, f)} \leq \ln \rho^{-4}(\mu, \nu). \tag{4.6}$$

The inequalities (4.5) and (4.6) evidently imply conditions (1) and (2). Since S is continuous (4.5) implies that

$$\inf\left\{ (Sh|h)_\mu \; : \; h \in H_\mu, (h|h)_\mu = 1 \right\}$$
$$= \inf\left\{ r_\nu(f, f) : f \in G, r_\mu(f, f) = 1 \right\}$$
$$\geq \frac{1}{4} \rho^4(\mu, \nu) > 0.$$

This inequality implies that S is even an invertible operator and thus condition (3) also holds.

Let us show now that (4.4) implies condition (4). We shall use Proposition 1.3. We take $H = H_\mu$ and $H_0 = A_\mu(G)$. Fix a finite-dimensional subspace $F \subset G$. Consider the σ-algebra $\hat{\mathcal{C}}(X, F)$, the restrictions μ_F, ν_F of μ, ν to $\hat{\mathcal{C}}(X, F)$, and the operator $S_F = P_{A_\mu(F)} S P_{A_\mu(F)} : A_\mu(F) \to A_\mu(F)$. It is easy to see that

$$r_\nu(f_1, f_2) = (S_F A_\mu f_1 | A_\mu f_2)_\mu. \qquad \forall f_1, f_2 \in F \tag{4.7}$$

Let now $h_1, ..., h_n$ be an orthonormal basis of $A_\mu(F)$ consisting of eigenvectors of R_F, and $s_1, ..., s_n$ be the corresponding eigenvalues. Then equality (4.7) implies that $h_1, ..., h_n$ are independent standard Gaussian random variables with respect to the measure μ and $h_1, ..., h_n$ are independent Gaussian random variables with parameters

$$(m(h_1), s_1), \ldots, (m(h_n), s_n)$$

with respect to the measure ν. Hence, by (4.4) and Proposition 1.1 we can write

$$0 < \rho(\mu, \nu) \leq \rho(\mu_F, \nu_F) = \prod_{k=1}^{n} \left(\frac{1}{2} s_k^{-\frac{1}{2}} + \frac{1}{2} s_k^{\frac{1}{2}} \right)^{-\frac{1}{2}} \exp\left(-\frac{1}{4} \frac{m^2(h_k)}{1 + s_k} \right).$$

In particular,

$$\sum_{k=1}^{n} \ln\left(\frac{1}{2} s_k^{-\frac{1}{2}} + \frac{1}{2} s_k^{\frac{1}{2}}\right) \leq \ln \rho^{-2}(\mu, \nu).$$

From here, taking into account that $s_k \leq ||S_F|| \leq ||S||, k = 1, ..., n$, and using the elementary inequality

$$(s-1)^2 \leq (\sqrt{s}+1)^2 (s+1) \ln\left(\frac{1}{2} s^{-\frac{1}{2}} + \frac{1}{2} s^{\frac{1}{2}}\right) \quad \forall s > 0,$$

we obtain

$$||S_F - I_F||_2^2 = \sum_{k=1}^{n} (s_k - 1)^2 \leq (\sqrt{||S||} + 1)^2 (||S|| + 1) \ln \rho^{-2}(\mu, \nu).$$

Since in this inequality the right-hand side does not depend on F, Proposition 1.3 is applicable and we conclude that $D = S - I$ is a Hilbert-Schmidt operator.

Let us show that if conditions (1),...,(4) are satisfied then $\nu << \mu$. Since, according to condition (4), $S - I$ is a symmetric compact operator, there is an orthonormal basis $(e_\lambda)_{\lambda \in \Lambda}$ of H_μ consisting of eigenvectors of S; denoted by $(s_\lambda)_{\lambda \in \Lambda}$ the corresponding family of eigenvalues. Obviously, condition (3) gives

$$0 < s_\lambda \leq ||S|| \qquad \forall \lambda \in \Lambda$$

and, by condition (4),

$$\sum_{\lambda \in \Lambda} (s_\lambda - 1)^2 < \infty. \tag{4.8}$$

This implies that the set

$$\Lambda_+ := \{\lambda \in \Lambda : |s_\lambda - 1| + |(\eta_0|e_\lambda)_\mu| > 0\}$$

is at most countable. Put

$$\xi_\lambda = \exp[-\frac{(e_\lambda - (e_\lambda|\eta_0)_\mu)^2}{2s_\lambda} - e_\lambda^2 + \ln s_\lambda], \qquad \lambda \in \Lambda.$$

Using formula (3.3) and relation (4.8) it is not hard to see that

$$\prod_{\lambda \in \Lambda} \int_X \xi_\lambda^{\frac{1}{2}} d\mu > 0.$$

Therfore by Proposition 1.2 we can conclude that the family $\xi_\lambda, \lambda \in \Lambda$, is $(\mu, 1)$-productable and

$$\xi := \prod_{\lambda \in \Lambda} \xi_\lambda > 0 \quad \mu\text{-a.e.} \tag{4.9}$$

Consider now the measure ν_1 on $\hat{\mathcal{C}}(X, G)$ such that

$$\frac{d\nu_1}{d\mu} = \xi$$

and let us show that $\nu = \nu_1$. This will imply $\nu \ll \mu$, $\frac{d\nu}{d\mu} = \xi$ and so, according to (4.9), we shall have that $\nu \sim \mu$.

To prove the equality $\nu_1 = \nu$ it is enough to verify that for an arbitrary $f \in G$ we have

$$\hat{\nu}_1(f) := \int_X \exp(if)\xi d\mu = \exp[im_\nu(f) - \frac{1}{2}r_\nu(f, f)](= \hat{\nu}(f)). \tag{4.10}$$

Condition (1) implies the equality

$$m_\nu(f) = m_\mu(f) + (A_\mu f | S^{-1} h_0)_\mu$$

$$= m_\mu(f) + \sum_{\lambda \in \Lambda} (f | e_\lambda)_\mu (S^{-1} h_0 | e_\lambda)_\mu. \tag{4.11}$$

We can write also

$$r_\nu(f, f) = (SA_\mu f | A_\mu f)_\mu = \sum_{\lambda \in \Lambda} s_\lambda (A_\mu f | e_\lambda)_\mu^2 \tag{4.12}$$

and

$$f = m_\mu(f) + \sum_{\lambda \in \Lambda} (A_\mu f | e_\lambda)_\mu e_\lambda. \tag{4.13}$$

By putting the expressions (4.13) and (4.9) in the second term of equality (4.10), taking into account that $e_\lambda, \lambda \in \Lambda$, are independent standard Gaussian random variables on (X, μ) and using equalities (4.11) and (4.12), we obtain equality (4.10).

Thus we have proved that (4.3) implies $\nu \ll \mu$ and

$$\frac{d\nu}{d\mu} = \xi.$$

Since, evidently, $\xi_\lambda > 0$ for all $\lambda \in \Lambda$, by Proposition 1.2 we have $\xi > 0$ μ-a.e., i.e., in fact we obtain $\nu \sim \mu$. So (4.3) implies that the measures are equivalent and the first assertion of the theorem is proved. It is easy to see that ξ equals the term in the left-hand side of (4.1), and therefore (4.1) is also proved.

Let us now prove (4.2). Fix $p \in]1, r'[$. Since $p < r' \leq s'_\lambda$, formula (3.2) implies

$$\int_X \xi_\lambda^p d\mu < \infty \quad \forall \lambda \in \Lambda. \tag{4.14}$$

Evidently

$$\xi = \prod_{\lambda \in \Lambda_+} \xi_\lambda.$$

If now Λ_+ is a finite set, then (4.2) follows at once from the independence of ξ_λ's and (4.14). So we can suppose that Λ_+ is a countable set and put $\Lambda_+ = \mathbb{B}$. We can write, according to Fatou's Lemma, independence of ξ_n's and relations (4.14) and (3.2), that

$$\int_X \xi^p d\mu \leq \liminf_n \int_X (\prod_{k=1}^n \xi_k)^p d\mu$$

$$\leq \prod_n \int_X \xi_n^p d\mu$$

$$= \left(\prod_n (p s_n^{p-1} - (p-1) s_n^p)\right)^{-\frac{1}{2}} \times$$

$$\exp[\frac{p(p-1)}{2} \sum_n \frac{(\eta_0|e_n)_\mu^2}{p - (p-1)s_n}]. \tag{4.15}$$

Since by (4.8) $\lim_n s_n = 1$, we have

$$\sum_n \frac{(\eta_0|e_n)_\mu^2}{p - (p-1)s_n} < \infty. \tag{4.16}$$

The elementary inequality

$$1 - (p s^{p-1} - (p-1)s^p) \leq c(s-1)^2,$$

which is valid for all s in a neighborhood of 1 for an appropriately chosen constant $c > 0$, and relation (4.8) shows that

$$\prod_n (p s_n^{p-1} - (p-1) s_n^p) < \infty. \tag{4.17}$$

Clearly (4.17), (4.16), and (4.15) imply (4.2) and the theorem is proved.

Remark. It is easy to see that the conditions (3) and (4) in Theorem 4.1 imply that S is an invertible operator. Also it is easy to show that these two conditions together are equivalent to the following condition: the operator S in condition (2) is such that $S - I$ is a Hilbert-Schmidt operator all eigenvalues of which are strictly greater than -1.

Thus we have proved dichotomy for Gaussian measures. At the same time we have obtained the necessary and sufficient conditions for equivalence and also the form of the density in the case of equivalence. In the next section we shall give an adaptation of this result to topological vector spaces. As we have noted above, the proof given here is similar to that of Theorem 2.1, being however a bit more complicated. The following statement explains a connection between Theorem 2.1 and Theorem 4.1.

Theorem 4.2. *Let, in notations of Theorem 4.1, μ and ν be two arbitrary Gaussian measures on $\hat{\mathcal{C}}(X, G)$. Assume that at least one $f \in G$ is not a constant function μ-a.e. Then the following assertions are equivalent:*

(a) $\mu \sim \nu$;

(b) There is a family $\eta_\lambda : X \to \mathbb{R}$, $\lambda \in \Lambda$, of $\hat{\mathcal{C}}(X, G)$-measurable functions with the properties:

(b_1) The mappings $\eta_\lambda, \lambda \in \Lambda$, generate the σ-algebra $\hat{\mathcal{C}}(X, G)$ both in modulus μ and ν (i.e., any $E \in \hat{\mathcal{C}}(X, G)$ is equal to some $E' \in \hat{\mathcal{C}}(X, \{\eta_\lambda : \lambda \in \Lambda\})$ μ-a.e. and the same is true for ν);

(b_2) The mappings $\eta_\lambda, \lambda \in \Lambda$, are independent standard Gaussian random variables with respect to μ and the same mappings are independent Gaussian random variables with respect to ν;

(b_3) The conditions

$$\sum_{\lambda \in \Lambda} m_\lambda^2 < \infty, \quad s_\lambda > 0 \quad \forall \lambda \in \Lambda \quad and \quad \sum_{\lambda \in \Lambda} (s_\lambda - 1)^2 < \infty,$$

are satisfied, where

$$m_\lambda = \mathbf{E}_\nu \eta_\lambda, \qquad s_\lambda = \mathbf{E}_\nu (\eta_\lambda - m_\lambda)^2 \qquad \forall \lambda \in \Lambda.$$

Proof. The implication (b)$\Rightarrow$ (a) follows from Theorem 2.1; it can be derived from Theorem 4.1 as well. To show the implication (a)$\Rightarrow$ (b) first of all we remark that (a) implies $H_\mu = H_\nu$ algebraically. Consider the identity map $\mathcal{I} : H_\mu \to H_\nu$. Evidently $S = \mathcal{I}^*\mathcal{I}$, where S is the operator in condition (2) of Theorem 4.1. By this theorem $S - I$ is a Hilbert-Schmidt operator. Hence, as in the proof of Theorem 4.1, we can find an orthonormal basis in H_μ consisting of eigenvectors of S. Select from equivalence class e_λ a $\hat{\mathcal{C}}(X, G)$-measurable function $\eta_\lambda : X \to \mathbb{R}$. Now it is routine to show that $\eta_\lambda, \lambda \in \Lambda$, has all the required properties. QED

Remark. This theorem shows that the problem of equivalence of two arbitrary Gaussian measures can be studied in the framework of Theorem

2.1. In this connection a natural question arises whether the assertions (b_1) and (b_2) of Theorem 4.2 are valid for any Gaussian measures (without the assumptions of their equivalence). The affirmative answer would imply that the Feldman-Hajek theorem, in principle, can be reduced to Kakutani's theorem.

The following result, that is a slight refinement of a result from [23], says that the dichotomy takes place for Gaussian Radon measures as well.

Theorem 4.3. *Let X be a Hausdorff topological space, G be a real vector space of functions $f : X \to \mathbb{R}$, which separates points of X, μ and ν be Radon probability measures on the Borel σ-algebra $\mathcal{B}(X)$ such that they are Gaussian with respect to G. Then μ and ν are either singular or equivalent. In case of equivalence $\frac{d\nu}{d\mu}$ is measurable with respect to the σ-algebra $\hat{\mathcal{C}}(X,G)$.*

Proof. Let μ_G, ν_G be the restrictions of μ and ν, respectively, to $\hat{\mathcal{C}}(X,G)$. Suppose $\mu \not\perp \nu$, then $\mu_G \not\perp \nu_G$. By Theorem 4.1 we have $\mu_G \sim \nu_G$. Let us show that this relation implies $\nu \sim \mu$. Evidently it is enough to see that $\nu \ll \mu$. For this, since our measures are Radon measures, it is sufficient to verify that for any compact $K \subset X$ with $\mu(K) = 0$ we have $\nu(K) = 0$.

Fix a compact $K \subset X$. Since G separates points of X, the set K is the intersection of the family $\mathcal{C}_K$ of all closed sets $B \in \hat{\mathcal{C}}(X,G)$ with $K \subset B$. So, since μ is a Radon measure, we can write

$$\mu(K) = \inf\{\mu(B) : B \in \mathcal{C}_K\}.$$

This implies the existence of a set $B_K \in \hat{\mathcal{C}}(X,G)$ such that $K \subset B_K$ and $\mu(K) = \mu(B_K)$. If now K is a compact subset of X with $\mu(K) = 0$, then we have $\mu(B_K) = 0$. Since $\nu_G \ll \mu_G$, we obtain $\nu(B_K) = 0$. Therefore $\nu(K) = 0$, and so $\nu \ll \mu$. Similarily we can show that $\mu \ll \nu$, and the dichotomy is proved.

Let us prove the last assertion of the theorem. Fix again a compact set $K \subset X$. Using the above notation, we can write

$$\nu(K) = \int_K \frac{d\nu}{d\mu}d\mu = \int_{B_K} \frac{d\nu}{d\mu}d\mu = \int_{B_K} \frac{d\nu_G}{d\mu_G}d\mu = \int_K \frac{d\nu_G}{d\mu_G}d\mu.$$

Since ν and μ are Radon measures this equality implies $\frac{d\nu}{d\mu} = \frac{d\nu_G}{d\mu_G}$ μ-a.e. QED

6.6   The Case of Topological Vector Spaces

In this section X denotes a real topological vector space, X^* denotes its topological dual space, and $G \subset X^*$ is a vector subspace, which separates points of X. Of course, Theorem 4.1 is applicable in this situation. But in this case the parameters of Gaussian measure can be described more concretely and conditions for the equivalence can be given in new terms.

Let μ be a Gaussian measure on $\hat{\mathcal{C}}(X, G)$. An element $b_\mu \in X$ (if it exists) is called the *mean* or the *baricenter* of μ if

$$m_\mu(f) = f(b_\mu) \qquad \forall f \in G.$$

Similarly, a linear operator $R_\mu : G \to X$ (if it exists) is called the *covariance operator* of μ if

$$r_\mu(f_1, f_2) = f_1(R_\mu f_2) \qquad \forall f_1, f_2 \in G.$$

Note that if X is a complete separable locally convex space and $G = X^*$, then the mean and the covariance operator exist for any Gaussian measure μ on $\hat{\mathcal{C}}(X, X^*)$ (see [26] or [27] for the case of Banach spaces); the same is true if X is an arbitrary (i.e., not necessarily complete or separable) locally convex space and μ is a Gaussian Radon measure (see [1]).

Recall that an operator $R : G \to X$ is called *positive* if $f(Rf) \geq 0$ for all $f \in G$ and is called *symmetric* if $f_1(Rf_2) = f_2(Rf_1)$ for all $f_1, f_2 \in G$. Clearly, any covariance operator is positive and symmetric.

A linear operator $R : G \to X$ is said to be *factorable* through a Hilbert space H if there is a linear operator $A : G \to H$ such that the dual operator A^* maps H into X and $R = A^*A$. If R is a factorable operator through a Hilbert space H and A is a corresponding operator, then we say that H is *a factorizing* Hilbert space and A is *a factorizing* operator for R. It is easy to see that any factorable operator is positive and symmetric. In many cases the converse assertion (*factorization lemma*) is also valid. We refer interested readers to [27] and [26] for more information. Recall also that if H is a real Hilbert space and $R : H \to H$ is a positive and symmetric operator in the ordinary sense, then R is factorable through H in the canonical way: there is the unique positive and symmetric *square root of R*, i.e., the symmetric and positive operator $R^{1/2} : H \to H$ such that $R^{1/2}R^{1/2} = R$. By the analogy, any factorizing operator A for a given symmetric and positive operator $R : G \to X$ can be called a square root of R.

Proposition 5.1. *Let μ be a Gaussian measure on $\hat{\mathcal{C}}(X, G)$ with a factorable covariance operator $R_\mu : G \to X$ and H and A be a factorizing*

Hilbert space and operator for R_μ, respectively. Then there is an isometric linear operator $U : H_\mu \to H$ such that $UA_\mu f = Af$ for all $f \in G$.

The proof is easy and we omit it.

Theorem 5.1. *Let X be a real topological vector space, $G \subset X^*$ be a separating vector subspace and μ and ν be Gaussian measures with respect to G on $\hat{C}(X, G)$. Then μ and ν are either singular or equivalent.*

Moreover, suppose that μ has the mean $b_\mu \in X$ and the covariance operator $R_\mu : G \to X$ that is factorable through a Hilbert space H with the factorizing operator $A : G \to H$. Then $\mu \sim \nu$ if and only if the following conditions are satisfied:

(i) ν has the mean $b_\nu \in X$ and $b_\nu - b_\mu \in A^*(H)$;

(ii) ν has covariance operator $R_\nu : G \to X$ factorable through H and there is an injective positive symmetric operator $T : H \to H$ such that $T - I$ is a Hilbert-Schmidt operator, and $R_\nu = A^*TA$.

Let (i) and (ii) be satisfied. Choose an element $h_{\mu,\nu} \in H$ such that $A^*h_{\mu,\nu} = b_\nu - b_\mu$, consider the (finite or infinite) sequence $s_1, \dots$ of nonzero eigenvalues of the symmetric Hilbert-Schmidt operator $T - I$ listed according to their multiplicities, and the (finite or infinite) complete orthonormal sequence $h_1, \dots$ of the corresponding eigenvectors. Consider also the isometry $U : H_\mu \to H$ with the property: $UA_\mu = A$ and put $U^*T^{-1}h_{\mu,\nu} = h_0$, $U^*h_n = e_n$, $n = 1, \dots$. Then the relations (4.1) and (4.2) remain valid.

Proof. Dichotomy is a direct consequence of Theorem 4.1. Suppose now that $\mu \sim \nu$ and verify (i) and (ii). According to Theorem 4.1 it is enough to show that the conditons (1),...,(4) of this theorem imply (i) and (ii). The condition (1) easily implies that the element $A^*U\eta_0$ is the mean of ν, i.e., (i) holds. Let now $P_U : H \to U(H_\mu)$ be the orthogonal projection. Denote $T = USU^* + I - P_U$. Then a direct verification shows that the operator A^*TA is the covariance operator of ν, i.e., (ii) also holds. Suppose now that the conditions (i) and (ii) hold. Then it can be shown in a similar manner that the conditions (1),...,(4) of Theorem 4.1 are satisfied and by this theorem we obtain that $\nu \sim \mu$. The other assertions of Theorem 5.1 follow from similar assertions of Theorem 4.1 in the same way. QED

Remark. Theorem 5.1 is a slight modification of a similar assertion announced in [25] for the case of Banach spaces.

Corollary 1. *Let H be a real Hilbert space and μ and ν be Gaussian measures on H. Then μ and ν are either singular or equivalent. Moreover, $\mu \sim \nu$ if and only if the following conditions are satisfied:*

(i) $b_\nu - b_\mu \in R_\mu^{1/2}(H)$;

(ii) There exists an injective positive symmetric operator $T : H \to H$

such that $T - I$ is a Hilbert-Schmidt operator and $R_\nu = R_\mu^{1/2} T R_\mu^{1/2}$.

If (i) and (ii) are satisfied then the relations (4.1) and (4.2) also are valid, where the parameters are defined as in Theorem 5.1 with only A replaced by $R_\mu^{1/2}$.

Proof. Put in Theorem 5.1 $X = H$, identify H^* with H and take $G = H$. QED

Remark. This corollary solves in principle the equivalence (or singularity) problems of two arbitrary Gaussian measures on Hilbert space. It is an important problem to find more observable conditions of equivalence or singularity for concretely given Gaussian measures on concrete spaces and the expression of densities in case of equivalence directly in terms of parameters of the measures (i.e. without using eigenvalues and eigenvectors of related operators). These questions for Hilbert spaces are well studied, see [6], [22].

Corollary 2. *Let $\mathbb{R}^{\mathbb{B}}$ be the space of all sequences $x : \mathbb{B} \to \mathbb{R}$ with the product topology and μ and ν be Gaussian measures on it. Then μ and ν are either singular or equivalent.*

Moreover, if μ is the direct product of standard Gaussian measures and ν is an arbitrary Gaussian measure with mean b_ν and covariance matrix t_ν, then $\mu \sim \nu$ if and only if the following conditions are satisfied:

(i) $b_\nu \in l_2$;
(ii) t_ν is the matrix of an injective operator in l_2;
(iii) $\sum_{k,j} [t_\nu(k,j) - \delta(k,j)]^2 < \infty$.

Proof. The space $(\mathbb{R}^{\mathbb{B}})^*$ can be identified with the space $\mathbb{R}_0^{\mathbb{B}}$ of all finitary sequences, and the covariance operator R_μ is the natural embedding of $\mathbb{R}_0^{\mathbb{B}}$ into $\mathbb{R}^{\mathbb{B}}$. Therefore it remains to apply Theorem 5.1 with $H = l_2$, $G = \mathbb{R}_0^{\mathbb{B}}$ and with the natural embedding $A : \mathbb{R}_0^{\mathbb{B}} \to l_2$ as a factorizing operator. QED

Remark. We see that even in the simple situation of Corollary 2 it is not easy to express equivalence conditions, namely condition (ii), explicitly in terms of the covariance matrix.

We want to formulate one more assertion, which is important for the Constructive Field theory. We need some notations for that.

Let H be an infinite-dimensional Hilbert space, X be the space of all (not necessarily continuous) linear functionals $x : H \to \mathbb{R}$ with the topology of pointwise convergence. Then X^* can be identified algebraically with H. Let us consider the σ-algebra $\hat{\mathcal{C}}(X, H)$. Fix a continuous linear operator $B : H \to H$. Then there exists a Gaussian (with respect to H) measure μ_B

on X that has the properties:

$$\int_X x(h) d\mu_B(x) = 0, \int_X |x(h)|^2 d\mu_B(x) = ||Bh||^2 \qquad \forall h \in H;$$

this follows from Bochner-Kolmogorov's theorem, see [27], p. 235. The measure μ_I is (the distribution of) the white noise.

Corollary 3. *Let $B : H \to H$ be a continuous linear operator. Then the measures μ_I and μ_B are equivalent if and only if B is an injective operator such that the operator $|B| - I$ (or, equivalently, the operator $B^*B - I$) is a Hilbert-Schmidt operator (here $|B| = (B^*B)^{1/2}$).*

In case of equivalence the density

$$\frac{d\mu_B}{d\mu_I}$$

is of the form (4.1), where $h_0 = 0, e_n(x) = x(h_n), n = 1, \ldots$ for μ_I-almost every $x \in X$ and $h_1, \ldots$ and $s_1, \ldots$ are related to the operator $T = |B|$ as in Theorem 5.1. Also

$$\int_X \left(\frac{d\mu_B}{d\mu_I} \right)^p d\mu_I < \infty$$

for all $p \in]1, r'[, r = ||B||^2$.

Proof. Take in Theorem 5.1 $\mu = \mu_I, \nu = \mu_B, G = H$. In this case R_μ and R_ν can be identified with the identity operator in H and with the operator B^*B, respectively. Then the required assertions follow easily from Theorem 5.1. QED

Remark. The last assertion of Corollary 3 is a slight refinement of a similar assertion, namely Theorem 1.23 in [21].

Bibliography

[1] Borel, C., Gaussian Radon measures on locally convex spaces, *Math. Scand.*, 1976, v.38, 265-284.

[2] Brody, E. J., An elementary proof of the Gaussian dichotomy theorem, *Z. Wahrsch. verv. Geb.*, 1971, v. 20, 217-226.

[3] Chatterji, S. D., and Mandrekar, V., Singularity and absolute continuity of measures. In: K.-D. Bierstedt, D. Fuchsteiner (Eds.) *Functional Analysis: Survey and Recent Results*, North-Holland, Amsterdam, 1977, 247-257.

[4] Chatterji, S. D. and Mandrekar, V., Equivalence and singularity of Gaussian measures and applications. In: *Probabilistic Analysis and Related Topics*, V.1, Ed., A. T. Bharucha-Reid, Academic Press, 1978, 169-197.

[5] Feldman, J., Equivalence and perpendicularity of Gaussian processes, *Pac. J. Math.*, 1958, v.8, 699-708; correction: ibid., 1959 v.9, 1295-1296.

[6] Gihman, I. I. and Skorohod, A. V., *Theory of Random Processes*, v.I, Nauka, Moscow, 1971.

[7] Hajek, J., On a property of normal distributions of any stochastic process, *Czech. Math. J.*, 1958, v.8, 610-618.

[8] Haliullin, S. G., On the dichotomy of ergodic measures on linear spaces, preprint, 1995.

[9] Ibragimov, I. A. and Rosanov, Iu. A., *Gaussian Random Processes*, Nauka, Moscow, 1970.

[10] Jacod, J. and Shiryaev, A. N., *Limit Theorems for Stochastic Processes*, Springer-Verlag, Berlin, 1987.

[11] Kakutani, S., On equivalence of infinite product measures, *Ann. Math.*, 1948, v.49, 214-224.

[12] Klein, A., Quadratic expressions in a free Boson fields, *Trans. Am. Math. Soc.*, 1973, v.181, 439-456.

[13] Kuo, H. H., Gaussian Measures on Banach Spaces. *Lecture notes in Math.*, 1975, v.463.

[14] Kuhn, T. and Liese, F., A short proof of the Hajek-Feldman theorem, *Teor. Verojat. Prim.*, 1978, v.23, 448-450.

[15] Neveu, J., Processus Aleatoires Gaussiens, Montreal, 1968.

[16] Rozanov, Iu. A., Gaussian Infinite Dimensional Distributions, *Trans. Steklov Inst. Math.*, 1968, v.CVIII, 1-136.

[17] Rao, M. M., *Foundations of Stochastic Analysis*, Academic Press, New York, 1981.

[18] Sato, H., On the convergence of products of independent random variables, *J. Math. Kyoto Univ.*, 1987, v.27, 381-385.

[19] Shale, D., Linear symmetries of free Boson fields, *Trans. Am. Math. Soc.*, 1962, v.103, 149-167.

[20] Shepp, L. A., Gaussian measures in function space, *Pac. J. Math.*, 1966, v.17, 167-173.

[21] Simon, B., The $P(\varphi)_2$ Euclidean (Quantum) Field Theory, Princeton University Press, Princeton, NJ., 1974.

[22] Skorohod, A. V., *Integration in Hilbert Space*, Springr-Verlag, Berlin, 1974.

[23] Sato, H. and Okazaki, Y., Separabilities of Gaussian Radon measures, *Ann. Inst. H. Poincare*, 1975, t.XI, 287-298.

[24] Talagrand, M., Mesures Gaussienes sur un espace localement convexe, *Z. Wahrsch. verv. Geb.*, 1983, v.64, 181-209.

[25] Tarieladze, V. I., On equivalence of Gaussian measures on Banach spaces, *Bull. Acad. Sci. Georgian SSR*, 1974, v.73, 529-532.

[26] Vakhania, N. N. and Tarieladze, V. I., Covariance operators of probability distributions in locally convex spaces, *Teor. Veroyat. Prim.*, 1978, v.23, 3-26.

[27] Vakhania, N. N., Tarieladze, V. I., and Chobanyan, S. A., *Probability Distributions on Banach Spaces*, D. Reidel, Dordrecht, 1987.

[28] Xia, Dao-Xing, *Measures and Integration on Infinite Dimensional Spaces*, Academic Press, New York, 1972.

INDEX